Calculus and Mathematical Reasoning for Social and Life Sciences

Daryl Cooper

University of California, Santa Barbara

KENDALL/HUNT PUBLISHING COMPANY
4050 Westmark Drive   Dubuque, Iowa 52002

Introduction. The text introduces some mathematics which is useful in the social and life sciences. The author has consulted with many people in various departments at UCSB to try to ensure the material covered is as relevant as possible to these studies.

The mathematics you will learn in this class is really quite sophisticated. It will enable you to do things that the smartest people in the world were incapable of doing for a thousand years, from the time of the Greeks until the Renaissance. You will learn more in 10 short weeks of study than many learned in a lifetime.

Calculus is the mathematics used to describe how things **change.** For example, consider the problem of describing the weather using mathematics. We might give the temperature and wind speed at this moment, but what we really want to know is what the weather will be like tomorrow. The laws governing the weather, at least as much as we understand of them, involve **differential equations.** These express how quickly certain things, such as temperature and pressure, change in terms of other things such as cloud cover. Once we have such a mathematical description of how quickly things change, we may attempt to calculate what the temperature will be tomorrow, by starting out from the temperature today. If the equations say something like "the temperature will drop by 2 degrees per hour" then we discover that 24 hours from now the temperature will have fallen 48 degrees. This process is called **integrating** the differential equations. Of course, the weather is very complicated. There are many equations. It is difficult to analyze because, for example, cloud cover affects temperature but temperature also affects cloud cover.

There are two fundamental processes in calculus. The first is to calculate how quickly something changes, and it is called **differentiation.** The second is the reverse; that is, if we know how quickly a quantity changes then we want to calculate what that quantity will be 24 hours from now, or whenever. This is called **integration.** These two processes are the reverse of each other.

Now it is a remarkable thing that differentiation turns out to basically involve determining the **slope** of a graph. For example, if we look at a graph showing the temperature throughout the day, then how quickly the temperature increases is given by how quickly the graph moves upwards. In other words by the **slope** of the graph.

A second remarkable thing is that integration involves finding out **how much area** there is underneath a graph. For example, if we look at a graph which shows the speed of a car at various times during a journey, then the distance the car travels is represented by the area under this graph.

In conclusion, the mathematics of change that is used to describe so much in science boils down to solving two **geometric** problems. Find the slope of a curve. Find the area underneath a curve.

I thank all those in various departments at UCSB who responded to my emails and phone calls during the planning phase of this book. Special thanks to **Ermila Moodley** for reading and giving invaluable feedback on early versions of this book. Also for the patience and support she has shown during the weekends and evenings that have been swallowed by this project. It would not have been possible without the assistance of **Matt White** who greatly improved the quality of the text, and provided solutions to the problems. Thanks to numerous people who have given me feedback, in particular to **Lily Cockerill, Eric Lichtenstein** and **Monika White** for supplying a list of errors. Also to **Roger Nisbet** for many discussions and for contributing a chapter to this text.

please email any comments to: cooper@math.ucsb.edu

How to use this book

Chapters 1-4 and **6** are mostly review. The instructor will do selected topics and problems from these chapters but you are expected to know all this material. **Chapter 10** is a refresher for students who are taking 34B but either did not take 34A or took it a long time ago. You can find out if you are ready for 34A by taking the **entrance quiz** in section 17.1.

Student's Guide

The last chapter gives advice on good ways to study math and science.

Written Work. You are expected to produce high quality written work. This means that it should be possible for someone to be able to **read and understand** what you have written. It is not enough that you get the correct answer. A prospective employer will not be impressed if you can not express yourself clearly in writing. If you look in a scientific journal at a paper using mathematics, you will see there is a lot of writing explaining the mathematics. You do not need to write trivialities such as "using the commutative property of addition". You do need to say things like "we now use equation (1) to eliminate x from equation (2)." There will be **bonus points** on exams for well written work.

This book covers:
- algebra
- word problems
- graphs
- linearity (straight lines)
- logarithms and exponentials
- derivatives
- integrals
- differential equations

If you do not master algebra, no matter how well you "understand" the later material, you will not be able to do the questions. The heading of word problems is short-hand for reasoning, analytical thinking, and the ability to handle problems of a **type** you have not seen before by **thinking.** This skill is probably the most important skill you can gain from this class. It is usually the topic which students find hardest. Developing this skill will serve you well in many areas of life which have nothing to do with math. The ability to break down a complex problem into small manageable pieces which can be tackled one step at a time is extremely useful. Word problems bring out these issues most clearly. Often the hardest part of the problem is deciding what needs to be done, what the significance of the available information is, and how that information can be used. Then one must develop a plan to solve the problem, proceeding in small steps, without losing sight of the final goal. Solving such problems involves writing sentences in **English.** Many students resist doing this, and fail as a result. Do not be one of them. When you write the solution to a word problem, you should try to mimic what the instructor and TAs write when they present solutions.

Studying. Do not put off doing any work until just before the exams, unless you are really keen to repeat this class. It is the story of the hare and tortoise all over again. If you work steadily throughout the quarter, you will feel much happier during exam week. It does wonders for self-esteem to think "gee, I put in the time, and I know I am going to pass!"

Exams. Get a good nights rest before any exam. **Do not study late.** I have graded thousands of exams. I have had hundreds of students come to me and say "I knew how to do it, but I just made this really stupid mistake." You need to be very alert in a math exam. Tiredness is your worst enemy. **Never leave an exam early.** Find some scratch paper, or use the back of the blue-book, and work out the problems again to see if you get the same answer. Try to check your answers. Read over them to see if you can see anything wrong. To prepare for exams, do the practise exam. Select homework problems at random from those assigned and see if you can do them **without looking at the model solutions.**

Know your weaknesses. If you play a competitive sport, you know it is important to work on your weak points. When you get an exam back, go over it to see where your mistakes were. Look for a pattern. The most likely is certain kinds of algebraic error. Make a note of this. In future exams, be on the lookout for exactly this kind of error when you **check your work.**

Study Groups. Once you have decided you will work hard to pass this class, you should decide to work **efficiently.** Also you should try to make the process of working **fun.** Both of these goals can be accomplished by forming a study group. You should meet regularly with a group of students to work together on homework. If one of you gets stuck, then you talk to each other. You will learn far more from this than going to a TA who will probably just tell you the answer. You learn from finding your own mistakes. Students who work hard in study groups get better grades. Besides, it is more enjoyable to work long hours in the company of others.

Lectures. Do not skip class. You may be used to missing class occasionally, and in some subjects this is not so important, but in math it is disastrous. When you return to class after skipping one, you will find you don't understand what is going on. Learning math is like building a tower. If you leave some bricks out near the bottom, it falls over. Having decided to come to class **arrive on time.** If you arrive 10 minutes late, you won't understand what is happening, and you may as well not have bothered. **Sit near the front.** People who sit near the front tend to get better grades. You can hear and see better at the front. You can ask questions, and there is not as much to distract you.

Homework. Math is like riding a bicycle, you can't learn to do it by watching someone else. You need to keep up to date on the homework. Do the homework when it is assigned. You will find what is said in class makes much more sense if you are current on the homework. Some of the word problems are quite tricky, and it will save you a lot of time if you discuss such problems with others. The act of discussing math with other people is one of the best ways to learn. Compare our solutions to yours. Make sure you understand any mistakes you made.

Discussions. This is your opportunity to get your questions answered. Always have a question ready. Examples of a good question are "I don't understand logarithms" or "how do you go about calculating half-life?" or "When do I have to write parentheses."

Reading. Each day there is a lecture you should read over the notes you took afterwards. This will help you remember what was said. Also things that were confusing in lecture sometimes become clear the second time round. Do the assigned reading **before class.** Usually this means reading about 4 pages and takes very little time. Although you might not understand it all, it will help the lecture fall into place. After the lecture, if you did not understand everything you read, re-read it.

Do not wait. If you are having difficulty, if there was a homework assignment you did badly on, get help **immediately.** You might go to your TA, or to the math lab, or to CLAS, or get a private tutor, or find someone in the class who knows what is going on and can explain it to you. The last option is often the best. Another student in the class who explains things to you will reap benefits. One of the best ways to learn something really well is to explain it to someone else. In fact sometimes I teach classes on subjects just so that I can learn them! Understanding something so well that you can explain it clearly to someone else is the best way to learn it.

Understand in your own words. Only you know what goes on inside your head. You have your own unique way of thinking. Try to translate what you are learning into language that really makes sense to you. Not the words the professor uses, but your own words. Make the ideas belong to you.

Contents

Chapter 1

Review of Algebra.

Mathematicians have a weird sense of humor. For example they use the word "review" to mean "It is quite likely that you don't know what I am about to tell you. So you better learn it." There are four things you have to be good at:
(1) the distributive law (using parentheses correctly).
(2) solving equations
(3) handling fractions
(4) handling powers like x^3.

If you have not done any algebra for a long time, then you should spend a lot of time reviewing these topics during the first two weeks.

Algebra is about x's. Consider the problem: Find the two numbers which added together give 7 and which when multiplied together give 12. Our job here is to discover two numbers. We must use the information given about these numbers to find out what they are. The problem is how to use the given information. The solution is to give **names** to the numbers you wish to find. Thus we start by writing down on our paper

x and y are the names of the numbers we wish to find.

Failure to do this preliminary step is a frequent cause of **trouble** for students. Now we can use the information we are told. The numbers added together give 7 means

$$x + y = 7.$$

Also their product gives 12 means that

$$x \times y = 12.$$

We now have 2 equations for 2 unknowns so we can solve them. The technique for problems with more than one unknown is this: **get rid of one of the unknowns.** If we have one equation which only has an x in it, then we can probably solve it. When we have 2 or more unknowns, the thing to do is to try to get rid of one of the unknowns. In our example, we can use the equation $x + y = 7$ to express y in terms of x thus

$$y = 7 - x.$$

This is a formula which expresses y in terms of x. So wherever we see a y, for example in the second equation $x \times y = 12$, we can replace the y so that there are only x's. This gives

$$x \times (7 - x) = 12.$$

This is a quadratic equation

$$x^2 - 7x + 12 = 0.$$

We can solve this by factoring

$$0 = x^2 - 7x + 12 = (x - 3)(x - 4)$$

9

and see that $x = 3$ or $x = 4$. Or we could use the formula for solving the quadratic equation

$$ax^2 + bx + c = 0 \qquad \Rightarrow \qquad x = \frac{-b \pm \sqrt{b^2 - 4ac}}{2a}.$$

If you wish to succeed you will pay attention to the following golden rules.

---

(1) Be neat. Space out your work. Do not cramp it up in a corner. Work **down** the page.

(2) BRIEFLY explain what you are doing. For example, write brief descriptions of ALL the unknowns in your work.

(3) When dealing with equations ALWAYS write the = sign.

(4) You MUST do the same things to both sides of an equation ("Ted 'n Fred rule" below).

(5) Only do ONE simplifying step per line.

(6) Be VERY careful to use parentheses. Always use both (and).

(7) Use a sharp pencil and neatly erase mistakes.

(8) Use × or · **as appropriate** to multiply.

(9) Put a box around your final answer.

YOU WILL BE MARKED VERY HARSHLY FOR FAILING TO OBEY THESE SIMPLE RULES. Rule 4 is the "Ted and Fred" rule. If you have an equation

$$Ted \ = \ Fred$$

then whatever you do to Ted you must also do to Fred, it is only fair!. Thus if you add 3 to Ted, you must add 3 to Fred also

$$Ted + 3 \ = \ Fred + 3.$$

For rule (8) you should write 3.6×4.2 not $3.6 \cdot 4.2$ BUT if when you write x it looks like × write $x \cdot x$ instead of $x \times x$. The intent is to avoid mistakes and make it clearer to read. It takes almost no time at all to follow these simple rules. They will save you hours of frustration, not to mention the points!

1.1   Notation and how to write Mathematics.

Mathematical writing is made up of **expressions** or **formulae** like

$$x^3 - 7x + 3.$$

These expressions are put together either with symbols like = or with English words to produce understandable **statements** like $x = 3$. There are some special symbols that are used in math.

& means AND

∴ means THEREFORE

∵ means BECAUSE

⇒ means IMPLIES

≈ means APPROXIMATELY EQUAL

∝ means PROPORTIONAL TO

< means LESS THAN

≤ means LESS THAN OR EQUAL TO

> means GREATER THAN

≥ means GREATER THAN OR EQUAL TO

The word "implies" in mathematics means the same as "therefore". It refers to the notion of logical implication. Mathematical writing like any other form of writing is meant to convey information to the person reading it. It is not meant to be a scrawl understood only by the person writing it during some calculation. It should be possible for a person to READ what has been written and for it to make SENSE. Numbers are like nouns in

English. English writing is composed of sentences. A sentence is the smallest unit of writing that has meaning. A single noun is not a sentence. Thus "dog." is not a sentence. It does not convey information. (Although one might argue that if you are walking in snow capped mountains and scream "Landslide!" that is in fact a sentence. There will be no screaming in this class.) In a similar way "17" does not convey information. When you write math, ANY math, it should be a sequence of mathematical sentences. The simplest mathematical sentence is an equation

$$x = 7.$$

This says something. It says one quantity, x equals another quantity, 7. Equals is a verb. This mathematical sentence $x = 7$ is linguistically about as exciting as "The dog exploded." But it is grammatically correct. Too often a poor student writes something like

$$x^2 + 5x + 6 = 0$$

$$(x + 2)(x + 3)$$

$$2, 3$$

or worse. Often it is crossed out, written again, crossed out again, and then written sideways or upside-down with a pathetic circle drawn around it. It is garbage grammatically speaking. It is also wrong. The first line starts out fine. But the second line is a noun, not a verb. This should be written:

$$
\begin{aligned}
&\textit{want to solve} \quad && x^2 + 5x + 6 = 0 \\
&\textit{factoring gives} \quad && (x + 2)(x + 3) = 0 \\
&\textit{solving gives } x + 2 = 0 \quad && \textit{or} \quad x + 3 = 0 \\
&\textit{and so} \quad x = -2 \quad && \textit{or} \quad x = -3 \quad \textit{are the solutions.}
\end{aligned}
$$

Another way to write this is

$$
\begin{aligned}
& x^2 + 5x + 6 = 0 \\
\Rightarrow\,& (x + 2)(x + 3) = 0 \\
\Rightarrow\,& x + 2 = 0 \quad \textit{or} \quad x + 3 = 0 \\
\Rightarrow\,& x = -2 \quad \textit{or} \quad x = -3.
\end{aligned}
$$

The first way of writing it is better. It helps the reader understand what you are doing. More importantly it helps YOU understand what you are doing. If you get into the habit of writing brief explanations of what you are doing, over a period of time you will find math gets easier. The repetition of explaining what is going on results in your brain working more efficiently. **This is probably the single simplest thing which you can do to get a better grade.**

The **implies** sign $\Rightarrow$ means the same thing as the **therefore** sign $\therefore$. What these signs mean is the following: if the first statement is assumed to be true then the second statement is also true.
Thus $x = 2 \Rightarrow x^2 = 4$ means that "if we know that $x = 2$ then it is true that $x^2 = 4$."
This is a completely different thing to the equals sign. The equals sign is used to say that two numbers are the same. The implies sign is used to say that if the first statement is true then the second statement is true. Thus equals signs are used to connect numbers. Implies signs are used to connect statements.

If you decide to use the $\Rightarrow$ sign remember that one statement can imply another statement, but a number cannot imply or be implied by anything. Thus

$$x^2 = 4 \Rightarrow \pm 2 \qquad \text{is wrong.}$$

$$x^2 = 4 \Rightarrow x = \pm 2 \qquad \text{is correct.}$$

1.2 Distributivity: *FOIL*ed again.

Algebra involves "letters" such as $x, y, A, B, v, t, \ldots$ These represent some number and so enjoy the same rules that numbers do, such as the distributive property. One sometimes imagines an envelope marked x containing a slip of paper inside with a number written on it that one can not see. In a particular situation, one might decide to call the letter x a **variable** and the letter A a **constant**. This is really a psychological thing. By saying that x is a variable we are imagining plugging in lots of different numbers for x. By saying A is a constant we imagine plugging in one number for A and never changing it. Sometimes we have to solve an equation for an unknown such as x, but there are other unknowns in the equation, such as A and these will appear in our solution for example

$$x = \frac{3 + A}{\sqrt{2}}.$$

However the distinction between variables and constants is more imaginary than real, it is a bit like the distinction between terrorists and freedom fighters: it all depends which side you are on.

You must be able to do symbolic calculations which involve several letters. Such calculations may involve many steps and you must be able to do them without error. There are just a few basic ideas to know. By far the most important is the **distributive** property. Many of you have learned the **FOIL** method of multiplying out. You should try to forget it!

Parentheses are instructions to do what is inside them **before even thinking about what is outside**. So $3 + (4 \times 5) = 23$ means something quite different to $(3 + 4) \times 5 = 35$. The order that you calculate things is given by the acronym **PEMDAS** which stands for **P**arentheses **E**xponentiation **M**ultiplication **D**ivision **A**ddition **S**ubtraction. So you do exponentiation first, then multiplication and divisions and finally addition and subtraction is the last thing you do. **But** parentheses are the first thing you do. In other words you work out what is inside parentheses before anything else.

Sometimes parentheses are not necessary, for example

$$(3 + 4) + 5 \;=\; 3 + (4 + 5)$$

so we do not normally bother with parentheses in such a situation because it does not matter what order you add things. If you add the 3 and 4 first, or if you add the 4 and 5 first makes no difference. The final result is to add up the three numbers $3, 4$ and 5 and it makes no difference what order you do it. But subtraction is a different matter:

$$(3 - 4) - 5 \;=\; -6 \qquad \text{but} \qquad 3 - (4 - 5) \;=\; 4$$

so if you have several subtraction signs it is important what order you do them. It is safest to use parentheses so there can be no doubt. But if there are no parentheses such as

$$3 - 4 - 5$$

the rule is to work left to right in the same way that we read. So

$$3 - 4 - 5 \;=\; (3 - 4) - 5.$$

Parentheses are **vital** when combining multiplication and addition. If there are no parentheses PEMDAS tells us we must multiply before adding. Thus

$$3 \times 4 + 5 \;=\; 12 + 5 \;=\; 17.$$

If you desire to multiply $4 + 5$ by 3 you **must** use parentheses:

$$3 \times (4 + 5) \;=\; 27.$$

That

$$3 \times (4 + 2) \;=\; (3 \times 4) + (3 \times 2)$$

goes by the phrase **multiplication distributes over addition.** Or the **distributive property** for short. The phrase is meant to convey the idea the multiplications symbol × has to be distributed: it goes to both the 4 and the 2. One of the **most common mistakes** on exams is that students "forget" this and write instead

$$3 \times (4 + 2) \ = \ 3 \times 4 \ + \ 2$$

often because they can't be bothered to write parentheses. When you read aloud $3 \times (4 + 2)$ it is wrong to say "three times four plus two" because the listener will quite properly assume that what you mean is to multiply the three and four together and then add two onto the answer. The correct way to read aloud $3 \times (4+2)$ is "three times the quantity four plus two." Another is "three times the result of adding four and two." In both cases you emphasize that you **first** add two and four **then** multiply by three. Now you might claim that you are not in the habit of going up to people and saying this kind of stuff anyway. That is not the point. The point is that **you say it to yourself** when working a problem and then go and make exactly the mistake I am describing. Sloppy expression of ideas produces sloppy thinking which produces mistakes. The parentheses are saying the same thing using symbols. Here are two ways to think about distributivity.

The first is just that $3 \times (4 + 2)$ means take three lots of $4 + 2$ and add them up. So you get

$$(4 + 2) \ + \ (4 + 2) \ + \ (4 + 2)$$

which is the same things as three lots of 4 plus three lots of 2

$$(4 + 4 + 4) \ + \ (2 + 2 + 2).$$

The second way is to think of multiplication in terms of the **area** of a rectangle.

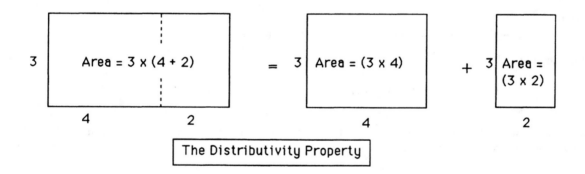

The Distributivity Property

Then $3 \times (4 + 2)$ is represented by a rectangle of height 3 and width $4 + 2$. Now this rectangle can be split apart into two rectangles of width 4 and width 2. The area of the big rectangle equals the sum of the areas of the two small rectangles. Therefore $3 \times (4 + 2) \ = \ (3 \times 4) \ + \ (3 \times 2)$.

Now this may seem a long discussion of the obvious. But this simple **idea** can be extended. First of all:

$$p \times (q + r) \ = \ (p \times q) \ + \ (p \times r).$$

Here the letters p, q, r stand for certain numbers, but you don't know which numbers they are. They could be anything. Even $3, 4$ and 2. It does not matter what they are because the formula above always works. It works for exactly the same reasons that we talked about before. You can think of it as saying "p lots of $(q + r)$ is the same as p lots of q added to p lots of r." Or you can think of a rectangle of height p and width $(q + r)$ divided into two smaller rectangles of width q and r. Then the area of the big rectangle is $p \times (q + r)$ and the areas of the two smaller rectangles are $p \times q$ and $p \times r$.

Now the **beauty** of this formula is that it works **no matter what** we choose for p, q and r and therefore spawns an infinite number of formulae. For example, if the number r happens to equal two other numbers added together, say $r = (s + t)$, then distributivity tells us

$$p \times (q + (s + t)) = (p \times q) + (p \times (s + t)).$$

We can simplify this (using distributivity again on the term $p \times (s + t)$) to get

$$p \times (q + s + t) = (p \times q) + (p \times s) + (p \times t).$$

This is NOT a new rule. It is simply using the distributive rule twice. You can imagine how you would use distributivity 3 times to show

$$p \times (q + r + s + t) = (p \times q) + (p \times r) + (p \times s) + (p \times t).$$

And so on.

Similarly, the above formula works **no matter what** we choose for p. If the number p just happens to be the sum of two other numbers, for example $p = x + y$, distributivity **still** says the same thing.

Example 1.2.1 *Multiply out* $(x + y) \times (q + r)$.

By distributivity,

$$(x + y) \times (q + r) = ((x + y) \times q) + ((x + y) \times r).$$

Now this is not the end of the story, because we can expand $(x + y) \times q$ and $(x + y) \times r$ using distributivity yet again:

$$((x + y) \times q) + ((x + y) \times r) = ((x \times q) + (y \times q)) + ((x \times r) + (y \times r)).$$

Thus, we get that

$$(x + y) \times (q + r) = ((x \times q) + (y \times q)) + ((x \times r) + (y \times r)).$$

Dropping redundant parentheses gives

$$(x + y) \times (q + r) = x \times q + y \times q + x \times r + y \times r.$$

Of course, this is what FOIL tells you. But distributivity has more to say since we can replace p in $p \times (q + r)$ by even more complicated expressions:

Example 1.2.2 *Multiply out* $(x + y + z) \times (q + r)$.

By distributivity,

$$(x + y + z) \times (q + r) = ((x + y + z) \times q) + ((x + y + z) \times r).$$

Again, we can expand $(x + y + z) \times q$ and $(x + y + z) \times r$ using distributivity to get:

$$((x + y + z) \times q) + ((x + y + z) \times r) = (x \times q + y \times q + z \times q) + (x \times r + y \times r + z \times r).$$

Dropping redundant parentheses we obtain

$$(x + y + z) \times (q + r) = x \times q + y \times q + z \times q + x \times r + y \times r + z \times r.$$

This can go on forever. More and more complicated expressions can be multiplied out **one step at a time.** It would be hopeless to try to memorize something like FOIL to do each of these problems. The great secret is to just remember two things:

$$\boxed{a \times (b + c) = (a \times b) + (a \times c) \qquad \text{Do one step at a time.}}$$

If you have to multiply a lot of things added together by a lot of things added together

> multiply everything in the first group by everything in the second group and add up

for example

$$(a + b + c) \times (x + y + z) = a \times x + a \times y + a \times z + b \times x + b \times y + b \times z + c \times x + c \times y + c \times z.$$

Problems

Multiply out each of the following, one step at a time, writing all the parentheses.

(1.2.1) (a) $(x - y) \times (x + y)$ (b) $(x + y) \times (x + y)$ (c) $(x + x^2) \times (1 - x)$

(1.2.2) (a) $(3x - 4y)(2x + 5y)$ (b) $(2x^2 - 3x + 1)(x - 1)$

(1.2.3) (a) $(x + y) \times (x + y) \times (x + y)$ (b) $(1 + p + p^2) \times (1 - p)$

(1.2.4) (a) $(1 + x)^4$ (b) $(1 - x + x^2)(1 + x + x^2)$

(1.2.5) (a) $(a + b)(2a + b)(a - b)$ (b) $(1 + x)(1 - x)(1 + x^2)$

(1.2.6) $(x^2 - 3x + 2)(x^3 - x^2 + 1)$

(1.2.7) Simplify (a) $(a - (b - c)) - ((a - b) - c)$ (b) $a - (a - (b - (a - b)))$ (c) $a - ((a - (a - b)) - (b - (a - b)))$

1.3   Revenge of the Fractions.

A fraction is

$$\frac{numerator \text{ is the thing on top}}{denominator}$$

You might remember the phrase "lowest common denominator." The word "lowest" is a reminder that the denominator is the bottom of a fraction. The first thing to understand about fractions is **canceling top and bottom:**

$$\frac{6}{2} = \frac{2 \times 3}{2 \times 1} = \frac{2}{2} \frac{3}{1} = \frac{3}{1} \quad \text{etc.}$$

The point here is that **if you multiply (or divide) the top and bottom of a fraction by the same thing, the value does not change.** Usually, you want **to simplify** a fraction. To do this, you have to **divide top and bottom by the same thing.** Again, this does not change the value of the fraction.

Now this all works when "letters" are involved. That is,

$$\frac{a \times b}{a \times c} = \frac{a}{a} \frac{b}{c} = \frac{b}{c}$$

and

$$\frac{a^3}{a} = \frac{a}{a} \frac{a^2}{1} = \frac{a^2}{1} = a^2$$

Sometimes though, things are not quite so simple:

$$\frac{x^2 + 2x}{2x^3 + x} = \frac{x(x + 2)}{x(2x^2 + 1)} = \frac{x + 2}{2x^2 + 1}.$$

Here the fraction we started with did not "obviously" have a factor on the top that we could cancel with a factor on the bottom. We first had to **factorize** the top and bottom. Here is an even trickier example

$$\frac{x^2 - 1}{x^2 + 3x + 2} = \frac{(x - 1)(x + 1)}{(x + 1)(x + 2)} = \frac{x - 1}{x + 2}.$$

Now a famous mistake:

$$\frac{2 + x^2}{3x} = \frac{x + 2}{3} \quad \text{cancelling an } x \text{ top and bottom!}$$

The x on the bottom "canceled" the x on the top in x^2 but unfortunately that is not right!! The numerator is $x^2 + 2$ so you have to divide **everything** in the numerator by x, not just part of the numerator. Think about this! In fact, a very common way for this mistake to happen is to write

$$2 + x^2/3x \qquad \text{instead of} \qquad (2 + x^2)/(3x).$$

The point is that if you have left out the parentheses, it certainly **looks** like

$$2 + x^2/3x = 2 + x/3.$$

The other two important ideas to understand are multiplication and addition. To multiply two fractions:

$$\frac{a}{b} \times \frac{c}{d} = \frac{ac}{bd}.$$

The rule is that you multiply what is on top and you multiply what is on the bottom.

Adding is stranger

$$\frac{a}{b} + \frac{c}{d} = \frac{ad + bc}{bd}.$$

Here is a way to see why this is true. Ask yourself "how much is 3 quarters plus 2 dimes". The answer is not 5 anything. When you add coins you must **add like to like.** But, we can figure out how much money this is by converting both quarters and dimes into cents. 3 quarters is 75 cents and 2 dimes is 20 cents so the correct answer is 95 cents. Before we could add dimes and quarters we had to convert both of them to cents. The same is true with fractions. What is

$$\frac{1}{3} + \frac{1}{4}?$$

Well, we can't directly add thirds and quarters. We must first convert them into a "common currency".

$$\frac{1}{3} \text{ is the same as } \frac{4}{12}$$

and

$$\frac{1}{4} \text{ is the same as } \frac{3}{12}.$$

Here, we convert both into $\frac{1}{12}$'s. Now we can add them to get $\frac{7}{12}$. So how do we convert $\frac{a}{b}$ and $\frac{c}{d}$ into a "common currency"? We use $\frac{1}{bd}$'s. This is good because $\frac{a}{b}$ equals $\frac{ad}{bd}$ and $\frac{c}{d}$ equals $\frac{bc}{bd}$. So now we can add:

$$\frac{a}{b} + \frac{c}{d} = \frac{ad}{bd} + \frac{bc}{bd} = \frac{ad + bc}{bd}.$$

Also remember that:

$$\frac{1}{(a/b)} = \frac{b}{a}.$$

Here is a more complicated situation:

Example 1.3.1 *Put over a common denominator and simplify:* $\frac{1}{2+x} + \frac{-1}{2-x}$

To do this, we convert the fractions to $\frac{1}{(2+x)(2-x)}$'s.

$$\frac{1}{2+x} + \frac{-1}{2-x} = \frac{1(2-x)}{(2+x)(2-x)} + \frac{(2+x)(-1)}{(2+x)(2-x)}$$
$$= \frac{(2-x) - (2+x)}{(2+x)(2-x)}$$
$$= \frac{-2x}{4 - x^2}$$
$$= \frac{2x}{x^2 - 4}.$$

1.3.1 Hidden Parentheses

Can you find the mistake in the following:

Example 1.3.2 *Solve*

$$\frac{x+2}{2x+3} = 4.$$

$$
\begin{array}{rcll}
\frac{x+2}{2x+3} & = & 4 & \\
\Rightarrow \quad x+2 & = & 2x+3.4 & \text{multiply both sides by } 2x+3 \\
\Rightarrow \quad x+2 & = & 2x+12 & \text{simplify} \\
\Rightarrow \quad -10 & = & x & \text{collect } x\text{'s.}
\end{array}
$$

If you cant see what is wrong, you are in **big trouble!**. There **must** be something wrong because plugging $x = -10$ into the original equation gives

$$\frac{-10+2}{2(-10)+3} = \frac{-8}{-17} \neq 4.$$

Clue: what is the heading of this subsection? If you still don't get it, ask someone **soon**.

Problems

For (1)-(3) put over a common denominator and simplify the following fractions:

(1.3.1) $\frac{x}{y} + \frac{x}{z}$

(1.3.2) $\frac{1}{x} + \frac{1}{y} + \frac{1}{z}$

(1.3.3) $\frac{ab}{a^2+b^2}\left(\frac{a}{b} + \frac{b}{a}\right)$

(1.3.4) Solve $\frac{a+2}{3} = 2a - 4$

1.3.2 Percentages

If someone eats 2 out of a total of 5 apples then they have eaten 40% of the apples. To convert a fraction into a percentage, multiply by 100%. In this example this gives

$$(2/5) \times 100\% = 40\%.$$

(1.3.5) Convert the the following fractions into percentages (*a*) 1/4 (*b*) 2/3 (*c*) x/y.

(1.3.6) Convert the following percentages into fractions and simplify (*a*) 20% (*b*) 85% (*c*) 300/x%.

(1.3.7) (a) What is 35% of \$400 ? (b) What is x% of 25 ? (c) What is 30% of 50% as a percentage ?

(1.3.8) (a) What is 17% of 3 added to 28% of 5 ?
(b) What is x% of 4 plus 7% of y ?
(c) What is x% of y% as a percentage ?

(1.3.9) Initially there were 3 liters of blue paint and 8 liters of crimson paint. A paint job uses 20% of the blue and 80% of the crimson paint.
(a) What percentage of the total combined amount of paint is used during the job ?
(b) What percentage of the total combined amount of paint remains after the job is finished ?

(1.3.10) Express x% of 3 plus y% of 8 as a percentage of 11.

(1.3.11) A manager starts with a salary of \$100,000. After one year he received a 10% pay rise. After another year his pay is cut by 10%. What is his salary after this ?

1.3.3   Exponents

The expression x^3 is called a **power** of x. It means $x \times x \times x$. The number 3 is, called an **exponent.** It tells us how many x's to multiply together. If you look at $x^2 \times x^3$ you see that it is $(x \times x)$ multiplied by $(x \times x \times x)$. The end result is $(x \times x) \times (x \times x \times x) = x^5$. More generally:

$$\boxed{x^m x^n \;=\; x^{m+n}.}$$

This formula, called **the first law of exponents,** expresses the simple idea that if you take m lots of x multiplied together and multiply it by n lots of x multiplied together then the end result is $(m + n)$ lots of x multiplied together. There is one more rule for exponents:

$$x^{-1} = \frac{1}{x} \qquad x^{-n} = \frac{1}{x^n}.$$

This slightly odd looking rule is in fact designed so that the first law of exponents works with **negative** exponents as well as positive ones. For example

$$\frac{1}{x^2}\, x^3 \;=\; x^{-2} \times x^3 \;=\; x^{-2+3} \;=\; x.$$

The point is that x's in the denominator of a fraction cancel with x's on top. So x's in the denominator should *count negative* in terms of exponents.

(1.3.12) Express the following as a power of a

$$(a)\ \ a^3 \times a^4 \times a^6 \quad (b)\ \ (a^4 \times a^{-3})^7 \quad (c)\ \ (1/a^{-2})^{-3}$$

(1.3.13) Calculate (without a calculator!)

$$(a)\ \ 10^5/10^3 \quad (b)\ \ (10^3)^2 \quad (c)\ \ (10^{473} \times 10^{-300} \times 10^{-173})^{5716}$$

1.4   Solving Equations.

If you have an equation with only one unknown x in it, then to solve the equation you do the same thing (Ted 'n Fred rule) to both sides of the equation until you get it into the form

$$x \;=\; \text{something with no } x\text{'s in it.}$$

Example 1.4.1 *Solve* $3x + 5 = 7x + 2$.

$$
\begin{aligned}
3x + 5 &= 7x + 2 \\
\Rightarrow \qquad 5 &= 4x + 2 \qquad &\text{subtract } 3x \text{ from both sides} \\
\Rightarrow \qquad 4x &= 3 \qquad &\text{subtract 2 from both sides} \\
\Rightarrow \qquad x &= \tfrac{3}{4} \qquad &\text{divide both sides by 4.}
\end{aligned}
$$

Example 1.4.2 *Solve the equation below for* x.

$$\frac{2x - 1}{3x + 4} \;=\; 2 - A$$

Here, the letter "A" represents a "constant" while "x" is the "unknown." It is important to realize there is a distinction. We think of a "constant" as a fixed number. Although we may not know its value, it isn't our job to find it. On the other hand, it **IS** our mission to find the "unknown." In this example, we do not know what either x or A actually is, but we are told to solve for x. Of course, our answer will have A in it; that is, our answer will be a formula for x which involves A. Later, if we should manage to discover a value for A, we can plug it into our formula for x and immediately find the value for x. But, the strategy in this example is still to get x onto one side and everything else on the other side.

$$
\begin{aligned}
\frac{2x-1}{3x+4} &= 2 - A \\
(2x-1) &= (3x+4) \times (2-A) && \text{multiply both sides by } (3x+4) \\
2x-1 &= 3x(2-A) + 4(2-A) && \text{multiply out using distributivity} \\
2x - 3x(2-A) &= 1 + 4(2-A) && \text{get } x\text{'s on one side: add 1 and subtract } 3x(2-A) \\
2x - 6x + 3Ax &= 1 + 8 - 4A && \text{multiply out} \\
x(-4+3A) &= 9 - 4A && \text{pull out a factor of } x \text{ and simplify} \\
x &= \frac{9-4A}{-4+3A} && \text{divide by } (-4+3A)
\end{aligned}
$$

Notice that the solution for x does indeed have A's in it. In other words what x is **depends** on A. This is also described by saying that x is a **function** of A. Saying that x is a **function** of A just means that whatever x actually IS depends on what A is. For example, if I am told that $A = 3$ then I discover that $x = -3/5$ from the formula.

The summary so far is:

> Get x on one side. Do one step at a time. Watch out for parentheses.

The next step is to be able to deal with **several** unknowns and equations at the same time.

Example 1.4.3 *Solve the pair of equations*

$$2x + 4y = 26 \qquad 4x - y = 7.$$

The strategy when there is more than one unknown:

> Use one of the equations to eliminate one of the unknowns from the other equations.

It is **vital** to **number** the equations so you can refer to them (and so you do not get confused!).

$$(1) \quad 2x + 4y = 26 \qquad\qquad (2) \quad 4x - y = 7$$

Use equation (2) to express y in terms of x

$$(3) \quad y = 4x - 7.$$

Now use this to **eliminate** y's in equation (1) leaving an equation with only x's

$$(4) \quad 2x + 4(4x - 7) = 26.$$

Now we solve equation (4).

$$
\begin{aligned}
2x + 4(4x-7) &= 26 \\
18x - 28 &= 26 && \text{simplify} \\
18x &= 26 + 28 = 54 && \text{add 28} \\
x &= \frac{54}{18} = 3. && \text{divide by 18}
\end{aligned}
$$

Now we know that $x = 3$. So, we can go back to equation (3) which tells us y in terms of x. Substituting $x = 3$ into (3) gives

$$y = 4(3) - 7 = 12 - 7 = 5.$$

Our solution is

> $x = 3 \quad y = 5.$

It is always a good idea (which often saves points on exams) to check the solution by substituting it into both equations. In (1),

$$2x + 4y = 2(3) + 4(5) = 6 + 20 = 26$$

which checks. Substituting into (2) gives

$$4x - y = 4(3) - 5 = 12 - 5 = 7$$

which also checks.

Example 1.4.4 *A rectangular box has volume* 40 *cm*^{3} *and the base has area* 10 *cm*^{2}. *Also the sum of length, width and height is* 11 *cm. What are the dimensions of the box?*

The first thing to do is decide to **name** the unknowns:

$$L = \text{length of bottom of box}$$
$$W = \text{width of bottom of box}$$
$$H = \text{height of box}$$

all measured in cm. Next we are told three pieces of information from which we produce three equations

(1)   volume $= 40 = L \times W \times H$

(2)   area of bottom $= 10 = L \times W$

(3) $L + W + H = 11.$

Now, here is a neat trick: equation (2) tells us that $L \times W$ is 10. There is an $L \times W$ appearing in equation (1). So we can simplify equation (1) by replacing $L \times W$ in equation (1) by 10. Now we get

(4) $40 = 10 \times H$

so $H = 4$. Now that we know what H is we can eliminate H from all the equations leaving us with only two unknowns. Substitute $H = 4$ into (3) gives:

(3) $L + W + 4 = 11$ $\Rightarrow$ (5) $L + W = 7$

Since H is not in (2) we now have only two equations [namely (2) and (5)] in two unknowns [namely L and W] and this is progress. Now use (5) to express L in terms of W

(6) $L = 7 - W.$

We can now eliminate L from (2) to get

(7) $10 = (7 - W) \times W.$

This is a quadratic equation which we must solve. Getting everything onto one side gives

(7) $W^2 - 7W + 10 = 0.$

This factors as

$$(W - 5)(W - 2) = 0$$

so that the solutions are

$$W = 2 \quad \text{or} \quad W = 5.$$

In the case that $W = 2$ then from (2) we see that $L = 5$ and we already know $H = 4$. The other possibility is $W = 5$ and $L = 2$. We should have **anticipated** this. The problem gives no information which allows us to distinguish between length and width, so our final answer reflects this fact: either $W = 2$ and $L = 5$ or vice-versa.

Example 1.4.5 *Solve the system of simultaneous equations*

(1) $2x + 3y - z = 3$
(2) $x - y + z = 2$
(3) $x + y + 2z = -1$

I decide, for no particular reason, that I will eliminate x first. To do this I will use (2) to express x in terms of y and z

(4) $x = 2 + y - z$ add y and subtract z from (2).

Now I will use this to eliminate x from the other two equations. (If I eliminate x from the equation (2) using this I will end up with the useless piece of information $0 = 0$.)

Equation (1) becomes:

$$2(2 + y - z) + 3y - z \ = 3$$
$$4 + 5y - 3z \ = 3 \qquad \text{simplify by collecting like terms}$$
$$(5) \qquad 5y - 3z \ = -1 \qquad \text{subtract 4}$$

Equation (3) becomes

$$(2 + y - z) + y + 2z \ = -1$$
$$2 + 2y + z \ = -1 \qquad \text{simplify by collecting like terms}$$
$$(6) \qquad 2y + z \ = -3 \qquad \text{subtract 2}$$

Now we have two equations, (5) and (6), which only involve two unknowns, namely y and z. This is progress. We use (6) to express z in terms of y

$$(7) \qquad z = -3 - 2y \qquad \text{subtract 2y from (6)}$$

Now substitute for z into (5) to get:

$$5y - 3(-3 - 2y) \ = -1$$
$$11y + 9 \ = -1 \qquad \text{simplify}$$
$$11y \ = -1 - 9 \qquad \text{subtract 9}$$
$$(8) \qquad y \ = \frac{-10}{11} \qquad \text{divide by 11.}$$

Now we have a value for y. So, we can substitute for y into (7) to get

$$(9) \qquad z = -3 - 2(-10/11)$$
$$= -3 + (20/11)$$
$$= (-33 + 20)/11 \qquad \text{put over a common denominator}$$
$$= -13/11.$$

Now substitute for z and y back into (4)

$$x = 2 + (-10/11) - (-13/11) = 2 + (3/11) = (22 + 3)/11 = 25/11.$$

A few remarks:

(1) Leave everything as fractions. Do not start working with decimals!

(2) Try to avoid fractions as much as possible. When I use an equation to eliminate an unknown, I try to choose one which will not introduce fractions. But that is not always possible.

(3) Be **very careful** with minus signs (I made a sign mistake the first time I typed this!)

Generally speaking in order to find solutions for some equations you need

$$\boxed{\text{the same number of equations as unknowns.}}$$

This is because each unknown is eliminated using one equation. You need enough equations to eliminate all the unknowns. If you have more unknowns than equations, typically you will end up with a solution that tells you some of the unknowns (eg x) in terms of others of the unknowns (eg C.) If you have **more** equations than unknowns there is a great danger that there are **no solutions** to your equations, though there might be.

Problems

(1.4.1) Solve
$$x + y \ = 3$$
$$x + 2y \ = 4$$

(1.4.2) Solve
$$x + y - 2z \ = 0$$
$$3x + y \ = 1$$
$$5x + 3y + 7z \ = 2$$

(1.4.3) Solve
$$2x + y = 3$$
$$6x + 4y = 9$$

(1.4.4) Why doesn't
$$-6x + y = 4$$
$$-12x + 2y = 7$$
have a solution?

(1.4.5) Give two different solutions (there are infinitely many) to $2x + 4y = 6$.

(1.4.6) (a) Solve for x the equation $2x + 4y = 6$.
(b) What does this solution give you when $y = 5$.

(1.4.7) Which of the following is a solution of $2x - 3y = 1$.

$$(a)\ x = 1\ y = 3 \quad (b)\ x = 2\ y = 1 \quad (c)\ x = 5\ y = 3$$

1.5   Substitution.

Given some formula, we often must perform a substitution. For example, if we substitute $x = 3$ into the formula

$$\frac{x^2 + 1}{x + 2}$$

then we get

$$\frac{3^2 + 1}{3 + 2} = 2.$$

When **substituting** 3 for x we must carefully replace **every** occurrence of x by 3. The same idea applies if we substitute something more complicated for x. For example if we substitute $x = y + 1$ we get

$$\frac{(y + 1)^2 + 1}{(y + 1) + 2} = \frac{y^2 + 2y + 2}{y + 3}.$$

It is **very important** to use parentheses carefully here. For example the numerator has an x^2 term. This means that **whatever** x is **it must be squared**. We are replacing x, **wherever we see it**, by $y + 1$. In other words, instead of squaring x we are going to square $y + 1$. We do **not** write $y + 1^2$. This is wrong because we must square the entire quantity that replaces x. The correct answer is $(y + 1)^2$. This is **very important**. The safe thing to do when you must substitute for x is to put parentheses around the thing you are going to plug in for x **before** doing anything else. To be sure you understand, think of doing the substitution $(y + 1)$ for x. In this case, every x is replaced $(y + 1)$ **including the parentheses**.

Sometimes when you make a substitution, the result is simpler than what you started with. For example if we substitute $y = x + 3$ into

$$5(x + 3)^4 + \frac{1}{(x + 3)^2}$$

we get

$$5y^4 + \frac{1}{y^2}.$$

In fact, often the reason for doing a substitution in the first place is to simplify something.

Substitution is a very useful technique for checking if an equation is correct. For example is the equation

$$\frac{x^2 + 2x + 1}{x + 1} = x + 2$$

correct? We can check by plugging in $x = 0$ and we find

$$\frac{0 + 0 + 1}{0 + 1} = 0 + 2$$

which is wrong. This is a very good thing to try when you have done a long piece of algebra to see if you have made a mistake.

Problems

In problems 1-4 simplify as much as possible

(1.5.1)

$$(a)\ \frac{x^3}{x^2}, \qquad (b)\ \frac{x^4 y^2}{xy}, \qquad (c)\ \frac{abc^2}{ac} \qquad (d)\ \frac{a^2}{a^3/a^5} \qquad (e)\ (1/x)^{-2} \qquad (f)\ (t^2 \times t^3)^2/t^{-3}$$

(1.5.2)

$$(a)\ \frac{5x + 10x^2}{3x + 6x^2}, \qquad (b)\ \frac{x^2 + 2x + 1}{x^2 + 4x + 3}, \qquad (c)\ \frac{\sqrt{x^2 + 4x + 4}}{4x + 8}$$

(1.5.3)

$$(a)\ \frac{\sqrt{x^{16}}}{\sqrt{4x^8}}, \qquad (b)\ \sqrt{\frac{x^2 + 6x + 3}{(x + 3)^2}}$$

(1.5.4)

$$\frac{ABC^3 + AB^3C + A^3BC}{A^2 + B^2 + C^2}$$

(1.5.5) Solve $2x - 3y = 5 \qquad -x + 2y = 4$

(1.5.6) Solve $\quad a \times x + b = c \times x + d \qquad$ for x.

(1.5.7) Solve

$$\frac{2}{3 + 2x} = 5$$

(1.5.8) Solve $(2x + 1)^{-1} = (x - 3)^{-1}$

(1.5.9) Solve $(x - 1)(x - 2)(x - 3) = 0$ and explain how you did it. [hint: there is a very easy way]

(1.5.10) Solve $(2x - 1)(4x + 2) = (8x + 3)(x - 4)$

(1.5.11) Find R such that $\pi R^2 = 16\pi$.

(1.5.12) A circular can has height h and the base is a circle with radius R. The volume is $\pi R^2 h$. If the volume must be 16π express R in terms of h.

(1.5.13) Solve $xy = 12 \quad \& \quad 2x^2 + 4xy = 66$

(1.5.14) Express x in terms of s and t if $\quad (s^2 - t^2)x = (s + t)(x + 1) \qquad$ and simplify your answer [hint: factor $s^2 - t^2$].

(1.5.15) Solve

$$2x - y + z = 3 \quad \& \quad 2y - 3z = 1 \quad \& \quad 5z = 10.$$

(1.5.16) A gas at temperature T, pressure P with volume V satisfies the gas equation

$$\boxed{PV = KT}$$

where K is a constant that depends on the gas. Express
(a) volume in terms of pressure and temperature
(b) pressure in terms of volume and temperature.

(1.5.17) Newton's law of gravity says that the gravitational force, F, between two objects distance R apart, one with mass M and the other with mass m is

$$F = \frac{GMm}{R^2}.$$

Here G is the universal constant of gravity. Express the distance in terms of the other quantities.

(1.5.18) (a) Use the quadratic formula to solve the distance formula

$$s = ut + \frac{1}{2}at^2$$

for t in terms of the other quantities. (you will see this formula later)

(b) Substitute $a = 6$, $t = 2$, and $u = 5$ into the above formula and find s.

(c) Now check your answer to (a) is correct by plugging in the values for s, a and u from (b) and seeing if your formula gives $t = 2$.

(1.5.19) Find two numbers so that twice their sum equals their product and one number is twice the other number.

(1.5.20) The perimeter of a rectangle equals one and a half times its area. Express the length of the rectangle in terms of the width.

(1.5.21) Put over a common denominator and simplify

$$\frac{x+1}{2x+1} + \frac{2x-1}{x-1}$$

(1.5.22) Solve [hint: put over a common denominator]

$$-x + \frac{x^2 - 2}{x+1} = 2.$$

(1.5.23) Multiply out and solve for x the equation $(1 + x)(1 - x + x^2 - x^3 + x^4) = 1 + a^5$.

(1.5.24) A certain circle and a square have the same area. Find the length of the diagonal of the square divided by the radius of the circle. [hint: draw a diagram and label unknowns]

(1.5.25) Work out **without** a calculator [hint: one step at a time!]

$$\left(1 + \left(\frac{2}{3} - \frac{3}{4}\right)^{-1}\right)^{-1}.$$

(1.5.26) Work out **without** a calculator

$$\left(\left(\frac{5}{4} - \frac{3}{4}\right) \times \left(\frac{13}{8} - \frac{9}{8}\right)^{-1}\right)^{-1}.$$

(1.5.27) Work out **without** a calculator

$$\left(1 + \frac{-1}{\sqrt{3 \times (8 - 5) - 4 \times (21 - 25)}}\right)^{-1}.$$

(1.5.28) Work out 25% of $(0.9333 - (1/3))$ to 3 decimal places.

(1.5.29) According to Einstein's Theory of Relativity, if an object has mass m when it is stationary (this is called the **rest mass**) and moves at a velocity of v relative to you (an observer) then the **observed mass** M that you measure the object as having is

$$M = \frac{m}{\sqrt{1 - v^2/c^2}}$$

where c is the speed of light. (a) What is the observed mass if the velocity is 99% of the speed of light? Find out what velocity the object must have if the observed mass is (b) twice the rest mass, (c) 100 times the rest mass. Your answer for v will be in terms of c.

(1.5.30) Ohm's law states that if a voltage of V volts is applied across a resistance of R ohms (the unit that **electrical resistance** is measured in) and a current of I amps flows then

$$V = IR$$

(a) Express resistance in terms of voltage and current
(b) Express current in terms of voltage and resistance.

(1.5.31) When an object of mass m_1 moving with velocity v_1 collides with an object of mass m_2 moving with velocity v_2 and sticks to it, then the law of **conservation of momentum** states that

$$m_1 v_1 + m_2 v_2 = (m_1 + m_2)u$$

where u is the final velocity of the combined object. Solve this equation for m_1 in terms of the other quantities. (This equation gives the recoil when you fire a gun.)

(1.5.32) If f is the focal length of a lens and u is the distance of an object from the lens and v is the distance of the image from the lens then

$$\frac{1}{u} + \frac{1}{v} = \frac{1}{f}.$$

(a) Solve this equation for the distance of the image in terms of the other two quantities. Simplify your answer.
(b) If the focal length is $10cm$ and the distance of the object from the lens is $50cm$ what is the distance of the image from the lens? (c) If the focal length is $10cm$ and the distance of the object from the lens is very large, for example $1,000,000cm$ what is the distance of the image from the lens approximately? (this explains the name focal length)

(1.5.33) Solve for x and y the equations

$$2x + 3y = a \qquad x + y = b.$$

(1.5.34) Make the following substitutions for x in the expression

$$\sqrt{x} + \frac{1}{x} + 3x^2.$$

$(a) x = 2 \quad (b)\ x = a \quad (c) x = c^2 \quad (d) x = a + b \quad (e) x = y + y^{-1}.$

(1.5.35) Make the substitution $x = y - 1$ into $x^3 + 3x^2 + 3x + 1$, and simplify your answer as much as possible.

(1.5.36) Make the substitution $x = a + a^{-1}$ into $x + 2x^{-1}$. Put your answer over a common denominator and simplify.

(1.5.37) One of the following equations is wrong, use substitution to find which one:

$$(a)\ \frac{2x + 1}{x - 2} + \frac{x - 2}{x + 3} = \frac{7 + 3x + 3x^2}{x^2 + x - 6} \qquad (b)\frac{x + 3}{2x - 1} + \frac{x - 2}{x + 4} = \frac{14 + 2x + 3x^2}{2x^2 + 7x - 4}$$

$$(c)\frac{x + 2}{x - 1} + \frac{x + 5}{2x + 3} = \frac{1 + 12x + 3x^2}{2x^2 + x - 3}.$$

(1.5.38) Is the following correct ? explain!

$$(1+x)(2+x)(3+x)(4+x) \; = \; 24 + 50\,x + 36\,x^2 + 10\,x^3 + x^4.$$

(1.5.39) Use substitution to determine which of the following are wrong. Show how to obtain those which you think are correct.

(a) $\dfrac{1}{x+y} \; = \; \dfrac{1}{x} + \dfrac{1}{y}$ (b) $\sqrt{x+y} \; = \; \sqrt{x} + \sqrt{y}$ (c) $(a+b)^3 \; = \; a^3 + b^3$

(d) $(x - x^{-1})^3 \; = \; x^3 - 3x + 3x^{-1} - x^{-3}$ (e) $\sqrt{x^4 y^8} \; = \; x^2 y^4$ (f) $\dfrac{1 + x + x^2}{1 + y + y^2} \; = \; 1 + (x/y) + (x/y)^2.$

1.6   Functions and Inverse Functions.

Functions are how we describe mathematically the idea that one quantity depends on another quantity. A function is a **rule** that converts one number called the **input** into another number **called the output**. It is helpful to imagine that a function is a machine like a calculator. You type a number into the machine, push a button, and then it prints out an output. An example of a function is

$$f(x) \; = \; x^2 + 1.$$

This particular function is given by a **formula**, namely $x^2 + 1$. It converts the input x into the output $x^2 + 1$. A word about this notation: $f(x)$ does **not** mean "multiply f by x." Instead, the letter f is really just a name

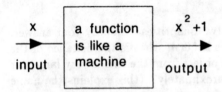

for the function. We pronounce $f(x)$ using the phrase "*eff of x*" to remind ourselves that this stands for a function. Another point: many important functions are not given by a formula. Instead, they are described by some other means such as a table of values or a graph. In fact, all we really need is a series of instructions that enables us to find the **value** of the function $f(x)$ once we are told the value of x.

In common speech we say things like "How long I go on holiday is a function of how much work I have." The phrase "is a function of" means the same thing as "depends on." This usage agrees with the mathematical idea of function. To say that y is a function of x just means that the value of y can be worked out from the value of x by some rule.

Often one quantity depends on several quantities. For example, the temperature, T, of a mass of gas depends on both the pressure, P, and the volume, V, of the gas. The formula is

$$T = \frac{PV}{K}$$

where K is a constant. We express this by saying that T is a **function of two variables**, namely P and V. This is written mathematically as $T(P, V)$ to indicate that the value of T depends on both P and V. Similarly one has functions of $3, 4, \cdots$ variables. Most of the time in this class we will study functions of a single variable.

Here is a straightforward example. An inch is 2.54 centimeters. So x inches equals $2.54x$ centimeters. The function $f(x) = 2.54x$ converts inches into centimeters. Similarly, the function $A(r) \; = \; \pi r^2$ converts an input that is the radius of a circle into an output which is the area of the circle.

WARNING: some students have an irresistible urge to always set things equal to zero. If you are asked for a function that does so and so, the answer is often a formula. But you should not just automatically set it equal to zero!

Here is another example: the graph shown determines a function $f(x)$. The graph gives us a way to convert one number into another. We can simply look up on the graph the value of the function for any particular input. For example, we see from the graph below that $f(2.6) = 18.5$.

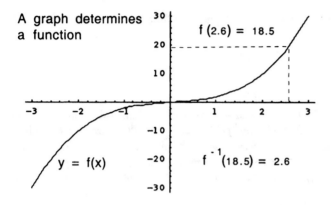

A graph determines a function

$f(2.6) = 18.5$

$y = f(x)$

$f^{-1}(18.5) = 2.6$

This next example uses the imaginary table of taxes below. To work out your taxes, take the amount of money you earn during the year and look up in the tax table how much tax you owe. This is a function. The input

income	10,000 to 14,999	15,000 to 19,999	20,000 and over
tax	600+	1000+	1,500+
	8% of the amount over 10,000	10% of the amount over 15,000	12% of the amount over 20,000

Table 1.1: Tax Table.

x is how much money you earn. The output is $f(x)$ is how much tax you pay. We see that if you earn $20000 then the tax you pay is $1,500. This means $f(20,000) = 1,500$. So the function f converts income, x, into tax, $f(x)$. Now suppose you know someone paid $1,300 tax last year. You can use the tax table to find out that their income was $18,000$. You look through the table until you see a tax amount of $1,300. Then you look to see how much income is needed to give this amount. What you are doing here is using the tax table backwards. Mathematically, we say that the **inverse function** f^{-1} is the function which takes as input the amount of tax paid and produces as output the income. It does the opposite of the function $f(x)$. The notation $f^{-1}(x)$ is read "eff-inverse of x." Be very careful: $f^{-1}(x)$ **does not mean**

$$\frac{1}{f(x)}.$$

This is an unfortunate case of ambiguous notation. Make sure you understand the difference between $f^{-1}(x)$ and $(f(x))^{-1}$.

We can find other examples of inverse functions. Using the graph of the function $f(x)$ above, we see that $f^{-1}(18.5) = 2.6$. This is just the same thing as saying that $f(2.6) = 18.5$

The inverse function $f^{-1}(x)$ is the function $f(x)$ **backwards.**

We know that a function converts an input to an output. The inverse function just does the **reverse**. It converts the output back to the input. So, if a certain function $f(x)$ converts an input of 2 into an output of 7 then the inverse function $f^{-1}(x)$ converts 7 back into 2. Mathematically this is expressed by saying

$$\text{if} \quad f(2) = 7 \quad \text{then} \quad f^{-1}(7) = 2.$$

The "black box" picture also suggests how the inverse function just "undoes" the original function.

The inverse function is the reverse of the original function

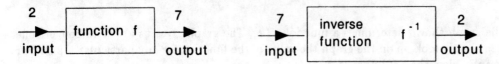

Definition 1.6.1 *The inverse of the function $f(x)$ is the function $f^{-1}(y)$ given as follows. The value of $f^{-1}(y)$ is the number x with the property that $f(x) = y$. Thus*

$$y = f(x) \qquad \Rightarrow \qquad x = f^{-1}(y).$$

Make sure that you see how this definition ties in with the tax example.

Another example: in the back of this text, you will find a table of square roots. This table allows us to evaluate the function $f(x) = \sqrt{x}$. So, using the table we can find that $f(37) = 6.083$. We now ask ourselves "what does it mean to do this backwards?" Well,"squaring" is the reverse of "square-rooting." But we can also use the definition above to figure this out. It says that $f^{-1}(6.083)$ is the number x with the property that $\sqrt{x} = 6.083$. By using the table backwards, we find that 37 has this property. So $f^{-1}(6.083) = 37$. We shall see more examples like this later.

Sometimes it's possible to figure out a function's inverse by using the following analogy. You are at home and must leave to shop for groceries. To do this, you turn off the light, lock the door, and then drive to the store. When you finish shopping, you drive back from the store, unlock the door, and finally turn on the light. In other words, to **undo** a sequence of actions, you do the reverse actions (like unlock instead of lock) **in the reverse order.** Let's apply this idea to the following example.

Example 1.6.2 *Find the inverse of $f(x) = 3x + 2$.*

FIrst solution: To compute $f(x)$, we first MULTIPLY x by 3 and then ADD 2 to this result. To find $f^{-1}(x)$ we do the inverse procedures backwards. Thus to find $f^{-1}(x)$, we first SUBTRACT 2 from x (which undoes adding 2) and then DIVIDE the result by 3 (which undoes multiplying by 3.) This gives us $f^{-1}(x) = \frac{1}{3}(x - 2)$.

Second solution: Put $y = 3x + 2$. We want to find x in terms of y. In other words we want a formula which you plug y into and out pops x. To get this formula we start with $y = 3x + 2$ and solve for x. Thus $3x = y - 2$ so $x = (y - 2)/3$. It is very easy to get confused about this !

Problems

(1.6.1) If $f(x) = 2x + 3$ what is (a) $f^{-1}(5)$ (b) $f^{-1}(y)$ [hint: what does x have to be to give an answer of 5?, of y?]

(1.6.2) If $f(x) = x^3$ what is (a) $f^{-1}(8)$ (b) $f^{-1}(a)$

(1.6.3) If $f(x) = \sqrt{x}$ what is (a) $f^{-1}(9)$ (b) $f^{-1}(t)$ (c) $f^{-1}(-4)$

(1.6.4) The function $f(x)$ converts US dollars into Japanese Yen. What does the inverse function do?

(1.6.5) What is the inverse of the function $f(x) = 8x^3$. [hint: put $y = 8x^3$ and solve for x. What does this have to do with the question?]

(1.6.6) Use the graph for this question to sketch the graph of the inverse function for $0 \leq x \leq 1$. [hint: first make a table of values] What do you notice?

(1.6.7) Use the graph for this question to find
(a) $f^{-1}(0.4)$ (b) $f(0.4)$ (c) $f\left(0.1 + f^{-1}(0.4)\right)$ (d) $f\left(f^{-1}(0.4)\right)$ (e) $f(f^{-1}(f(0.5)))$

(1.6.8) Find a formula for the inverse of $f(x) = 1 + x^{-1}$. Use your formula to find $f^{-1}(2)$ and $f^{-1}(1)$. What goes wrong?.

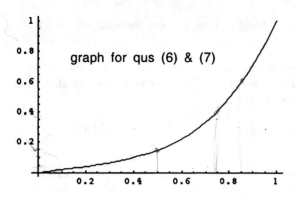

graph for qus (6) & (7)

(1.6.9)

$$\text{IF} \quad f(x) = \sqrt{x} + \frac{1}{x} + 3x^2 \quad \text{Find} \quad (a) \; f(2) \quad (b) \; f(a) \quad (c) \; f(c^2) \quad (d) \; f(a+b) \quad (e) \; f(y+y^{-1}).$$

(1.6.10) Find the inverse function of $f(x) = 4x + 6$.

(1.6.11) Use the tax table above to (a) find $f^{-1}(1860)$ (b) If someone paid \$3000 what was their income?

1.7 Pythagoras' Theorem

There is an important formula for right angled triangles. "The length of the hypotenuse squared equals the sums of the squares of the other two sides"

$$\boxed{a^2 + b^2 = c^2.}$$

The diagram shows a proof. You start with a right angled triangle and then cleverly arrange four of them in

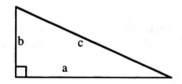

area of triangle = ab/2

area of big square = $(a+b)^2 = a^2 + 2ab + b^2$

= 4x(area of triangle) + (area of small square)

= $4(ab/2) + c^2 = 2ab + c^2$

subtracting 2ab from both sides gives:

$$\boxed{c^2 = a^2 + b^2}$$

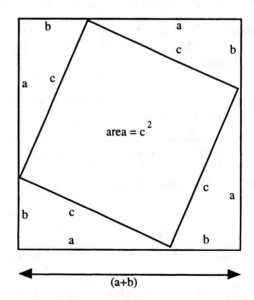

area = c^2

(a+b)

the corners of a large square (which has each side of length $a + b$). The large square is then divided up into 4 copies of the triangle plus a smaller square in the middle which is "tilted". If you now add up the areas of the four triangles plus tilted square and set this equal to the area of the big square, you get an equation, and when you simplify it, out pops Pythagoras' theorem by magic!

You might enjoy this proof, or you might not. But the result is very useful, even if you are not interested in right-angled triangles. The reason this formula is so useful is that it provides a way to **calculate the distance** between two points in the plane. This is something one often needs to do.

Example 1.7.1 *An airplane P is 4 miles north of Santa Barbara and plane Q is 3 miles east of Santa Barbara. How far apart are the two planes?.*

We draw a diagram, showing Santa Barbara as one point marked S, then a second point marked P 4 miles north and a third point marked Q 3 miles west. These points form a right angled triangle. If we set C to equal the

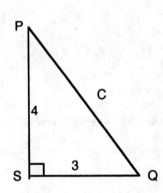

Distance between airplanes

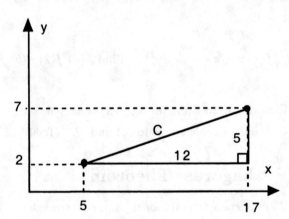

Distance between two points in the plane

distance between the P and Q then

$$C^2 \;=\; 3^2 \;+\; 4^2 \;=\; 9 + 16 \;=\; 25$$

so that $C = 5$. Thus the planes are 5 miles apart.

Example 1.7.2 *Find the distance between the points* $(5, 2)$ *and* $(17, 7)$.

Look at the diagram. We see that the line connecting the two points is the diagonal of a right angled triangle, the other two sides of which are parallel to the x and y-axes. Set C to equal to the distance between the two points, then by Pythagoras:

$$C^2 \;=\; 5^2 \;+\; 12^2 \;=\; 25 + 144 \;=\; 169 \;=\; 13^2$$

so $C = 13$. This example should make clear why the following formula gives the distance between two points (x_1, y_1) and (x_2, y_2) in the plane:

$$\boxed{(\text{distance between } (x_1, y_1) \text{ and } (x_2, y_2) \,) \;=\; \sqrt{(x_2 - x_1)^2 \;+\; (y_2 - y_1)^2}}$$

You will find it useful at exam time to remember

$$\boxed{3^2 + 4^2 = 5^2 \qquad 5^2 + 12^2 = 13^2.}$$

By using Pythagoras' theorem **twice** we get a formula for the distance between two points (x_1, y_1, z_1) and (x_2, y_2, z_2) in space:

$$\boxed{\sqrt{(x_2 - x_1)^2 \;+\; (y_2 - y_1)^2 \;+\; (z_2 - z_1)^2}}$$

This is probably one of the most frequently used formulae in the world. Just about any time an object is built, this formula gets used. However we will not have much cause to use this formula in this class.

Problems

(1.7.1) Find the length of the unknown sides in the triangles shown.

(1.7.2) Find x in the diagram. [hint: you will need to find other things first]

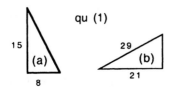

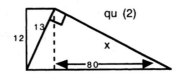

(1.7.3) What is the distance between the points $(-3, 7)$ and $(37, 49)$

(1.7.4) Car A and Car B leave the origin at noon. Car A travels north at 33 mph and car B travels east at 56 mph.
(a) How far apart are they after 1 hour?
(b) How far apart are they after t hours?
(c) When are they 195 miles apart?

(1.7.5) The diagram shows a rectangular box of dimensions 12 by 16 by 21. Find the length of the diagonal, shown dotted, connecting a pair of opposite corners. [hint: first find the lengths of the heavy lines. You will use Pythagoras twice.]

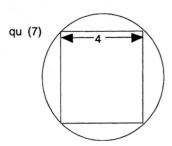

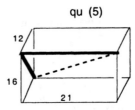

(1.7.6) Generalize your solution to the previous problem as follows. If a rectangular box has dimensions x by y by z show that the length of the diagonal connecting a pair of opposite corners is $\sqrt{x^2 + y^2 + z^2}$. Draw a diagram and briefly explain what you are doing.

(1.7.7) The circle shown contains a square of side length 4. What is the area of the circle ?

(1.7.8) Two cars leave the origin at noon. One goes north at 60 mph and the other goes east at 40 mph. Express the distance between in terms of the number of hours they have been driving.

1.8   My Biggest Mistake was....

The single mistake which will cause students in this class to lose the most points on the final is **untidiness**. Is it really worth repeating this class just because you can't be bothered to produce neat work? Untidiness causes errors. Would you feel safe if the pilot of the plane you are in has junk scattered all round the cockpit? Would you like to be operated on by a surgeon who can't find the scalpel because it is with all that other stuff somewhere under the operating table? The best way to get the grader on your side is to be tidy. Tidiness is next to mathliness. Here are some of the most common mistakes. If you follow the golden rules you will avoid many of them.

Multiplying out. The expression $2 \times (3x + 5)$ gives $6x + 10$. The most common mistake caused by laziness during a longer calculation is to write

$$2 \times 3x + 5$$

by just leaving out the parentheses. Then the next line as often as not incorrectly becomes $6x + 5$. Another common mistake is with

$$(2x - 1)(3x - 2).$$

To multiply this out you MUST go one step at a time and use parentheses. Thus

$$(2x - 1)(3x - 2) = 2x(3x - 2) - 1(3x - 2)$$
$$= 6x^2 - 4x - 3x + 2$$
$$= 6x^2 - 7x + 2.$$

Notice that it would be very easy to mess up by writing $-1 \cdot 3x - 2$ instead of $-1(3x - 2)$ and then writing $-3x - 2$.

Another common mistake is to write down an equation and then start simplifying *on the same line*. For example you *should* write:

$$L \times w = 5000$$
$$\Rightarrow \quad \frac{L \times w}{L} = \frac{5000}{L} \quad \text{dividing both sides by L}$$
$$\Rightarrow \quad w = 5000/L.$$

But, alas, many people write a single line

$$\frac{\not{L} \times w}{\not{L}} = \frac{5000}{L}$$

This is a **great** way to make mistakes because when you look back at it you cant see what you did!

Remember:

$$\frac{1}{x + y} \neq \frac{1}{x} + \frac{1}{y}$$

If you are tempted to write this just think, does

$$\frac{1}{1 + 1} = \frac{1}{1} + \frac{1}{1} \qquad ????$$

Instead of carefully writing $1/(x + y)$ you leave out the parentheses and put $1/x + y$. Then in the next line you think this means $(1/x) + y$. So, you do something like subtract y to leave x. This in turn leaves you with a completely wrong answer. The ONLY way to avoid this kind of error is to use PARENTHESES. Another important point:

$$(a + b)^2 \neq a^2 + b^2 \qquad \textit{you MUST multiply out parentheses using distributivity}$$

A related error is to leave out the parentheses so that although you mean $(a + b)^2$ you write $a + b^2$. I wish I had a dollar for every student in this class who will do this on the final. You should also realize that

$$\sqrt{x + y} \neq \sqrt{x} + \sqrt{y}.$$

You are STUCK with something like $\sqrt{x + y}$. There is nothing you can do to simplify it. Do not even try! If you remember logarithms, you might note that

$$log(x + y) \neq log(x) + log(y).$$

If you're not careful, you could fall into this trap by confusing this with the correct formula $log(x \times y) = log(x) + log(y)$.

1.9 A Note on Homework and Exam Solutions.

Your failure or success in this course depends to a very great extent on the **quality** of your work. There is a bonus for high quality work on exams. It is very, very important for you to write clear, neat, logical, and readable answers which make sense in ENGLISH. Here is a good test which you can use to judge the quality of your written solutions. After you have completed writing, read you answer out loud (ignore any funny looks from friends) without inserting any words or phrases; that is, just read exactly what is present on your page. Now ask yourself, "would a person who listened to my reading understand the solution?" Your answer to this question must be an unqualified "Yes!" If not, you should ask more questions to help you improve the work. Did you feel you needed to add words to make the meaning clear? Perhaps the steps were not in a logical order so that the chain of reasoning seemed disconnected. Was your handwriting messy? Here is a problem which appeared in a previous class. Don't worry if you do not understand the problem.

Problem: The volume of a sphere of radius r is $\frac{4}{3}\pi r^3$ and its surface area is $4\pi r^2$. If the volume is increasing at a rate of $8\pi \ cm^3/min$, how fast is the surface area increasing when the volume is $36 \ cm^3$? Give units.

Here are two solutions that were handed in by students. An example of a GOOD solution is the following:

Rate of Ch. of surface area: $\frac{dA}{dt}$

Volume $= V = \frac{4}{3}\pi R^3$ where $R = R(t)$ is radius.

Area $= A = 4\pi R^2$

$\frac{dV}{dt} = 8\pi \ cm^3/min.$

Differentiating A:

$$\frac{dA}{dt} = \frac{d}{dt}(4\pi R^2) = 4\pi \, 2R \, \frac{dR}{dT} \qquad (*)$$

To find $\frac{dA}{dt}$, differentiate V:

$$\frac{dV}{dt} = \frac{d}{dt}\left(\frac{4}{3}\pi R^3\right) = \frac{4}{3}\pi \, 3R^2 \frac{dR}{dt}$$

$$\frac{dV}{dt} = 8\pi \Rightarrow \frac{dR}{dt} = \frac{2}{R^2}$$

Substituting into $(*)$:

$$\frac{dA}{dt} = \frac{4\pi 4R}{R^2} = \frac{16\pi}{R}$$

When $V = 36\pi$, $R = 3$

So

$$A'(3) = \frac{16\pi}{3} \ cm^2 \min.$$

There difference here is quite evident. You should try to read each of these solutions aloud and note the vast improvement in the good answer. Try to be like the good student in your own work.

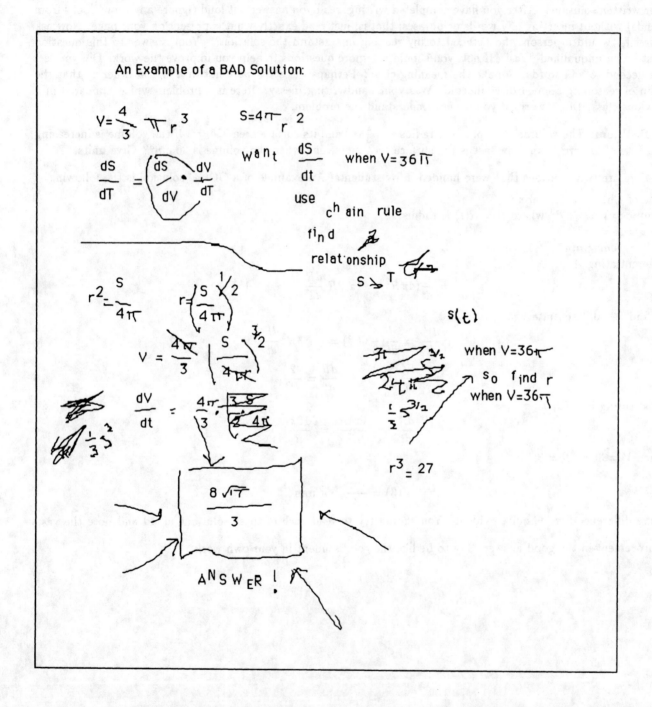

An Example of a BAD Solution:

Chapter 2

Graphs.

We human beings are not computers. We have to be able to understand things. We are not very good at looking at long lists of numbers and understanding what they mean. But we often must understand the important things the numbers tell us. For example, someone who has a heart attack may be connected to a machine that measures electrical signals coming from the heart. The machine measures the number of **volts** (this is an electrical measurement) many times a second. If these numbers were printed out, it would be very hard for a doctor to follow the heart's condition. By themselves, the numbers are hard to interpret. Instead, the machine draws a graph of these numbers. It is much easier for us to look at the graph to see what is happening.

Usually, a hospital patient has a chart which describes the patient's condition to the hospital staff. Among other things, this chart will include a graph of the person's temperature. This picture tells something about the patient's health. As another example, consider the price of a certain stock. Instead of just presenting the

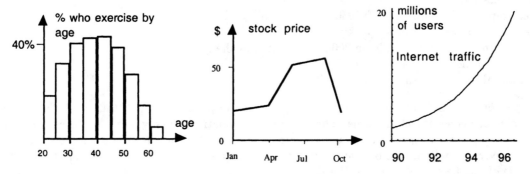

numbers, it is much better to make a graph. It's easy to use the graph to measure stock performance: if the stock-price graph is going upwards, the stock is doing well. The example graph below shows that things started going wrong for our stock in September.

There are different ways to graph data. Some may be better than others depending on what data is available. Here are some of the most common types of graph:
(1) A bar chart.
(2) A graph made up of several lines of varying slopes.
(3) A graph of some smooth curve.

Problem 2.0.1 *Use the three graphs above to answer the following questions.*

(1) Half a group of people is aged 20-25, the other half aged 25-30. What percentage exercise?
(2) Starting in 1992, how long did it take for the number of internet users to double?
(3) What is the ratio of how fast the stock price fell in september to how fast it increased in february?
(4) When did the stock price rise most quickly?
(5) Half a group of people is aged 55-60, the other half aged 60-65. Does this group exercise more or less than the group in question (1)?

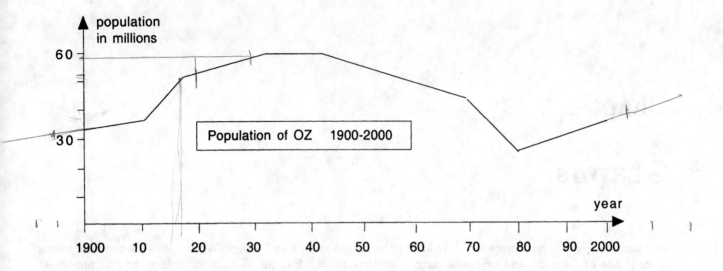

Problem 2.0.2 *A graph contains a lot of information. The graph shows the population of the country of Oz during the years 1900-2000. Using this graph answer the following questions*

(1) How many people lived in Oz in 1940?
(2) In what year did the population of Oz first reach 50 million?
(3) How much did the population increase by between 1920 and 1930?
(4) When was the population of Oz greatest during this time?
(5) When was the population least?
(6) What happened to population during 1970-1980?
(7) When was the population growing most quickly?
(8) What do you guess the population will be in 2005? [hint: extend the graph]
(9) What would you guess the population was in 1895?
(10) What can you say about how many people died during the 70's ?

You are able to figure out all kinds of things from the graph. The **slope** of the graph tells you whether the population is increasing or decreasing at any time. The maximum population is easily seen. You used the graph "backwards" to answer (2). In question (1), you use the graph "forwards". Hidden in this simple example is the idea of function, inverse function, rate of change (meaning derivative) and much more. Trying to find this information in a table of numbers would be much harder.

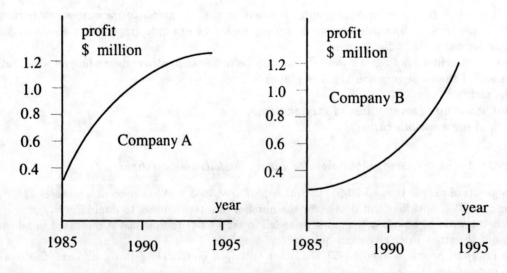

Example 2.0.3 *Here are graphs for two bio-technology companies. The graphs show the annual profits. Notice*

that both companies had the same profits in the years 1985 and 1995. Which one did better in 1990?. Which do you guess will do better in 1996?

Solution: If you just look at the numbers, all you see is that in 1995 both companies made the same profit. But looking at the graphs you quickly see that company B is likely to do better than company A in 1996. Not because it is making more money than company A in 1995. In fact company B has NEVER made more money than company A. So in particular, Company A was making more profit in 1990. On the other hand, we clearly see that if things go on the way the graphs suggest then company B will overtake company A. The important thing with these graphs is how steeply they **slope**. In 1995 the graph of company B is sloped more steeply than that of company A. This means that if this continues then company B will earn more profits than company A. How can we explain this briefly? In 1995 both companies make the same profit. But the profits of company B are **increasing more quickly** than those of company A. The slopes of the graphs tell us how **quickly the profits are increasing.**

Suppose we want to communicate how much rain falls in Seattle throughout the year. A good way to do this is to show the average amount of rain that falls in each month. This comes down to 12 numbers. We might

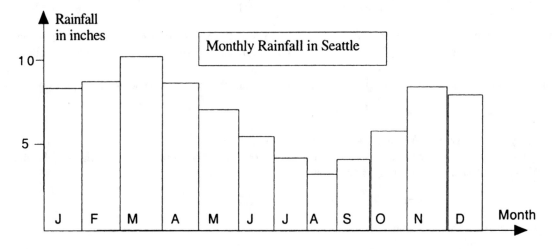

then make a bar chart, so that the height of the box for June represents how many inches of rain falls in June. You can look at this graph and quickly see when it is wettest and when it is driest. There is something else you can do. You can use the graph to figure out how much rain falls in the entire year. Just add up the amounts that fall each month. The amount that falls in June is represented by the height the box for June. If we decide that the width of each box is one unit, then the **area** of each box equals its height. So adding up the heights of all 12 boxes is the same as adding up the **areas** of all 12 boxes. This gives the total area under the graph. So we have discovered

(total rainfall in one year) = (total area under graph).

This is something that is perhaps a little strange. The **area under the graph** tells us something useful. Question: using this, if you want to know the total rainfall between January and June inclusive, what area is this represented by? How can you easily tell, just by looking at the graph, whether more rain falls in the period January to June or July to December?

2.1 Solving equations graphically.

The great thing about this method is that we can actually use it in two ways:

(1) We can use graphs to help us understand algebraic equations.

(2) We can use algebra to help us understand graphs.

Here is an example of the first type. The equations

$$y = 3x - 2 \qquad y = x + 4$$

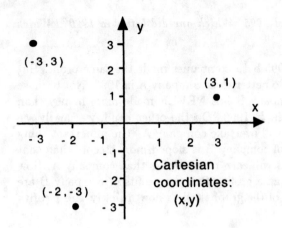

The French mathematican Descartes popularized Cartesian coordinates.

The point (3,1) is 3 units to the right of the origin and 1 unit upwards.

A function f can be described by a picture showing for each x value the corresponding value f(x)

graph as straight lines. If we want to know where the two lines meet then we solve these two equations to get

$$3x - 2 = x + 4 \quad \Rightarrow \quad x = 3.$$

Once we know the x-coordinate of where the two lines meet we can find the y coordinate using either equation: $y = 3 + 4 = 7$. So, the two lines meet at $(x, y) = (3, 7)$. This is an example of using algebra to solve a problem about graphs.

On the other hand, a graph can be used to solve a complicated equation. In order to solve the equation

$$x^3 - 3x + 1 = 0$$

we draw the graph of $y = x^3 - 3x + 1$ We then use the graph to see which value of x gives zero. This is easy to see because it is just where the y coordinate is zero. In other words, it is the point at which the graph cuts across the x-axis. We see that there are actually 3 solutions

$$x \approx -1.9 \qquad x \approx 0.3 \qquad x \approx 1.5.$$

Similarly if we want to solve the equation

$$x^3 - 3x + 1 = 5$$

we can use the same graph. This time we want to know what value of x gives $y = 5$. So to make things simple

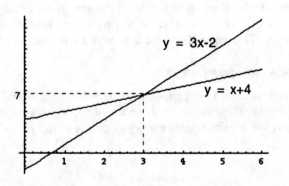

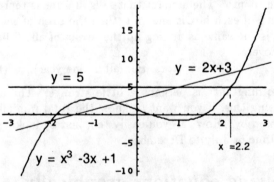

we draw the line $y = 5$ on the graph and see where it hits the graph of $y = x^3 - 3x + 1$. Actually, there is another way we could have done this problem. If we did not already have the graph of $y = x^3 - 3x + 1$ drawn, then in order to solve

$$x^3 - 3x + 1 = 5$$

we might take the 5 to the other side and solve the equivalent equation

$$x^3 - 3x - 4 = 0.$$

We could do this by drawing the graph of $y = x^3 - 3x - 4$ and seeing where it crosses the x-axis.

To solve the equation $x^3 - 3x + 1 = 2x + 3$ we can use the graph of $y = x^3 - 3x + 1$ we already have. This time we also draw the graph of $y = 2x + 3$. We must find an x value such that $x^3 - 3x + 1 = 2x + 3$. This happens where the two graphs meet. Where they meet gives a number x for which $x^3 - 3x + 1$ and $2x + 3$ are equal. From the diagram we see there are 3 solutions.

Example 2.1.1 *Suppose that phone company NDJ offers a monthly fee of $5 then 18 cents per minute. Phone company BU&U offers a monthly fee of $18, your first hour of calls free, and then 10 cents per minute. Which company should you choose?*

Solution: We can draw graphs which show how much we pay each month plotted against how many minutes of calls we make. If we plot the information for both companies on a graph, we will be able to see the answer. The

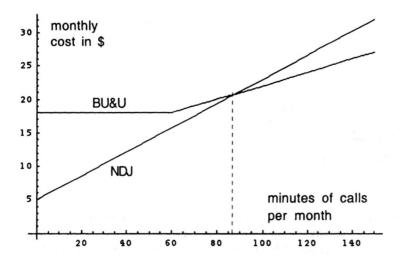

graphs cross at $t = 87$. If you make less than 87 minutes of calls per month then NDJ is less expensive. If you make more calls, BU&U is the best buy.

2.2 Graphing with graph paper.

Much of calculus is developed using graphs. Graphs convey a lot of information once you learn to read them. You will be drawing the graph of $y = x^2$ for homework. It is shown below. You have seen this graph many times. The graph is a picture of the function $f(x) = x^2$. The picture simply shows for each number x the corresponding value of the function x^2. So, if you want to use the graph to estimate $(1.35)^2$, then you look along the x-axis to find 1.35 and then look to the point on the graph above to obtain $y = 1.8225$ (although you will be doing well to read an answer of 1.82 or 1.83 from the graph). You can also "read the graph backwards." If you want to know "what value of x gives $x^2 = 3$", you look for 3 on the y-axis. Then look sideways to find the point $(1.73, 3)$ on the graph. This means that $(1.73)^2 \approx 3$. We talk about the **function** $f(x) = x^2$ and the **graph** of this function $y = x^2$. This might seem a bit confusing. The **name** of the function is $f(x)$. The graph of $y = x^2$ shows x-values and y-values. The graph of $y = f(x)$ is the same as the graph $y = x^2$. The point is that by saying "the graph $y = x^2$," you inform the reader that the axes for your graph are labeled x and y. This is important; sometimes the axes might have different labels, such as t (for time) and m (for mass).

A **linear plus constant** function has the form $f(x) = mx + b$. Sometimes it is just called a **linear function** but this is not really correct, as will be explained in chapter (6.2). Linear-plus-constant functions are found in many situations. The first point about having a graph which is a **straight line** is that it allows you to easily find the formula for the function (which you will not know unless the data came from a formula in the first place). Second, if the graph is a straight line, you can extrapolate it with some confidence to make predictions. That is, suppose your data only cover the x values between 0 and 4, but the graph is a straight line. You can continue that straight line to predict the y value for $x = 6$.

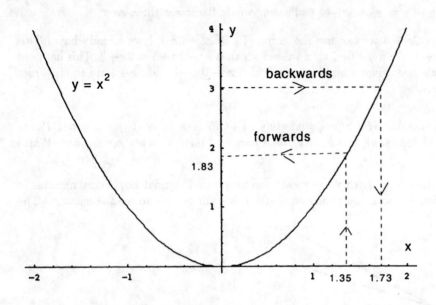

x	0	.25	0.5	0.75	1	1.25	1.5	1.75	2
x^2	0	0.0625	0.25	0.5625	1	1.5625	2.25	3.0625	4

Table 2.1: Table for graphing $y = x^2$ for Qu (1)

Sometimes you are asked to **draw** a graph (this should be very neat and accurate and done on graph paper), and sometimes to **sketch** a graph (this should be very neat, but not as accurate and done on regular paper.) When asked to **sketch** a graph, draw the axes with a ruler and clearly label them. Then neatly sketch the graph, being careful to emphasize any important features. You do not need to plot tables of values. In fact often you do not have precise data from which to sketch the graph.

Here are some general guidelines for **drawing** graphs. You are expected to produce neat high quality work. Many, many points will be taken off for careless or sloppy drawings! The first step is to make a **table of values** to plot. The table should show the x and y coordinate of each point you plot. Sometimes you are given this data; other times you should **calculate it before doing anything else**. Use a **sharp** pencil (preferably a #2 pencil.) Decide on the range of values for the x and y-axes. Draw the axes with a ruler, label them x and y, and write the units if appropriate. Make **tick** marks on the axis next to numerical labels (eg at $x = 1, 2, 3, 4, 5$ put small marks on the x-axis.) Next, go through the table of values and plot small crosses, one for each data point. You might be told to draw the best straight line you can through these points. In this case, manipulate a ruler until it looks to be in the best position then draw the line. On the other hand, if you are told to draw a smooth curve through the points it is easier to do a good job if you use a **flexible curve**. Gently bend it into a shape which goes over each of the data points. Be sure to even out any bends in the flexible curve that should not be there. Now draw the curve. If the curve you drew does not go through all the data points, erase it and try again.

Problems

There is graph paper at the end of the textbook

(2.2.1) Draw the graph $y = x^2$ for $-2 \le x \le 2$. Use the table of values above and the fact that the graph is **symmetric** across the y-axis. Mathematically this means $(-x)^2 = x^2$. What this means when you look at the graph is that the parts of the graph on either side of the y-axis look the same, except that one is the mirror image of the other. This graph will often be used later in the course. So keep it safe.

(2.2.2) The graph for this question shows the position on a line of an object plotted against time.

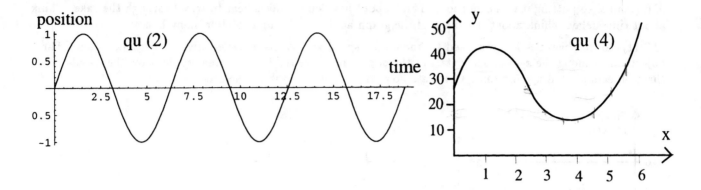

(a) Describe the motion of the object. (b) When is it moving most quickly?
(c) When is it stationary? (d) When is it moving backwards?

(2.2.3) Sketch a plausible graph showing the temperature during the following time period. Indicate on your graph where each piece of information is used. Mark appropriate points on the axes. "I awoke around 3am feeling cold and went to look for another blanket. It had not been cold when I went to bed at midnight, but a chill had developed since then. When I got up at 8am, there was frost on the window panes and grass, but that all disappeared as the day rapidly warmed up. By noon, I was in shirt-sleeves until around 2pm a sudden gust of cold air off the lake sent me looking for my sweater. That did not last long and the temperature gradually increased until sunset. It is amazing how fast the temperature drops when the sun goes down. I expect tonight will be similar to last night."

(2.2.4) Make a table of values from the graph shown, spaced at x-intervals of 0.5.

(2.2.5) Newton's law of cooling says "hot things cool faster than warm things." To be more precise, it says that how quickly a thing cools is proportional to how much hotter it is than the surroundings. Use this information and sketch, on the same graph, temperature versus time for a hot cup of coffee and of a cup of water which is colder than the air, when both are sitting in a room at a temperature of $60°$ F. Briefly explain any relevant features on your graph. [hint: what is it about your graph that shows how quickly the temperature of the coffee is changing?]

(2.2.6) A pumpkin is dropped from the top of Cheadle hall at a height of 30 meters above ground level. It takes 2.5 seconds to hit the ground.
(a) Sketch a graph showing the height of the object during the descent.
(b) Sketch a graph showing the speed of the object during its descent [it speeds up steadily at the rate of 10 ms^{-1} every second.]

(2.2.7) When an object falls the air streaming past it as it falls pushes upwards on it. The faster it falls the greater is this force of air resistance. If the object falls from a great enough height, after a few seconds it is going so fast that the air resistance balances the pull of gravity and the object does not speed up any more. It has reached a limiting speed called the **terminal velocity.** The terminal velocity of a feather is very slow, that for an elephant is much greater. Draw a graph showing the speed of an object as it falls from an airplane high above the earth's surface.

(2.2.8) A batter hits a home run. The ball is in the air for 3 seconds.
(a) Draw a graph showing the course of the baseball through the air. This is a graph with height on the y-axis and distance along the ground on the x-axis.
(b) Draw a graph showing the height of the ball plotted against time.
(c) Draw a graph showing distance traveled along the ground plotted against time.

(2.2.9) Briefly describe how you would use a graphical method to solve the pair of equations $y = x^4 - 3x + 7$ and $y = f(x)$ where $f(x)$ is a function whose graph you already have. Illustrate your answer with sketch graphs.

(2.2.10) A cold cake is placed into a hot oven. Sketch 3 graphs showing temperature versus time, one each for a point near the surface of the cake, a point at the center of the cake, and a point midway between the center and the surface. To help you think about this problem you may find it useful to think about how a block of ice cream melts. Another thing to think about is insulation. Imagine wrapping a hot cup of coffee in a giant bundle of fur and then putting it in a cool room. Think about how long it takes heat to travel through the cake. Think about time delay. Think about the shape of the graph at the start and also after many hours.

(2.2.11) To do this question you need to know that speed is distance traveled divided by time taken. Three objects move along the x-axis. They start at the origin and move right. The three graphs show the speeds of the three objects plotted against time. Calculate how far each object is from the origin at $t = 4$.

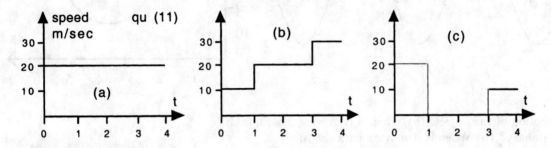

(2.2.12) The graph shows the distance in meters of an object from the origin after t seconds. What was the speed of the object at (a) $t = 0.5$ (b) $t = 1.5$ (c) $t = 3.5$
(d) what is the significance of the fact that the graph slopes downwards between $t = 2$ and $t = 4$? [hint: speed = distance gone divided by time taken]

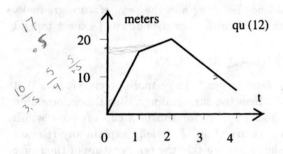

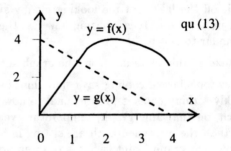

(2.2.13) The figure shows the graphs of the function $f(x)$ as a solid line and the graph of $g(x)$ as a dotted line. Make a table of values spaced 0.5 apart for the function $k(x)$ given by $f(g(x))$ and use this to sketch the graph of $k(x)$ for $0 \le x \le 4$. If you have done this carefully you should notice something about the graph you have just drawn. What?

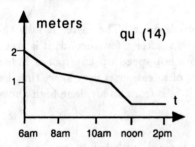

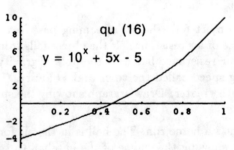

(2.2.14) The graph for this question shows the height of water in a water tank as a function of time. The tank is cylindrical. The base is a circle of radius 2 meters. How much water was used between 6am and noon? On the average how many liters per hour were used during this time [a liter is 1000 cm^3.]. During what time period was water leaving the tank most rapidly?

(2.2.15) A clean lake contains 10 million cubic meters of water. A river enters the lake at the rate of 5 cubic meters per minute. Suppose that water evaporates from the lake at the same rate, so the volume of the lake does not change. One day the river becomes polluted with 100 grams of organic solvents per cubic meter of water.
(a) sketch a graph showing the amount of pollutant (in grams) in the lake during a time span of 100 days.

(b) Suppose now that there is another river which takes 5 cubic meters per minute out of the lake, and that there is no evaporation. Again sketch a graph showing the amount of pollution in the lake. [hint: assume that the concentration of pollution in the lake will never exceed that of the incoming water. This is because the two rivers and lake act like a pipe with a gigantic tank in the middle. The water in the tank starts out pure and gets progressively more polluted, but never more so than the incoming water]

(2.2.16) Use one step of linear interpolation to find an approximate solution of the equation

$$10^x + 5x - 5 = 0$$

using the graph shown on the previous page. Substitute your final answer for x into the equation to see how accurate it is.

(2.2.17) Sketch a graph of distance traveled against time from the following information. "I was stopped at the lights for a minute. When they turned green, I gently put my foot on the accelerator and pulled away. I was going about 30 when this guy in the far lane shoots by at around 80. I put my foot to the floor, turned on the flashing lights and sped after him. Within a minute I was doing 95. I caught up with the guy after a couple of minutes, and tailed him doing 80 for a couple more minutes before he thought to look in his rear-view mirror. He pretended I wasn't there for a minute or so then slammed on his brakes (as did I) and pulled over. Guess what. He had no insurance either."

(2.2.18) The table of data below shows the population (in thousands) of a city. Plot this data on semi-log graph paper, using one unit in the y direction for 1000 people. Draw the best straight line you can with this data.
(a) Find the exponential function that best fits this data.
(b) what was the population in 1989?
(c) what do you think the population will be in the year 2000?

year	1976	1978	1980	1982	1984	1986	1988	1990	1992	1994	1996
population	2.0	2.33	2.65	3.09	3.60	4.13	4.78	5.44	6.24	7.21	8.43

For questions 19 to 24 first complete the given table of values, then plot the graph on regular graph paper using this table of values.

(2.2.19) Plot the graph of $y = x^3$ for $-2 \le x \le 2$. The table of values shown is for positive values of x only. For negative x use that $(-x)^3 = -(x^3)$.

x	0	.25	.5	.75	1	1.25	1.5	1.75	2	2.25	2.5	2.75	3
x^3	0	.02	.12	.42	1		3.38	5.36	8	11.39	15.63	20.80	27

(2.2.20) Graph $y = (x+1)(x-1)(x-2) = x^3 - 2x^2 - x + 2$ for $-2 \le x \le 3$ using the table of values given.

x	-2	-1.5	-1	-0.5	0	0.5	1	1.5	2	2.5	3
$(x+1)(x-1)(x-2)$	-12		0	1.87		1.12		-0.62		2.62	8

(2.2.21) Plot the graph of $y = 2^x$ for $-3 \le x \le 3$.

x	-3.	-2.5	-2.	-1.5	-1.	-0.5	0.	0.5	1.	1.5	2.	2.5	3.
2^x	0.12	0.18	0.25	0.35		0.70	1.	1.41	2.	2.83		5.66	8.00

(2.2.22) Plot the graph of $y = 1/x$ for $-2 \le x \le 2$. Pay careful attention to what happens at $x = 0$. Notice once again that replacing x by $-x$ negates the value of $1/x$ also. What does this say about the graph?

x	0.25	0.5	0.75	1.	1.25	1.5	1.75	2.
$1/x$	4.	2.	1.33	1.		0.67	0.57	

(2.2.23) Find the area above the x-axis and under the graph of $y = x^2$ between $x = 0$ and $x = 1$ by counting small squares on the graph you made for (2.2.1).

(2.2.24) Draw a tangent line on the graph of $y = x^2$ and measure the slope of the tangent line at each of the points $x = -2, -1, 0, 1, 2$. on the graph you made for (2.2.1).

(2.2.25) Draw a secant line for the graph $y = x^2$ going through the points where $x = 0.8$ and $x = 1.2$ and measure it's slope on the graph you made for (2.2.1).

(2.2.26) With a little thought, you can use the table of values for x^3 given above to plot each of the following graphs. On the **same graph**, sketch the graphs $y = x^3$, $y = (x + 1)^3$, $y = (x - 1)^3$, $y = x^3 + 1$, and $y = x^3 - 1$ for $-2 \leq x \leq 2$. Clearly label each graph.

(2.2.27) Plot the graph for $-8 \leq x \leq 8$ of the **inverse** of the function $f(x) = x^3$. Use the table of values for x^3.

(2.2.28) Plot the graph of $ln(t)$ against $log(t)$ for $1 \leq t \leq 10$. This means you will make a table of values for $t = 1, 2, 3 \cdots, 9, 10$. For example $log(3) = .4771$ and $ln(3) = 1.099$, so plot the point $(.4771, 1.099)$. You can find these values, in the tables of logs and natural logs at the back of these notes. What kind of curve do you get?

(2.2.29) The graph shows the number of people infected with disease X. How long did it take for the number of people with this disease to double from the number who were infected in (a) 1980 (b) 1990
(c) How quickly were people becoming infected in 1995?

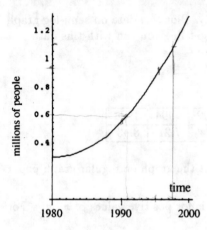

Chapter 3

Problem Solving and Logical Reasoning.

Many students think that mathematics is a collection of complicated formulae to be memorized. Actually the reverse is true. There are a few basic ideas that can be disguised in hundreds of ways. It is crazy to try to memorize methods to solve each individual problem. When we learn English, we memorize some words and then we put them together to say whatever we want. It would be crazy to memorize every sentence we wanted to say. The goal of this chapter is to help you to think in a logical step by step way. This is the most important skill that you can gain from this class. The ideas must make SENSE to YOU otherwise you will be unable to reason about the problems.

3.1 Mathematical Modeling and Problem Solving.

Mathematical modeling is the name given to the process of taking some situation and describing certain features of it using mathematics. For example, the number of people infected by a disease changes with time as people get infected, recover, or die. A mathematical model might attempt to give a formula to describe the progress of the infection. This model might provide a numerical answer to the question "if things go on the same way, how many sick people will there be in 3 years?" It might also answer this: "if we inoculate 50% of the population, how many sick people will there be in 3 years?"

The world is a complex place and a mathematical model must often be a gross simplification of a complicated situation. It takes skill and experience to decide which things to include in the model and which to leave out. However, even the best models may fail to provide very accurate answers. For example, mathematical models exist for predicting the weather. But, if we attempt to use such a model to forecast weather one month from today in a place like San Francisco (where weather frequently changes), then the answer we get will probably be wrong.

In this course you will be asked to model some very simple situations. In these cases there usually is one right answer, although there may be several ways to obtain it. The goal is not that you memorize how to solve each kind of problem that we do; instead, you should learn the problem solving skills that will help you to meet problems you have not seen before. It is very important that the steps used to solve each problem MAKE SENSE to you. You should not execute steps just because your friend (or the book) does them. If you have no idea why you are carrying out particular steps, then you cannot possibly understand the solution. The most essential thing to develop is the ability to create correct steps on your own.

Algebra is the only mathematical skill required for this section. The difficulty is in using the algebra to describe things. Once we have the right equations written down, the problem will usually be fairly easy to solve. However, if you just guess equations, you are most likely to be wrong. It is then completely pointless to spend time solving the wrong equations. This is not the way to get partial credit! Expend most of your effort in finding the right formulae.

Problem Solving Strategies.

Here are some things which will help you solve unfamiliar problems. See also sections 16.4 and 16.11.

(1) Read the question several times to make sure you really understand what is being asked. Make sure you take note of **all** the information you are given.

(2) Write down in **plain English** what you are trying to find; what your **goal** is. Sometimes students, in their haste, try to find the wrong thing.

(3) Give symbolic names (like w or x) to any quantities (unknowns) in the problem. Write down very clearly what each letter represents, even if it seems obvious to you that w is the **width**. In such a case you should draw a diagram, showing what the width is, so you do not confuse it with the length.

(4) Draw **neat** diagrams clearly showing as much information as possible.

(5) Write in **complete sentences**, not short-hand or key phrases, thoughts about the problem, in a "**thinks box**".

(6) Avoid pronouns: **it, this, that** try to be specific and clear.

(7) Try explaining to a friend in **very simple and precise** terms what you understand and where you are stuck.

(8) Never say stuff like "this is the y for x" or "you plug this in here and move that there."

(9) Try to imagine yourself **in the problem.** For example, if a problem involves two cars, imagine yourself in one of the cars.

(10) If you can not solve the given problem, try to make a simpler related problem that you can solve. If you can solve the simpler problem, perhaps you will get ideas for the original problem. I often do this when a problem has lots of unknowns in it. For example if a car is moving with speed v I might first work the problem with speed 10 and see if I can do that. If I can, then probably changing the 10 to v in a few places will solve the original.

(11) Try to use your **common sense** as much as possible. Everyone is a lot smarter than they think, if they use all their mental abilities. Common sense is far more useful than a formula.

(12) Make a **plan.** Examples of plans are given for the next 8 problems.

We will illustrate problem solving by giving some examples together with an **outline** or **plan** of the solution method. This outline will be at a higher level than "set A equals L times W and now plug in $W = 3$ and $A = 6$ and solve for L." These instructions do indeed tell us how to get to the answer $L = 2$. But, they do not help us understand the bigger picture. The following examples are intended to be **models** of how to think about problems. This is **high level** thinking. It is hard. You will find it easier at first to do this with the help of a friend or two.

Problem 3.1.1 *There are three consecutive numbers. The sum of these numbers is 300. Find the numbers.*

The first step is to give a name to the first number as it is the unknown in this problem. Once we have named the first number, the other two numbers can be expressed in terms of this single unknown. The condition that these three numbers add up to 300 can now be expressed as an equation which involves just this unknown. Solve the equation. Use the solution to write down the three numbers. Check the answer is correct.

Problem 3.1.2 *Liquid A has 10 grams of element X per liter. Liquid B has 15 milligrams of element Y per* cm^3. *It is desired to make 100 grams of molecule Z. To make 18 grams of Z requires 2 grams of X and 16 grams of Y. How much of the two liquids must be used?*

Convert everything into the same units. Figure out how many grams of X and Y are needed to make 100 grams of Z. Figure out how much each of the two liquids is needed to get these amounts of X and Y. Check you answer.

Problem 3.1.3 *A rectangle is twice as wide as it is long. If the area is 450 square meters, what are the dimensions of the rectangle?*

The length and width are the unknowns. We obtain one equation which says the width is twice the length. We get another equation using that the area is the product of length and width. This gives two equations involving two unknowns. Solve them. Check the answer is correct.

Problem 3.1.4 *Express the area of a square in terms of the perimeter.*

The goal here is to find a formula for the area of a square. We know that the area of a square is the square of the length of one side. So we need to find the length of one side of a square in terms of the length of the perimeter. A square has 4 sides of equal length, so we can find the length of one side from the length of the perimeter. Check the formula you end up with is correct by plugging in some numbers.

Problem 3.1.5 *Car A leaves Los Angeles at noon driving north to San Francisco at 60mph. Car B leaves at 1pm traveling the same route at 70mph. When does Car B catch up with Car A?.*

The unknown here is time. The idea is to work out how far each car travels in a certain (arbitrary) time and then set these two quantities equal to each other. How far each car has traveled by T hours after noon can be expressed in terms of T. This gives an equation with T in it and you can solve this equation to find T. In figuring out how far car B has traveled by the time T hours after noon, be careful not to forget that car B did not move during the first hour after noon. In other words, to figure out how far car B has traveled by time T first find out how many hours car B has been moving and then multiply by the speed. Check the answer by seeing how far each car has traveled in the time you obtain.

Here is another way to solve this problem. Car A has a 1 hour **head start.** Figure out how far ahead of Car B it is at the end of that hour. Now Car B chases after car A. Work out **how much faster** Car B is going than car A. This is the speed of car B **relative** to car A. It is how quickly B catches up with A. Next figure out how long it will take car B to make up the distance between it and Car A.

Problem 3.1.6 *A swimming pool is 2m deep, 5m wide and 60m long. Water is pumped into it at the rate of 100 liters per minute. How long will it take to fill the pool?*

The unknown is a time. A liter is a unit of volume. Find out how many liters are in a cubic meter. Figure out the volume of the pool in cubic meters. Convert this to liters. Calculate how many minutes it takes to pump this number of liters. Check the answer.

Problem 3.1.7 *A box has rectangular top, bottom and sides. The top and bottom are square. The volume must be $2m^3$. Express the total surface area of the box in terms of the height of the box.*

A box has 6 sides, so we need to add up the areas of all 6 sides. To work out the area of a side involves knowing the two dimensions of that side. So we need to know the length, width and height of the box. We can express the surface area in terms of these three quantities. **Name them.** But we need an answer that only involves the height. So we must somehow replace the width and length by something else. We have extra information. The base of the box is square so length and width are equal. This means we can express the area in terms of just two things: height and width. We are also told the volume is $2m^3$. This tells us something about length, width and height. This is another equation. We can replace length in this equation by width. It is now an equation involving only height and width. So we can solve it to find the width in terms of the height. The formula for total surface area can now be modified by replacing all occurrences of the width by this equation. Summary: express the surface area in terms of length, width and height. Get two more equations and use them to eliminate length and width from this formula.

Check you answer by plugging in some numbers.

Problem 3.1.8 *Oil is leaking from an oil tanker at the rate of 5000 liters per hour. Eight liters of oil spread out over 10 square meters of ocean surface. A circular oil slick forms. Express the radius of the oil slick as a function of the time the tanker has been leaking. When will the oil slick have a radius of $1Km$?.*

Decide on units [meters or kilometers.] We need to find a formula for the radius of the oil slick involving time.[You should give the time since the spill started a name!] The oil slick is a circle. The radius and area of a circle are related. We can figure out the area of the oil slick as a function of time [formula with time in it]. Then use the area to find the radius. This gives the formula for radius in terms of time. To find out when the radius is $1Km$ use the formula and solve for time.

3.2   Word Problems

Problems

(3.2.1) A rectangular field is to have an area of 1000 m^2 and is to be surrounded by a fence. The cost of the fence is \$20 per meter of length. Express the total cost of the fence in terms of the **width** of the field.

(3.2.2) I have three numbers. The biggest one is twice the middle one, and the biggest one plus the middle one is four times the smallest one. The smallest one plus the middle one is two less than the biggest one. What are the numbers? Explain.

(3.2.3) (a) Express the surface area (total of all six sides) of a cubical box (all sides have same length) in terms of the volume of the box. (b) Express the volume of this box in terms of the total surface area.

(3.2.4) To do this problem you need to know that $(distance) = (speed) \times (time)$.
(a) A car travels for 3 hours at 50 mph how far does it go?
(b) A car travels 150 miles in 2 hours at constant speed. What was the speed?
(c) A car travels at 40 mph. How long does it take to go 200 miles?
(d) A car travels for t hours at 50 mph how far does it go?
(e) A car travels at v mph. How long does it take to go 200 miles?
(f) A car travels for t hours at v mph how far does it go?
(g) A car travels x miles at v mph how long does it take?
(h) A car travels y miles in t hours, what is its speed?
(i) A car goes at v mph for a hours then goes in the opposite direction at u mph for b hours. How far is the car from where it started at the end?
(j) make up your own question along these lines and answer it

(3.2.5) (a) A liter of paint covers an area of 4 square meters. It takes 3 liters of paint to paint a rectangular wall that is 2 meters high. How wide is the wall ?
(b) A liter of paint covers an area of A square meters. It takes v liters of paint to paint a rectangular wall that is h meters high. How wide is the wall ?
(c) Use your answer from (b) to work out the answer to (a). Do you get the same answer as before?

(3.2.6) A painter is paid \$10 per hour and can paint $8m^2$ with a roller in this time. Renting a spray gun costs \$5 for each hour of use, and the painter can paint $20m^2$ per hour with the spray gun. It takes 1/2 an hour for the painter to clean up after using the roller, but 1 hour using the spray gun. You must pay the painter for her time and you must also pay for renting the spray gun. How much money do you save by using the spray gun on a paint job that involves painting $40m^2$.

(3.2.7) The first 7 **Fibonacci** numbers are $1, 1, 2, 3, 5, 8, 13, \ldots$ each number in the sequence is the sum of the previous two numbers. Thus $13 = 8 + 5$ and $8 = 5 + 3$.
(a) What are the 8'th and 9'th Fibonacci numbers?
(b) What do you get if you subtract the 8'th Fibonacci number from the 9'th Fibonacci number?
(c) Why did you get that answer for (b)?

(3.2.8) I have \$$x$. Veggie-burgers cost \$$v$ each and sodas costs \$$s$ each. If I buy y veggie-burgers how many sodas can I buy?

(3.2.9) Central park is a rectangle with an area of 840 acres. It is 1/2 a mile wide. One acre is 43, 560 square feet. Two people start at the south-west corner at 1:30 pm and start walking round the park in opposite directions. One walks at $2mph$ the other walks at $3mph$. When do they meet again?. A mile is 5280 feet.

(3.2.10) I have milk that contains 1% fat and milk that contains 4% fat. A customer wants a double latté made with 1/2 pint of 2% milk. How much of each type of milk should I use ?

(3.2.11) There are more than 1/4 million species of beetle. Assume the average length of a beetle is $1cm$ and the average walking speed is $10cm$ per second. These beetles walk up a gang-plank that is 5 meters long onto an ark. They do this in pairs, side by side, one male and one female from each species. The pairs of beetles are spaced $1cm$ apart. How many days will it take for 1/4 million species of beetles to embark onto the ark?

(3.2.12) The radius of the earth is 4000 miles. How fast is someone on the equator moving compared to someone at the north pole due to daily rotation of the Earth?

(3.2.13) When a car is driven at 55 mph or slower it goes 50 miles on one gallon of fuel. For every 5 mph faster than 55 mph that the car goes, the number of miles per gallon traveled is reduced by 3.
(a) How many gallons are used to go 200 miles at 80 mph?
(b) How many gallons are used to go x miles at 80 mph?
(c) How many gallons are used to go x miles at v mph (give one answer when $v \le 55$ and another answer for when $v > 55$ mph)

(3.2.14) An office block is 180 meters high and each floor has a height of 3 meters. The base of the office block is a square which is 50 meters on each side. Half way up, the office block becomes narrower, with a horizontal cross-section which is a square that is 30 meters on each side. Each office has floor area a square 5 meters by 5 meters. Make a sketch of the office block. How many offices are there?

(3.2.15) (a) A cube has surface area $96m^2$. What is the volume of the cube?
(b)[hard] Same question, but this time the surface area of the cube is $A \ m^2$.

(3.2.16) (a) A square has perimeter $36m$ what is the area of the square?
(b) Same question, but this time the perimeter of the square is L meters.

(3.2.17) (a) A right-angled triangle has a 45° angle. If the area is 8 cm^2 what is the length of the perimeter?
(b) Same question, but this time the area is $A \ cm^2$

(3.2.18) (a) A rectangle has the same area as a square. The long side of the rectangle is four times the length of the short side. The perimeter of the square has length $196mm$. How much longer is the perimeter of the rectangle than the perimeter of the square?
(b) Same question but this time the perimeter of the square is $L \ mm$ and the long side of the rectangle is x times as long as the short side.

(3.2.19) I have two cans of paint. Can A has 9 parts of blue paint to one part of yellow paint. Can B is 20% blue paint and the rest is yellow paint. How much paint should I use from each can to obtain 1 liter of paint which is half blue and half yellow.

(3.2.20) (a) A highway patrolman traveling at the speed limit is passed by a car going $15mph$ faster than he is. After one minute, the patrolman speeds up to $100mph$. How long after speeding up until the patrolman catches up with the speeding car. The speed limit is $55mph$.
(b) Same question, but this time the patrolman speeds up to a speed of v mph ($v > 70$).

(3.2.21) Car A leaves San Diego at noon driving at 60 mph along a route which is 400 miles long to San Francisco. Car B leaves San Francisco 2 hours later traveling along the same route at 80 mph. How far from the midpoint of the route are they when they meet? Are they closer to San Diego or to San Francisco?

(3.2.22) A fighter plane, F, is told to intercept plane X which is 100 miles in front of F flying directly away from F at $500mph$. Plane F is told it must intercept X in 6 minutes. How fast should F fly?

(3.2.23) (a) A shell fish absorbs 40% of the heavy metals in the water it ingests. The concentration of heavy metals is $.0002mg$ per cubic meter. The shellfish ingests 3 liters of water per hour. How much heavy metal does it absorb in 3 months? [there are 1000 liters in 1 cubic meter]
(b) same question but this time there are $x \ mg$ of heavy metal per cubic meter.

(3.2.24) A chess board has 64 squares. If you put one grain of rice on the first square, two grains on the second square, four grains on the third, eight grains on the fourth, and so on. Approximately how many grains are on the last square. No calculators.

(3.2.25) A glass aquarium has no top, and the bottom and sides are rectangular. The glass costs $3 $/m^2$. Express the total cost in terms of the **length, width** and **height** of the aquarium. Explain.

(3.2.26) A box, with rectangular sides, base and top is to have a volume of 2 cubic feet. It has a square base. Express the surface area of the box in terms of the width of the base. If the material for the base and top costs $10 $/ft^2$ and that for the sides costs $20 $/ft^2$ express the total cost as a function of the width.

(3.2.27) A window has the shape of a semi-circle placed on top of a rectangle as shown (on next page).
(a) Express the area of the window in terms of the width and height of the rectangle.

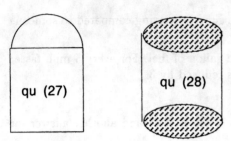

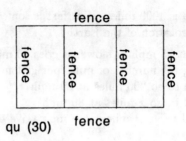

qu (27)

qu (28)

qu (30)

(b) If the area is 2, express the height in terms of the width.

(3.2.28) A cylindrical metal can is to be constructed with a volume of 24 cm^3. It has a base and top. Express the surface area in terms of the radius of the base.

(3.2.29) There are 40 animals in a farmyard. Some are cows and some are chickens. In total there are 104 legs. How many chickens are there?

(3.2.30) A farmer wants to make a rectangular field with a total area of 2000 m^2. It is surrounded by a fence. It is divided into 3 equal areas by fences as shown. Express the total length of all the fence required in terms of the length of one of the fences which divide the field.

(3.2.31) A jet airliner flies at 300 mph for the first half hour and last half hour of a flight. The rest of the time it flies at 600 mph. How long does it take to fly from LA to NY, a distance of 2100 miles?.

(3.2.32) The number of bacteria after t hours is $N(t) = A \cdot 2^t$. Initially there are 100 bacteria. How many are there after 4 hours?.

(3.2.33) (*) A manufacturer wishes to make tee-shirts for the band **Indigo Girls.** They sell for $20 each. She must deliver all the tee-shirts in 20 days time. The manufacturer will first set up some machines. Once all machines are set up she will turn them on. Each machine takes one day to set up. A machine produces 200 shirts in one day. All the machines must be set up before any of them are turned on. Express the amount of money she will receive for the shirts in terms of the number of machines she decides to set up.

(3.2.34) I have some nickels, dimes and quarters. In total I have 18 coins. I have twice as many nickels as dimes. If I multiply (*no.nickels*) × (*no.dimes*) × (*no.quarters*) express the result in terms of the number of nickels.

(3.2.35) As dry air rises, it expands and this causes it to cool about 1 degree celsius for every 100 meter rise, up to a height of about 12 Km. If the ground temperature is 22^oC write a formula for the temperature in terms of height above ground.

(3.2.36) (*) In a 6 acre apple orchard, it is decided to plant 20 or more trees per acre. If 20 trees are planted per acre then each mature tree will yield 600 apples. For each additional tree planted per acre, the number of apples produced by each tree decreases by 12 per year. Express the total number of apples produced per year in terms of the number of trees per acre.

(3.2.37) Which number gives the same result when you subtract 5 as when you divide by 5?

(3.2.38) The light beam from a lighthouse revolves 4 times per minute. How fast does the beam move past someone standing 200 meters from the lighthouse?

(3.2.39) An airplane departs from LA and flies to NY every 30 minutes. The trip takes 4 hours and 5 minutes. An airplane takes off from NY at the same time that one takes of from LA and flies to LA at the same speed. How many planes does it pass going in the opposite direction?.

(3.2.40) The fee charged for parking a car is $1.00 for the first half hour and then $0.50 for each additional half hour up to a maximum of $5.00. Draw a graph showing the parking fee against time.

(3.2.41) Express the surface area of a cube in terms of its volume.

(3.2.42) What is the area of the largest square you can fit inside a circle of radius 1? [hint: draw a diagram, think about the diagonal of the square]

(3.2.43) (*) An airline sells **all the tickets for a certain route at the same price.** If it charges $200 per

ticket it sells 10,000 tickets. For every \$10 the ticket price is reduced, an extra thousand tickets are sold. Thus if the tickets are sold for \$190 each then 11,000 tickets sell. It costs the airline \$100 to fly a person.
(a) Express the total profit in terms of the number of tickets sold.
(b) Express the total profit in terms of the price of one ticket.

qu (44)

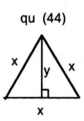

qu (45)

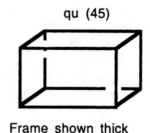

Frame shown thick

qu (47)

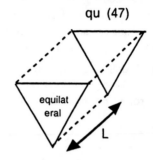

(3.2.44) The diagram shows an equilateral triangle (one with all sides of equal length).
(a) Find y in terms of x.
(b) Find the area in terms of x. [hint: Pythagoras.]

(3.2.45) An aquarium with a square base has no top. There is a metal frame. Glass costs \$3 /m^2 and the frame costs \$2 /m. The volume is to be 20 m^3. Express the total cost in terms of the height. [hint: work out the cost of the glass and frame separately]

(3.2.46) An empty swimming pool is in the shape of a rectangle 20 meters long, 5 meters wide and 2 meters deep. A child starts to fill the pool using a saucepan with a circular base having a diameter of 20 cm and a depth of 10 cm. The child can fill the saucepan and empty it into the pool twice per minute. How many days would it take the child (working 24 hours per day) to fill the pool ?

(3.2.47) (a) Use the answer to 44(b) to find the volume of the **triangular prism** shown. The ends of this prism are equilateral triangles.
(b) (**) Water is entering this prism at the rate of A m^3/hr. The prism is empty at time 0. Express the depth of the water in meters in terms of the length of time water has been entering the trough and the length of the prism. [hint: qu (44)].

(3.2.48) A commuter railway has 800 passengers per day and charges each one \$2. For each 2 cents that the fare is increased, 5 fewer people will go by train. Express the income from the train in terms of the ticket price.

(3.2.49) A poster is to have a total area of 500 cm^2. There is a margin round the edges of 6 cm at the top and 4 cm at the sides and bottom where nothing is printed. Express the printed area in terms of the width of the poster.

qu (50)

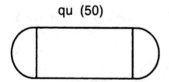

(3.2.50) A sports field is to have the shape shown: a rectangle with semi-circles put on the two ends. It must have a perimeter of 1000 m. Express the area enclosed in terms of the diameter of the semi-circular ends.

(3.2.51) When I ride my bike in still air, the air resistance exerts a force on me proportional to the square of my speed. The amount of energy I use to pedal to overcome this is proportional to the product of the force of air-resistance and the distance I travel. How much more energy do I use to bike for 1 hour at 20 mph instead of 1 hour at 10 mph.

(3.2.52) On the planet Golgafrincham there are two systems for measuring temperature. Using the Celia scale, water freezes at $0°$ C and boils at $50°$ C. Using the Furryhat scale, water freezes at $20°$ F and boils at $120°$ F.

Both systems divide the temperatures in between into equal increments. If x is the temperature in degrees Celia and y is the temperature in degrees Furryhat
(a) Find a formula which converts temperatures measured in Celia into Furryhat.
(b) Find a formula which does the reverse.
(c) What temperature is the same in both Celia and Furryhat ?

(3.2.53) A humming bird needs 10 grams of sugar and 8 grams of protein each day. One honeysuckle flower provides 20 mg (milligrams) of sugar and 10 mg of protein. One nasturtium flower provides 10 mg of sugar and 10 mg of protein. It takes 10 seconds to feed from a nasturtium and 20 seconds per honeysuckle. How many seconds does it take to get exactly the food it needs. (1 gram equals 1000 mg.)

(3.2.54) The radius of the Earth is about $4,000$ miles. One rope is laid on the ground all the way around the equator. A second rope is placed exactly 3 feet above the second rope all the way around the equator. How much longer is the second rope than the first rope ? (a) first **guess** an answer, then (b) work it out. (c) Are you surprised ?

3.3   Logic.

The basic idea behind logical reasoning is IF...THEN... . To illustrate, let's look at a simple game. You might have fun playing it with a friend who is NOT in this class and beating them (until they discover your secret). Put a lot of pennies on the table. The basic rule is **Last Player Wins: If you can't move, then you Lose!**. The players take turns. Each turn a player can remove one or two coins. That is it!

I will explain how to win by logical reasoning. Let us start by thinking about a simple game with not many coins. If there is exactly one or two coins then the first player wins by picking them both up.
Starting with 1 or 2 coins is a win for player number one.

But if there are three coins then the first player will lose because the first player must pick up either one or two coins. Whatever the first player does, this leaves the second player with one or two coins, so the second player picks them up and wins.
Starting with 3 coins is a win for player number two, no matter what player number one does.

The logical method is to continue working with ever more coins, seeing who wins with 4 coins, then with 5 coins and so on. Let's see what happens with a 4 coin game. Susan wants to win, so she would like to arrange that after her first more, Tom is left in a losing position. So Susan picks up one coin leaving 3 behind. Tom is now confronted with a pile of 3 coins. Tom has to do something. Tom reasons "My situation is exactly what I would face if we were just **starting** this game with 3 coins. It is my move. I think of myself as player number one now. There are 3 coins. When there are 3 coins it is a win for player number two (worked out above). This means whatever I do I lose." So Susan wins. The summary is:
Starting with 4 coins is a win for player number one. The winning move is to pick up just one coin.

Do you have any idea yet what to do if there are 17 coins and it is your move? Let's continue. There are 5 coins. Susan wants to win. If she picks up one coin, then there are 4 coins left and it is Tom's go. Tom would be starting out with four coins and that is a win for player number one (which is Tom now). So Susan does not pick up one coin. If she picks up two coins, that leaves Tom with 3 coins. Tom is now player number one with 3 coins and that is a win for player number two (Susan). So Susan picks up 2 coins and wins. Summary:
Starting with 5 coins is a win for player number one, who should pick up two coins.

Susan starts with 6 coins. If she picks up one coin that leaves 5 for Tom. Starting with 5 coins is a win for player number one (provided Tom knows what to do. With luck he is not in this class and just guesses what to do. If Tom picks up 1 coins, he will lose. This is a chance for Susan to win if Tom does not understand the game). If Susan picks up two coins that leaves 4 which is also a win for player number one. Either way Susan will lose (but only if Tom is lucky or knows what to do). Thus
Starting with 6 coins is a win for player number two, no matter what player number one does.

Susan starts with 7 coins. If she picks up just one coin, this leaves Tom to start with 6 coins. He will lose whatever he does. Thus

Starting with 7 coins is a win for player number one who picks up 1.

Susan starts with 8 coins. She picks up two leaving 6 for Tom. Tom loses.
Starting with 8 coins is a win for player number one who picks up 2.

Susan starts with 9 coins. If she picks up one, that leaves Tom to start with 8 and Tom wins (IF he knows what to do). If Susan picks up two coins, that leaves Tom to start with 7 coins and he wins (IF he knows what to do). Thus:
Starting with 9 coins is a win for player number two, no matter what player number one does.

So what we know so far is that starting with $1, 2, 4, 5, 7$ or 8 coins is a win for player number one. Starting with $3, 6$ or 9 coins is a win for player number two. This assumes both players know the best way to play. There is a pattern here. The second player wins with $3, 6, 9, 12, 15, \cdots$ coins otherwise the first player wins. So if Susan wants to win and Tom does not know what is written here, she should put down a large number (say 30) of pennies on the table. She then asks Tom if he wants to start. Susan's goal is to get to a situation Tom starts his move with a number of coins divisible by 3. If she can get to this situation she will win. If Tom removes 1 coin, she takes away 2. If Tom removes 2 coins she takes away 1. She does the opposite of whatever Tom does. This way, after Tom and Susan each make one move, exactly 3 coins are removed. The trick is to get to the situation where Tom starts with $3n$ coins. If Susan starts with 30 coins she is in danger. But Tom need only make one mistake. Susan can try removing 1 or 2 coins at random. If Tom always does the opposite of what Susan does, then she will lose. But if just once Tom makes the same move as Susan, then she wins. This happens because she can use her next move to leave Tom with $3n$ coins.

The above is an example of LOGICAL ANALYSIS. The method that we used over and over again was to say "IF this is the situation THEN this is what happens." We also used a standard technique

> *If you don't know how to solve the given problem, solve an easier one.*

We wanted to immediately solve the problem: how do I win this game? We started with the easier problem: how do I win when the game starts with 2 coins? Then a slightly harder problem: starting with 3 coins. This is one of the most powerful methods for solving problems: make up an easier problem that you can solve. Then gradually try to make the easier problem closer to the original problem. It is often a good technique with **story problems.**

Problems

(3.3.1) Logically analyze the game which is like the above except that each player is now allowed to remove $1, 2$ or 3 coins each time.

Chapter 4

Units.

4.1 Dimensions and Units

The **SI** (Système International) system of units used in science is shown in the table.

mass	length	time	energy	force
Kilogram	meter	second	Joule	Newton
Kg	m	s	J	N

Table 4.1: SI units

Length can be measured in inches, kilometers, or other units. We will concentrate on the metric unit called the meter. You need to be able to convert between different units of length. To do this, you need to know

$$1 \; foot = 12 \; inches$$
$$1 \; yard = 3 \; feet$$
$$1 \; mile = 1760 \; yards$$
$$1 \; centimeter = 10 \; millimeters$$
$$1 \; meter = 100 \; centimeters$$
$$1 \; kilometer = 1000 \; meters$$
$$1 \; mile \approx 8/5 \; kilometers$$
$$1 \; meter \approx 1.1 \; yards$$
$$1 \; inch \approx 2.5 \; centimeters.$$

Area is length × length. The unit for area is the unit for length squared. So if we measure length in feet, then area will be in square feet. For example, a rectangle that is 3 feet by 5 feet has an area of 15 square feet. This is abbreviated to 15 ft^2. Similarly, volume is length × length × length. Hence the unit for volume is the unit for length cubed. In particular, cubic feet = ft^3 and cubic centimeters = cm^3. There are other units of volume in common use. These are the pint, the gallon, and the liter.

$$1 \; ft^3 \approx 7.5 \; gallons$$
$$1 \; gallon = 8 \; pints$$
$$1 \; m^3 = 1000 \; liters$$
$$1 \; liter = 1000 \; cm^3$$
$$1 \; liter \approx 2 \; pints.$$

The next units are for weight and mass. The **weight** of an object is the amount of **force** it exerts when it is placed on a weighing machine. Thus the units of force and of weight are the same. Gravity pulls the object down.

The American unit of weight is the pound. If you buy 12 pounds of potatoes, then this means that the potatoes push down with a force of 12 pounds due to gravity. BUT if you take these same potatoes to the moon where gravity is only 1/6 as strong as on Earth, these potatoes will only weigh 2 pounds. Weight is a measurement of force.

Now **mass** is closely related to weight but different. For most everyday purposes, the distinction does not matter. A **kilogram** is a unit of mass. If you buy a kilogram of potatoes and take then to the moon you still have a kilogram of potatoes; the mass does not change. Roughly speaking, mass is "how much matter" is contained in an object. So in our last example, the mass indicates "how many atoms" are contained in the sack of potatoes. This does not depend on the strength of gravity. On the earth, a kilogram weighs about 2.2 pounds. On the moon a kilogram weighs only 1/6 as much. But, if we stick to the Earth then

$$1 \; Kilogram \; weighs \; 2.2 \; pounds.$$

It is not strictly correct to say that 1 *Kilogram* = 2.2 *pounds* but people say it anyway! (Unless they live on the moon in which case they say something else). Incidentally, in case you wonder what the American unit of mass is, it is the **slug.** It's true! One slug is a mass which weighs 32 pounds.

$$1 \; gram = 1000 \; milligrams$$
$$1 \; kilogram = 1000 \; grams$$
$$1 \; metric \; ton = 1000 \; kilograms$$
$$28 \; grams \; weighs \approx 1 \; ounce.$$

Time

$$1 \; minute = 60 \; seconds$$
$$1 \; hour = 60 \; minutes$$
$$1 \; day = 24 \; hours$$
$$1 \; year = 365 \; days.$$

When you do a calculation involving numbers with units some care is required. First, decide the system of units you are going to use. Then convert any information you have into those units. For example, if a problem uses both minutes and hours you must decide which unit of time to work with. If you choose to use hours and 15 minutes appears in the problem, then you must convert this to 1/4 of an hour. Failure to convert everything to one system of units leads to unbearable misery.

Next, the answer will often have to be given with units. There is big difference between "the sheep ate 3 tons of grass per minute" and "the sheep ate 3 kilograms of grass per hour." You may not know what units to expect in the answer. But you can work out the units for your answer as follows:

Whenever you multiply two quantities, multiply the units. Whenever you divide one quantity by another, divide the units. For example. Suppose we multiply 5 meters by 3 kilograms and divide by 7 seconds. This gives us this result:

$$\frac{5 \; m \times 3 \; kg}{7 \; seconds} = \frac{15}{7} \frac{m \; Kg}{sec}$$

That is, the units are meters kilograms per second.

A common example is speed. If you travel 100 miles in 2 hours, then your average speed is (100/2) *miles/hour* the units are "miles per hour." Another common example is acceleration. If you jump out of an airplane, you will fall slowly at first, but then faster and faster. At the end of one second your speed is 32 *ft/sec*. At the end of 2 seconds you speed is 64 *ft/sec*. For each extra second you are fall, your speed increases by 32 *ft/sec*. Now, acceleration is just how quickly speed increases. The rate your speed increases is 32 *ft/sec* every second, or 32 "feet per second per second" or just 32 *ft/sec*^{2}. Thus, the units of acceleration in this answer are "feet per second squared" or *ft sec*$^{-2}$. The **acceleration due to gravity** is 32 *ft/sec*^{2} (on Earth. On the moon it is about 5 *ft/sec*^{2}.). In the SI system it is 9.8 *ms*$^{-2}$. This is often approximated with 10 *ms*$^{-2}$.

You should NEVER add or subtract quantities that have different units. If you do, then you are almost certainly making a mistake. This can be a useful thing to check during calculations. You would never add 5 meters to 7 kilograms because the answer would not make any sense. It would be 12 of something, but what?.

Huntsville	
Population	2570
Founded	1827
Elevation	4850
Total	9247

sometimes a calculation
is meaningless because
the units do not match

Some people find it hard to remember which expression gives area and which gives circumference for a circle:

$$2\pi R \qquad\qquad \pi R^2.$$

There is a simple clue: the first formula has the units of length (because the radius R is a length) and so gives circumference (which is also a length). The second equation has the units of length squared or area, (because R^2 is length times length). The number π is a "pure" number which means that it does not have any units.

The SI unit of force is the **newton.** This is the force required to accelerate a mass of 1 Kg at a rate of 1 ms^{-2}. Now, the acceleration due to gravity is about 10 ms^{-2}. Therefore the force that Earth's gravity exerts on a mass of 1 Kg is about 10 Newtons. Since a kilogram weighs about 2.2 pounds, 1 Newton is about 0.22 pounds or 3 ounces.

The SI unit of **work** or **energy** (these are the same thing) is the Joule. One Joule is the work required to push against a force of 1 Newton a distance of 1 meter. Thus

$$1\ Joule\ =\ 1\ Newton\ times\ 1\ meter.$$

If you are not familiar with work and energy, here is a brief explanation. You must supply energy to do any of the following:
(a) Heat water.
(b) Lift an object.
(c) Push something along the ground against friction.
(d) Ride a bicycle against air resistance.
(e) Contract a muscle in your arm.

The **Law of conservation of energy** says that energy is neither created or destroyed, but just changes form. Thus energy cannot be manufactured out of nothing. A source of energy is required to do work. Now, energy can be changed into different forms. You change chemical energy in food to mechanical energy when you stand up. You convert mechanical energy into heat when you rub your hands together. A generator converts mechanical energy into electrical. A light bulb converts electrical energy into light and heat.

On a frictionless surface (eg smooth ice), it takes no work to move sideways: once an object is pushed it keeps going (until air resistance and friction slow it down). However, pushing an object uphill takes work. When you stop pushing, the object stops moving. It may be tiring to hold a heavy object in your out-stretched arms **even if they are not moving.** But unless you move the object, you are not doing any work on it. Thus, the definition of energy is the amount of work that is done when, for example, a certain mass is lifted a certain distance. In other words, when you push the mass a certain distance opposed by a certain force.

work = force × distance.

Another unit of energy is the **calorie.** It is the quantity of energy needed to raise 1 gram of water by 1 degree celsius. One calorie equals 4.2 Joules. **A big calorie** is 1000 calories. If you eat a candy bar with 220 calories, those are big calories.

One can sometimes check a formula to see if it is **dimensionally correct.** This means that the units on both sides of the formula agree. An example of a formula which is not dimensionally correct is

$$\pi R^2 = 3v$$

where v is a speed (units m/sec) and R is a radius (units meters). The units on the left hand side are *meters*^{2}. These are not the same as the units on the right hand side which are meters per second. So this formula is incorrect.

> Problems

In the following problems **show what numbers you multiply** to get the final answer. Briefly explain what each of these numbers does.

(4.1.1) A car travels at constant speed for 3 days and covers 1400 miles. What is the speed in centimeters per minute?

(4.1.2) If a car travels 35 miles per gallon of fuel (in other words the car does 35mpg) how many kilometers does it travel per liter of fuel?

(4.1.3) The Earth travels in a circle around the sun once every year. The radius of the circle is 98 million miles. What is the speed of the Earth in miles per hour? In centimeters per day?

(4.1.4) The Earth has a radius of 4000 miles. One cubic meter of rock has an average mass of 5 metric tons. What is the mass of the Earth in grams? (the volume of a sphere of radius R is $4\pi R^3/3$.)

(4.1.5) A triangle has an area of $24m^2$. What is its area in square millimeters?

(4.1.6) (a) In Canada some roads have a speed limit of 95 kmh. What is this limit in miles per hour?
(b) You are driving 75 mph in Nevada. How many miles do you travel in one minute?
You are driving on a two lane road that is 12 miles long. From the start, you have been stuck behind a car that is going 60 mph.
(c) How long will it take for you to reach the end of the road if you do not pass ?
(d) How fast must you travel to reduce this time by one minute if you pass right at the start ?

(4.1.7) The planet Earth is 5 billion years old (a billion is one thousand million). Life first evolved on Earth around 4 billion years ago. The species Homo Sapiens is around 1 million years old. Imagine that the 5 billion years of the Earth's existence is represented by one single year starting on January 1'st and ending December 31'st.
(a) On what day and roughly at what time on that day does Homo Sapiens first appear ?
(b) The earliest civilizations of which we have records are no more than 10,000 years old. What time does this correspond to ?
(c) Can you think of a better way to describe the relationship between the age of the Earth and the age of our species?

(4.1.8) Light travels 186, 282 miles in one second. A **light year** is the distance light travels in one year. It is about 4.3 light years to the nearest star, Proxima Centauri, (other than our sun !).
(a) How many miles away is Proxima Centauri ?
(b) The distance from Los Angeles to New York is about 2, 000 miles. If this distance is represented by the width of a grain of sand (say 1/100 of an inch) how many miles away would Proxima Centauri be ? [refer to the start of this chapter for converting units]
(c) The diameter of our galaxy, the Milky Way, is about 100,000 light years. With this representation what would the diameter of the Milky Way be ?
(d) Imagine we now represent the Milky Way by a single grain of sand. The observable universe is 15 billion light years across. How many feet would this correspond to ? [According to the HitchHiker's Guide to the Galaxy: "Space is big, I mean really big. You may think it's a long way to the chemist but that's just peanuts to space."]

4.2   Area and Volume.

We know the following:

$$\text{(area of a rectangle)} = \text{length} \times \text{width}.$$

$$\textit{(area of a circle of radius } R) = \pi R^2.$$

It's also important to know:

$$\text{(area of a triangle)} = (1/2)\text{base} \times \text{height}.$$

$$\text{(area of a trapezium)} = \text{(average height)} \times \text{width}.$$

The way to remember the formula for the area of a triangle is to think of a triangle as half of a rectangle, so it should have half the area of a rectangle with the same base and height. (Well some triangles ARE half a rectangle but others are not. A right angled triangle IS half a rectangle. The same formula works for all triangles. Thus, it certainly helps us to remember it if we just think of a triangle as half a rectangle.)

It is very useful to know how to work out the area of a triangle, because many more complicated shapes can be broken up into triangles. This means we can work out their area by adding up the areas of these triangles. It will turn out later in integration that we want to know the area of the **trapezium** shown above. This is just a rectangle with a triangle stuck on top. We can work out its area using this fact as follows:

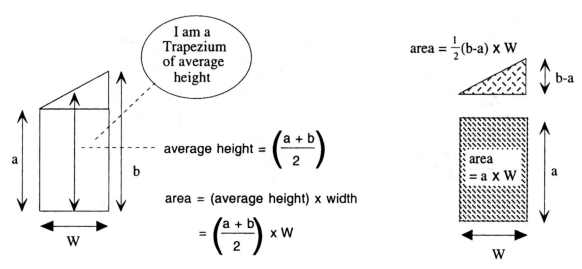

$$\text{(area of trapezium)} = \text{(area of rectangle)} + \text{(area of triangle)}$$

$$= (a \times W) + \frac{1}{2}(b - a) \times W$$

$$= (a \times W) + \frac{1}{2}(b \times W) - \frac{1}{2}(a \times W)$$

$$= \frac{1}{2}(a \times W) + \frac{1}{2}(b \times W)$$

$$= \frac{a}{2} \times W + \frac{b}{2} \times W$$

$$= \left(\frac{a+b}{2}\right) \times W.$$

The quantity $(a + b)/2$ is called the **average height** of the trapezium because the height on one side is a and the height on the other side is b. Hence, this is the average of the height on the two sides.

Here is another way to find the area of a trapezium. We can take two trapezia and turn them and then move so they fit side by side. The result is that the two trapezia fit together as shown to form a rectangle. The height

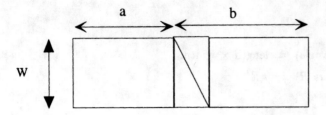

Two trapezia side by side form
a rectangle with area
(a+b) X W so half of this equals
the area of one trapezium

of the rectangle is $a + b$ so the area of the rectangle is $(a + b) \times w$. This gives the area of **two** trapezia. So half this is gives the area of one trapezium.

The volume of a rectangular box is *length* $\times$ *width* $\times$ *height*. You can see why this is true if you imagine a box which measures 5 inches long by 4 inches wide by 3 inches high. It is made up of one-inch cubes. The base of the box is a rectangle with 5 cubes along one side and 4 cubes along the other side. So there are 5×4 cubes in the bottom layer. The height of the box is 3 inches so there are 3 layers. Each layer has 5×4 cubes so the total number of cubes is $5 \times 4 \times 3$. Each cube has a volume of one cubic inch. So the total volume is $5 \times 4 \times 3$ cubic inches.

$$\boxed{\text{Volume} = \text{(area of base)} \times \text{height}}$$

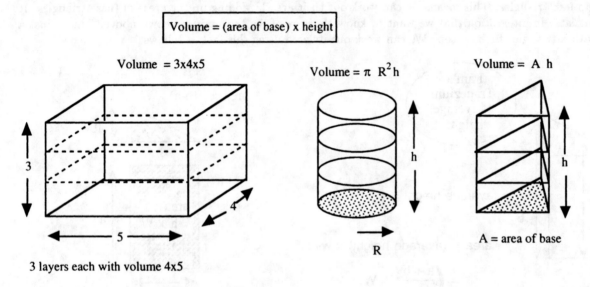

3 layers each with volume 4x5

Another way to think of the formula for the volume of a rectangular box is:

$$\boxed{\text{volume of box} = \text{(area of base)} \times \text{height}.}$$

This formula works for **any shape with a horizontal base and top and vertical sides.** For example a **circular cylinder** (like a soda can) with a circular bottom of radius r cm and height h cm has volume $(\pi r^2) \times h$. A *toblerone* candy bar is a **triangular cylinder**. The volume of this is (area of base triangle)$\times$(height).

If you think of these shapes as made up of horizontal **slices** in the same way that we sliced up the rectangular box you should be able to see why this formula works. The number of slices equals the height. The volume is the number of slices (= height) times the volume of one slice. If a slice is 1 inch thick it has 1 cubic inch of volume for each square inch of area in the base. So the volume of each 1 inch thick slice is the area of the base (in square inches) times 1 inch.

4.3   Growth of Area and Volume.

Suppose we have some shape on a piece of paper, perhaps a circle, a rectangle, a triangle, or even the outline of a moth. Now suppose we draw the same shape again but exactly three times as large. This means the width and height, in fact every length is exactly three times bigger. The larger shape has more area than the smaller shape, but how much more? Most people guess that the bigger shape has three times the area of the smaller shape because it is three times as big. This is a good guess, but very wrong. The large shape had 9 times as much area as the smaller shape. The number 9 is 3 squared; in other words, it is 3×3. Here is the explanation.

The area of a rectangle is width times height. If we make the rectangle three times bigger, then BOTH the width AND the height are three times larger. What happens to the area? Well, since area is width times height, the area is multiplied by three when we make the width three times bigger. Then the area is multiplied by another three when we make the height three times bigger. Thus, the area has been multiplied by three TWICE; that is, the area has been multiplied by three squared or 9.

If we make a rectangle 5 times bigger, the area is multiplied by 5^2. This is a general fact: if we make the rectangle x times larger (which means that both width and height are made x times bigger), then the area is multiplied by x^2.

So much for rectangles. But what about other shapes?. Well, they all work the same way. It does not matter if the shape is circle or a moth. If we enlarge any shape by a factor of x, the area is multiplied by x^2. We can ee this as follows: whatever the shape is, imagine drawing lots of rectangles inside it. Fill it with little rectangles. We won't quite succeed with something like a circle because it is round and can't be filled exactly with rectangles. But, never mind. We can get very close. We can use so many rectangles and make then so small that very little of the shape is not covered by one of the rectangles. Then the total area of all the rectangles will be so close to the area of the shape that no one would care about the difference. Now make everything three times bigger; that is, enlarge the shape and all the tiny rectangles inside it. What is the area of the big new circle? Well, it is almost exactly equal to the total area in all the enlarged rectangles. We know that when we make rectangles three times larger, their area is multiplied by 3^2. So the big rectangles in the big shape have 3^2 times the area of the small rectangles in the small shape. Thus the area of the big shape is 3^2 times the area of the small shape. We have discovered that for any shape, if we increase the linear dimensions by multiplying by a factor of x then the area is multiplied by x^2. In summary:

> Area grows as the square of the linear dimensions.

This all sounds simple enough, but maybe seems a bit dull. Well it is not! Quite soon we will see that these simple ideas have rather astonishing consequences. A mathematical biologist named Haldane pointed out, for example, that a cockroach 6 feet long is impossible on mathematical grounds (and I bet you are glad about that!).

Here is an application. Pretend that you do not know the formula for the area of a circle. From the above discussion, we now know that a circle of radius R must have R^2 times the area of a circle of radius 1. If we decide to call the area of the circle of radius 1 by the name of π then the area of the circle of radius R is πR^2. Similarly for spheres. The surface area of a sphere of radius R is R^2 times the area of a sphere of radius 1. Once you know that the area of a sphere of radius 1 is 4π then you know that the area of a sphere of radius R is $4\pi R^2$.

What we have done for area we can do for volume. If we have a solid object like a box, a football, an automobile, or a baby, we might imagine making it four times bigger. This means we make the object four times as wide, four times as high, and four times as long. In fact every direction is four times as big. The volume of the big object is $4 \times 4 \times 4$ times the volume of the original object. The reason can be seen as follows. If we have a box then the volume is length $\times$ width $\times$ height. It is the product of three dimensions all multiplied together. When we increase each dimension (namely length, width and height) by multiplying by four, the volume gets multiplied by four 3 times. Once because width is multiplied by four, once for length , and once for height. So the new box has $4^3 = 64$ times the volume of the original box. Now, what works for boxes works for any solid shape. Again, we can see this is true by imagining taking any solid object and dividing it up into lots of really tiny boxes.

Thus, we have discovered that if we make a solid object x times larger, then the volume is increased by x^3. This is what is meant by the phrase

> Volume grows as the cube of the linear dimensions.

The "linear dimensions" just means any length we care to measure: width, height, length of the diagonal, circumference, ANY length. Just as long as we make all directions larger by the same factor x then the volume is multiplied by x^3. The term "cube" in "cube of the linear dimensions" refers to cubing x as in x^3. It is no coincidence that the word "cube" refers to a box with sides of the same length, and also to the process multiplying a single number by itself three times.

The following essay shows how simple mathematical considerations of length, area and volume can be used to explain many things in biology. The same reasoning also explains many things in engineering, astronomy, and architecture to name but a few. The basic point is "you can't just take something that works, make it a lot bigger (or smaller) and expect it to still work." The reason is that length, area and volume grow at different rates.

4.4   On being the Right size: an essay by J.B.S. Haldane.

The most obvious differences between different animals are differences of size, but for some reason the zoologists have paid singularly little attention to them. In a large textbook of zoology before me I find no indication that the eagle is larger than the sparrow, or the hippopotamus bigger than the hare, though some grudging admissions are made in the case of the mouse and the whale. But yet it is easy to show that a hare could not be as large as a hippopotamus, or a whale as small as a herring. For every type of animal there is a most convenient size, and a large change in size invariably carries with it a change of form.

Let us take the most obvious of possible cases, and consider a giant man sixty feet high - about the height of Giant Pope and Giant Pagan in the illustrated *Pilgrim's Progress* of my childhood. These monsters were not only ten times as high as Christian, but ten times as wide and ten times as thick, so that their total weight was a thousand times his, or about eighty to ninety tons. Unfortunately, the cross-sections of their bones were only a hundred times those of Christian, so that every square inch of giant bone had to support ten times the weight borne by a square inch of human bone. As the human thigh-bone breaks under about ten times the human weight, Pope and Pagan would have broken their thighs every time they took a step. This was doubtless why they were sitting down in the picture I remember. But it lessens one's respect for Christian and Jack the Giant Killer.

To turn to zoology, suppose that a gazelle, a graceful little creature with long thin legs, is to become large; it will break its bones unless it does one of two things. It may make its legs short and thick, like the rhinoceros, so that every pound of weight has still about the same area of bone to support it. Or it can compress its body and stretch out its legs obliquely to gain stability, like the giraffe. I mention these two beasts because they happen to belong to the same order as the gazelle, and both are quite successful mechanically, being remarkably fast runners.

Gravity, a mere nuisance to Christian, was a terror to Pope, Pagan, and Despair. To the mouse and any smaller animal it presents practically no dangers. You can drop a mouse down a thousand-yard mine shaft; and, on arriving at the bottom, it gets a slight shock and walks away. A rat is killed, a man is broken, a horse splashes. For the resistance present to movement by air is proportional to the surface of the moving object. Divide an animal's length, breadth, and height each by ten; its weight is reduced to a thousandth, but its surface only a hundredth. So the resistance to falling in the case of a small animal is relatively ten times greater than the driving force.

An insect, therefore, is not afraid of gravity; it can fall without danger, and can cling to the ceiling with remarkably little thought. It can go in for elegant and fantastic forms of support like that of the daddy-long-legs. But there is a force which is as formidable to an insect as gravitation to a mammal. This is surface tension. A man coming out of a bath carries with him a film of water of about one fiftieth of an inch in thickness. This weighs roughly a pound. A wet mouse has to carry about its own weight of water. A wet fly has to lift many times its own weight and, as everyone knows, a fly once wetted by water or any other liquid is in a very serious position indeed. An insect going for a drink is in as great danger as a man leaning out over a precipice in search of food. If it once falls into the grip of the surface tension of the water - that is to say, gets wet - it is likely to remain so until it drowns. A few insects, such as water-beetles, contrive to be unwettable, the majority keep well away from their drink by means of a long proboscis.

Of course tall land animals have other difficulties. They have to pump their blood to greater heights than a man and , therefore, require a larger blood pressure and tougher blood-vessels. A great many men die from burst arteries, especially in the brain, and this danger is presumably still greater for an elephant or a giraffe. But animals of all kinds find difficulties in size for the following reason. A typical small animal, say a microscopic worm or rotifer, has a smooth skin through which all the oxygen it requires can soak in, a straight gut with sufficient surface to absorb its food, and a simple kidney. Increase its dimensions tenfold in every direction, and its weight is increased a thousand times, so that if it is to use its muscles as efficiently as its miniature counterpart, it will need a thousand times as much food and oxygen per day and will excrete a thousand times as much of waste products.

Now if its shape is unaltered its surface will be increased only a hundredfold, and ten times as much oxygen must enter per minute through each square millimeter of skin, ten times as much food through each intestine. When a limit is reached to their absorptive powers the surface has to be increased by some special device. For example, a part of the skin may be drawn out into tufts to make gills or pushed in to make lungs, thus increasing the oxygen-absorbing surface in proportion to the animals bulk. A man, for example, has a hundred square yards of lung. Similarly, the gut, instead of being smooth and straight, becomes coiled and develops a velvety surface, and other organs increase in complication. The higher animals are not larger than the lower because they are more complicated. They are more complicated because they are larger. Just the same is true of plants. The simplest plants, such as the green algae growing in stagnant water or on the bark of trees, are merely round cells. The higher plants increase their surface by putting out leaves and roots. Comparative anatomy is largely the story of the struggle to increase surface in proportion to volume.

Some of the methods of increasing the surface are useful up to a point, but not capable of a very wide adaptation. For example, while vertebrates carry the oxygen from the gills or lungs all over the body in the blood, insects take air directly to every part of the their body by tiny blind tubes called trachea which open to the surface at many different points. Now, although by their breathing movements they can renew the air in the outer part of the tracheal system, the oxygen has to penetrate the finer branches by means of diffusion. Gases can diffuse easily through very small distances, not many times larger than the average length traveled by a gas molecule between collisions with other molecules. But when such vast journeys - from the point of view of a molecule - as a quarter of an inch have to be made, the process becomes slow. So the portions of an insect's body more than a quarter of an inch from the air would always be short of oxygen. In consequence, hardly any insects are more than half an inch thick. Land crabs are built on the same general plan as insects, but are much clumsier. Yet like ourselves they carry oxygen around in their blood, and are therefore able to grow far larger than any insects. If the insects had hit on a plan for driving air through their tissues instead of letting it soak in, they might well have become as large as lobsters, though other considerations would have prevented them from becoming as large as man.

Exactly the same difficulties attach to flying. It is an elementary principle of aeronautics that the minimum speed needed to keep an aeroplane of a given shape in the air varies as the square root of its length. If its linear dimensions are increased four times, it must fly twice as fast. Now the power needed for the minimum speed increases more rapidly than the weight of the machine. So the larger aeroplane, which weighs 64 times as much as the smaller, needs 128 times its horsepower to keep it up. Applying the same principles to the birds, we find that the limit to their size is soon reached. An angel whose muscles developed no more power weight for weight than those of an eagle or a pigeon would require a breast projecting for four feet to house the muscles engaged in working its wings, while to economize in weight, its legs would have to be reduced to mere stilts. Actually a large bird such as an eagle or a kite does not keep in the air mainly by moving its wings. It is generally to be seen soaring, that is to say balanced on a rising column of air. And even soaring becomes more and more difficult with increasing size. Were this not the case eagles might be as large as tigers and as formidable to man as hostile aeroplanes.

But it is time we passed to some of the advantages of size. One of the most obvious is that it enables one to keep warm. All warm-blooded animals at rest lose the same amount of heat from a unit area of skin, for which purpose they need a food-supply proportional to their surface and not to their weight. Five thousand mice weigh as much as a man. Their combined surface and food or oxygen consumption are about seventeen times a man's. In fact a mouse eats about one quarter its own weight in food every day, which is mainly used in keeping it warm. For the same reason small animals cannot live in cold countries. In the arctic region there are no reptiles or amphibians, and no small mammals. The smallest mammal in Spitzbergen is the fox. The small birds fly away in winter, while the insects die, though their eggs can survive six months or more of frost. The most successful mammals are bears, seals, and walruses.

Similarly, the eye is a rather inefficient organ until it reaches a large size. The back of the human eye on which an image of the outside world is thrown, and which corresponds to the film of a camera, is composed of a mosaic of 'rods and cones' whose diameter is little more than the length of an average light wave. Each eye has about half a million, and for two objects to be distinguishable their images must fall on separate rods or cones. It is obvious that with fewer but larger rods and cones we should see less distinctly. If they were twice as broad two points would have to be twice as far apart before we could distinguish them at a given distance. But if their size were diminished and their number increased we should see no better. For it is impossible to form a definite image smaller than a wavelength of light. Hence a mouse's eye is not a small-scale model of a human eye. Its rods and cones are not much smaller than ours, and therefore there are far fewer of them. A mouse could not distinguish one human face from another six feet away. In order that they should be of any use at all the eyes of small animals have to be much larger in proportion to their bodies than our own. Large animals on the other hand only require relatively small eyes, and those of the whale and elephant are a little larger than our own.

For rather more recondite reasons the same general principle holds true of the brain. If we compare the brain weights of a set of very similar animals such as the cat, cheetah, leopard, and tiger, we find that as we quadruple the body weight the brain weight is only doubled. The larger animal with proportionately larger bones can economize on brain, eyes, and certain other organs.

Such are a very few of the considerations which show that for every type of animal there is an optimum size. Yet although Galileo demonstrated the contrary more than three hundred years ago, people still believe that if a flea were as large as a man it could jump a thousand feet into the air. As a matter of fact the height to which an animal can jump is more nearly independent of its size than proportionate to it. A flea can jump about two feet, a man about five. To jump a given height, if we neglect the resistance of the air, requires an expenditure of energy proportional to the jumper's weight. But if the jumping muscles form a constant fraction of the animal's body, and if the energy developed per ounce of muscle is the same, then the height that can be jumped is independent of the size, provided it can be developed quickly enough in the small animal. As a matter of fact an insect's muscles, although they can contract more quickly than our own, appear to be less efficient; as otherwise a flea or grasshopper could rise six feet into the air.

And just as there is a best size for every animal, so the same is true for every human institution. In the Greek type of democracy all citizens could listen to a series of orators and vote directly on questions of legislation. Hence their philosophers held that a small city was the largest possible democratic state. The English invention of representative government made a democratic nation possible, and the possibility was first realized in the United States, and later elsewhere. With the development of broadcasting it has once more become possible for every citizen to listen to the political views of representative orators, and the future may perhaps see the return of the national state to the Greek form of democracy. Even the referendum has been made possible only by the institution of daily newspapers.

To the biologist the problem of socialism appears largely a problem of size. The extreme socialists desire to run every nation as a single business concern. I do not suppose that Henry Ford would find much difficulty in running Andorra or Luxemburg on a socialistic basis. He has already more men on his payroll than their population. It is conceivable that a syndicate of Fords, if we could find them, would make Belgium Ltd. or Denmark Inc. pay their way. But while nationalization of certain industries is an obvious possibility in the largest of states, I find it no easier to picture a completely socialized British Empire or United States than an elephant turning somersaults or a hippopotamus jumping a hedge.

4.5   The Inverse Square Law.

Many laws of physics involve an "inverse square law". For example:
gravity, light, electric field, magnetic field, heat.

All these laws take the form that a certain quantity depends on the distance, R, between two objects in the following way: the quantity is inversely proportional to R^2. In other words,

$$\boxed{\text{quantity} \propto \frac{1}{R^2}.}$$

Now there is a very simple **geometric** reason for this:

$$\boxed{\text{The surface area of a sphere of radius } R \text{ is proportional to } R^2.}$$

Here is the explanation. First think about light coming from a light source at a distance R from an object. The closer the object is to the light source the brighter the light on it is. In fact:

$$(\text{intensity of light on object}) = \frac{K}{R^2}.$$

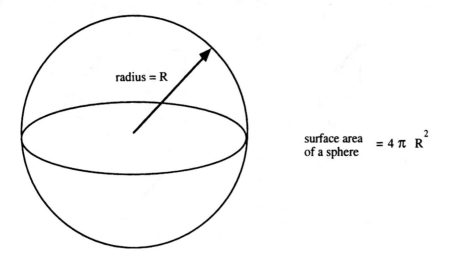

surface area
of a sphere $= 4\,\pi\,R^2$

Here K is a constant which depends on the light source. Imagine the light streaming out from the light source. As it travels further away, it has to spread out to cover a larger and larger area. Imagine the light source is at the center of a sphere. The light spreads out over the sphere. If we take a bigger sphere, then the light has to cover a bigger area. There is a fixed amount of light, so the amount of it that lands on any particular spot of the sphere depends on the size of the sphere. If a large sphere has 25 times the surface area of a small sphere (both of which have the light source at their center), then the light on the small sphere is 25 times stronger than the light on the big sphere. The radius of the big sphere is 5 times the radius of the small sphere **because the surface area of a sphere is proportional to the square of its radius.** So, a sphere 5 times as large has light intensity only 1/25 as much.

The same reasoning applies to heat, radio waves, radiation, gravity (you should imagine gravity as a bit like light, streaming out from the object), etc.

4.6   Hyperspace: the fourth dimension and beyond.

Mathematicians and people who use mathematics are familiar with the idea of "many dimensions." To people not in the know, this sounds exotic. But in fact it is a very simple idea. First: do not expect to find out the answer to the question "what is the fourth dimension?." What is about to be described is a **mathematical** idea, not a physical one. When we speak about length, width and height as being "the three dimensions" what it really means is this. If you want to specify where a thing is you need to specify three numbers. These numbers might be, for example, how far the object is in front of you, how far it is to the left of you, and how far it is above you. These three numbers determine the object's location. So everything that is anywhere (in this room, in this world, or even in the universe) is specified by these three numbers. In this sense we say that our universe has three dimensions. This just means that we need three numbers to locate any point in the universe. In many situations, it requires more than 3 numbers to describe a situation mathematically, and it is convenient to refer to these as **dimensions.** Thus if you are working with 4 unknowns you might talk about your problem being **4 dimensional.** I will describe to you one of the simplest 4-dimensional objects: a hyper-cube. Many books contain pictures of it. It is a four-dimensional analog of a cube. Let me describe a way to imagine it.

Start with a single point. Now move it sideways and the point sweeps out a line. Take the line and move it sideways and the line sweeps out a square. Take the square and move it sideways and the square sweeps out a cube. The diagram shows this. Now you must imagine taking the cube and "sweeping it sideways" into a fourth dimension we do not possess in this universe. However, the fact that something is not real has never stopped anyone from thinking about it (contemplate Darth Vader.) So, the diagram shows two solid cubes connected by dotted lines. The solid cubes are the "top" and "bottom" of the hypercube. The dotted lines show the paths traced out by the corners of the cube as they are swept in the fourth dimension.

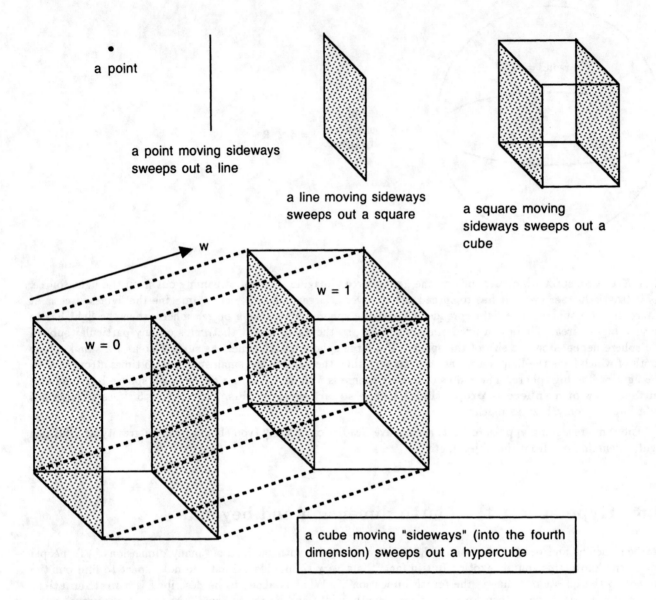

a point

a point moving sideways
sweeps out a line

a line moving sideways
sweeps out a square

a square moving
sideways sweeps out a
cube

w

w = 1

w = 0

a cube moving "sideways" (into the fourth
dimension) sweeps out a hypercube

There is a way to describe this with equations.

a line is $\qquad 0 \leq x \leq 1$

a square is $\qquad 0 \leq x \leq 1 \quad 0 \leq y \leq 1$

a cube is $\qquad 0 \leq x \leq 1 \quad 0 \leq y \leq 1 \quad 0 \leq z \leq 1$

a hypercube is $\quad 0 \leq x \leq 1 \quad 0 \leq y \leq 1 \quad 0 \leq z \leq 1 \quad 0 \leq w \leq 1.$

I am using (x, y, z) for coordinates in 3-space and the w coordinate for the fourth dimension so that the four
coordinates (x, y, z, w) are used for 4-space.

A square of side length x has area x^2, a cube has volume x^3, a hypercube has 4-dimensional hyper-volume x^4.
There is no reason to stop with four dimensions: there is a hyper-hyper-cube in 5 dimensions...

The area inside a circle is πR^2. The volume inside a sphere is $4\pi R^3/3$ the 4-dimensional hyper-volume inside a
4-dimensional hyper-sphere is $\pi^2 R^4/2$ ....

Problems

(4.6.1) If there are two identical pyramids except one is twice as large as the other (ie all its linear dimensions are twice those of the other pyramid), how much more does the larger one weigh?. Assume they are both solid and made of the same stone.

(4.6.2) Assume that the strength of a long object like a tree trunk or a steel beam is proportional to the cross-sectional area. If you take the plans for a sky-scraper and double all the dimensions how safe will the bigger building be? [hint: weight]

(4.6.3) Assume that the rate that a snowball melts is proportional to it's surface area. If one snowball is 3 times larger than another snowball, how much longer will it take to melt. Explain.

(4.6.4) If the volume of a certain pyramid is 5 and another pyramid is made which is x times larger, what is the volume of the new pyramid ?

(4.6.5) How much longer does it take to inflate a balloon to a diameter of 12 inches instead of a diameter of 4 inches. Assume the rate that air enters is constant. Explain your answer.

(4.6.6) What things in nature can you think of that are explained by different rates of growth of length, area and volume?

(4.6.7) A student wrote down the formula $\pi(a^2 + b^2 + c^2)$ for the **volume** of an ellipsoid where a, b, c are the lengths of the major axes of the ellipsoid. Without even knowing what an ellipsoid is, or what these lengths are, explain why this formula **must** be wrong.

(4.6.8) If the units of A is *meters/sec* and the units of B is seconds what are the units of A/B and of AB.

(4.6.9) If a is an acceleration and v is a velocity and s is a distance check whether or not all three terms in the equation

$$s = vt + \frac{1}{2}at^2$$

have the same dimensions.

(4.6.10) A concrete beam has a square cross-section. The weight that a beam can hold is proportional to the area of the cross-section and inversely proportional to the length of the beam. If two beams are to be able to hold the same weight, but one is 9 times longer than the other, how much wider should the cross-section be?

(4.6.11) If you double the length of every edge explain what happens to
(a) the area of a square (b) the volume of a cube (c) the hypervolume of a hypercube.

(4.6.12) Suppose a hose pipe supplies water at a fixed rate. There are two oil drums each in the shape of a circular cylinder. The larger drum is twice the height and the base has twice the diameter compared to the small drum. What is the ratio of the times taken to fill the two drums (ie time for bigger/time for smaller) ?

(4.6.13) The **density** of an object is its mass divided by its volume. Block A is the same shape as block B but twice the **size** (every dimension is twice as big). Block A weighs 3 times as much as block B. Which block is denser ? What is the ratio of the densities (ie more dense/less dense) ?

(4.6.14) A pyramid is upside down so the pointed end is at the bottom and the square base is at the top. It is full of water. The water drains out at a constant rate. A duck is floating on the water surface in the pyramid. As the water drains out the duck moves downwards. Would you expect the duck moves downwards
(a) at a constant rate ? (b) speeds up ? (c) slows down ? Explain your answer.

Chapter 5

Ideas behind Calculus

Calculus involves the ideas of **limit, derivative** and **integral**. A derivative describes how quickly something changes. An integral is a way of adding things up. And a limit is what the end result of doing infinitely many more and more accurate calculations gives. In this brief chapter we lay the ground work for the study of calculus later on.

5.1   Error and Relative Error and Limits.

Often we do not calculate an exact value, but instead an approximate answer. If the exact value is 10 and we find an approximate answer of 9.5, then the **error** in our approximate answer is $10 - 9.5 = 0.5$. How bad is this error? It depends on the situation. Sometimes such an error might not matter (for example if we want to know how long it takes to drive somewhere, the difference between 10 minutes and 9.5 minutes is probably not important). Other times, it does matter: if you think your flight leaves in 10 hours, but really it leaves in 9.5 hours, you are in trouble.

Frequently, the importance of an error of 0.5 depends on the size of the quantity. An error of 0.5 out of 100 is a small error. An error of 0.5 out of 2 is a large error. What is important is the size of the error compared to the number in question. If the exact answer is x and the error is 0.5 then the **relative error** is $0.5/x$. So, an error of 0.5 in 100 is a relative error of .005. But, an error of 0.5 in 2 is an relative error of 0.25. A relative error of 1 means that the error is as big as the exact answer since

$$(relative\ error)\ =\ \frac{error}{exact\ answer}.$$

A relative error bigger than 1 is really bad. Sometimes, we talk about **percentage error**. For example, if the exact value is 2, an error of 0.5 is an error of 25%. Thus:

$$(percentage\ error)\ =\ 100\%\ \times\ \frac{error}{exact\ answer}.$$

It's easy to think of this in terms of money. If you think you have \$19 in your pocket and you really have \$20, then the error is $20 - 19 = 1$. So the relative error is $1/20$ and the percentage error is 5%.

One of the main ideas in calculus is to find approximate answers where the relative (or percentage) error is very small. In fact, the idea of a "limit" is to take make the relative error become smaller and smaller until it vanishes!

Suppose that we have a method to calculate an approximate answer to a certain question and we know that as we repeat this method our approximate answer becomes more accurate though it will never be quite exact. Suppose we do 1 step and get the answer $x_1 = 2.1$. Then we do 2 steps and get the answer $x_2 = 2.01$. After three steps we get the answer $x_3 = 2.001$. Four steps gives us $x_4 = 2.0001$. At this point, we might guess that as we do more and more steps that the approximate answers $x_5, x_6, \cdots$ we get become closer and closer to 2. Then we would guess that the exact answer is 2.

If the approximate answers keep on getting closer and closer to 2, then we say that **the limit of the approximate answers is 2.** This is sometimes written in the following way.

$$\lim x_n = 2.$$

This should be read as "the limit of the numbers x_n is 2." Another way to write it is

$$\lim_{n \to \infty} x_n = 2.$$

This means exactly the same thing. Read it aloud as "the limit as enn goes to infinity of exx-sub-enn equals two." The notation $n \to \infty$ is read "as n goes to infinity." It means "as n gets bigger and bigger." The whole thing means when n is big the numbers x_n are very close to 2. They get closer and closer to 2 as you make n bigger and bigger.

The table above lists some values of a function f at x-values close to 3. It certainly looks like

$$\lim_{x \to 3} f(x) = 2.5.$$

But notice that sometimes $f(x)$ is bigger than 2.5 and sometimes smaller than 2.5.

x	3.1	3.01	3.001	3.0001	3.00001
$f(x)$	2.56	2.49	2.503	2.499	2.5003

Here are some additional examples.

$$\lim_{n \to \infty} \frac{1}{n} = 0.$$

This just says that as you make n bigger and bigger, the number $1/n$ gets smaller and smaller and gets closer and closer to 0. So "in the limit" $1/n$ "becomes" 0. Another example is

$$\lim_{n \to \infty} \frac{n+1}{n} = 1.$$

This is because

$$\frac{n+1}{n} = 1 + \frac{1}{n}.$$

Now, we saw that $1/n \to 0$ (read this as "$1/n$ goes to 0") as $n \to \infty$ (read this as "n goes to infinity," or what amounts to the same thing "as n gets bigger and bigger".)

Sometimes we want to talk about what happens to one quantity when we make x very close to 3. For example

$$\lim_{x \to 3} \frac{x^2 - 9}{x - 3} = 6.$$

One reads this aloud as "the limit as x goes to 3 of $(x^2 - 9)/(x - 3)$ is 6." This means that as x gets closer and closer to 3 the quantity $(x^2 - 9)/(x - 3)$ gets closer and closer to 6. This is a bit tricky. If we try to put $x = 3$, we end up with $0/0$ which is useless. The remarkable thing is though, that if we replace x by a number very close to, but not quite equal to, 3 then the result is very close to 6. For example if you put $x = 3.01$ into the formula it gives

$$\frac{(3.01)^2 - 9}{3.01 - 3} = 6.01.$$

There is a simple way to see what is going on in this particular example by doing a little algebra. The point is that $x^2 - 9$ factors as

$$x^2 - 9 = (x - 3)(x + 3).$$

So we can simplify

$$\frac{x^2 - 9}{x - 3} = \frac{(x - 3)(x + 3)}{x - 3} = x + 3.$$

This means that

$$\lim_{x \to 3} \left(\frac{x^2 - 9}{x - 3} \right) = \lim_{x \to 3} (x + 3).$$

Now the limit of $(x + 3)$ as x goes to 3 is clearly just $3 + 3 = 6$. This can also be expressed by writing $(x + 3) \to 6$ as $x \to 3$.

Many of the ideas in calculus involve using limits. In this course, we will not stress this very much. If you hear the phrase "in the limit as x goes to 0" just imagine that x is a very small number like .0000001. Or perhaps 10^{-100}.

Problems

(5.1.1) Write out in English that a child would understand the practical meaning of the following mathematical statements. Do not use any math jargon.

$$\lim_{n \to \infty} x_n = 0$$
$$\lim_{x \to 2} x^2 = 4$$
$$\lim_{x \to 1} (x + (1/x)) = 2$$
$$\lim_{x \to \infty} x^{-2} = 0$$
$$\lim_{n \to \infty} \left(\frac{x_n}{x_{n+1}} \right) = 1.$$

(5.1.2) Find the following limits

$$(a) \lim_{x \to 2} (x + 2) \quad (b) \lim_{x \to \infty} x^{-1} \quad (c) \lim_{x \to 0} 1/x \quad (d) \lim_{x \to \infty} (3 + x^{-1}) \quad (e) \lim_{x \to \infty} x/(2x + 1)$$

[hint: if $x \to \infty$ try plugging in a large number for x then guess the exact answer. If $x \to 0$ try plugging in a very small number for x.]

(5.1.3) If you have \$350 in the bank but your bank balance shows \$385 what is the percentage error?

(5.1.4) If you think the gas tank in your car is 1/3 full but it is really 1/4 full, what percentage error have you made?

(5.1.5) Find

$$\lim_{h \to 0} \left(\frac{(1 + h)^2 - 1}{h} \right).$$

[hint: simplify the fraction before thinking about limits.] What happens if you just try to plug in $h = 0$ before doing any simplifying?.

(5.1.6) Set

$$x_1 = \frac{1}{2} \quad x_2 = \frac{1}{2} + \frac{1}{4} \quad x_3 = \frac{1}{2} + \frac{1}{4} + \frac{1}{8} \quad x_4 = \frac{1}{2} + \frac{1}{4} + \frac{1}{8} + \frac{1}{16} \quad \cdots$$

Write out x_5. Work out what each of the numbers x_1 to x_5 is. What do you think the limit of x_n is as $n \to \infty$?

(5.1.7) A rectangle has width 3.1 and height 2.9 What is the percentage error if an approximate value of $3 \times 3 = 9$ is used for the area of this rectangle?

5.2 The Change in $f(x)$.

In calculus, you will often hear something like "if x is increased from 3 to 4 then x^2 increases by 7". All this means is $4^2 - 3^2 = 7$. The change in $f(x)$ between a and b means $f(b) - f(a)$. The question "if x is increased by 1 how much does $f(x)$ increase?" has the answer $f(x + 1) - f(x)$. It is a good idea to imagine the diagram

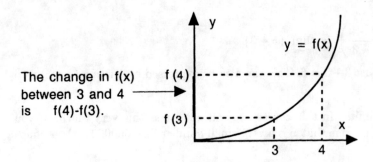

shown above which shows the increase in $f(x)$ between 3 and 4. The word "increase" should be understood to allow a **negative** increase, in other words a **decrease**. Mathematicians are funny this way. The word **change** means increase or decrease. However, it is again possible to have a change of 0, which is not really any change at all!

It is important to realize that when we talk about changing x in something like $f(x) = 3x^2 + 2x$ all we mean is the following: first imagine plugging in one number for x. Next, imagine plugging in a different number for x. Think about the two answers we get for $f(x)$. Now subtract one from the other.

Problems

(5.2.1) If x is increased from 4 to 5 how much does

$$\frac{1+x}{2+x}$$

change by? Increase or decrease?

(5.2.2) What is the change in $f(x) = x^3 + x$ when x is increased by 0.1 starting from $x = 2$?

(5.2.3) For each of the values of h given, when x is increased from 1 to $1 + h$, work out

$$\frac{\text{the change in } x^2}{h}$$

The values of h are $1, 0.1, 0.01, 0.001$

(5.2.4) If x is increased from 2 to $2 + h$ how much does $x^2 + x$ increase by?

5.3   Adding things up: Summation Notation.

Often we must add up a lot of numbers. To be more precise we must **imagine** adding up a lot of numbers. Suppose you want write down all the numbers from 1 to 20 added together. Then you write down

$$1 + 2 + 3 + 4 + 5 + 6 + 7 + 8 + 9 + 10 + 11 + 12 + 13 + 14 + 15 + 16 + 17 + 18 + 19 + 20.$$

There is a shorter way to write this down:

$$\sum_{n=1}^{20} n.$$

The symbol $\sum$ is a Greek letter "sigma" which is the equivalent of an English "S." The "S" stands for the word "sum" meaning "add up." Next, the $n = 1$ underneath and the 20 on top mean that the sum starts with $n = 1$ and progresses until $n = 20$. In between, n takes on each of the values $1, 2, 3, \cdots, 19, 20$ in turn. It acts like a machine. We tell the machine "You are going to add up twenty things. You will start with $n = 1$ and end up with $n = 20$." The actual thing that is going to be added up is the object that follows the Σ (n in this example). If we want to write down

$$1^2 + 2^2 + 3^2 + 4^2 + 5^2 + 6^2 + 7^2 + 8^2 + 9^2 + 10^2$$

using summation notation, we write

$$\sum_{n=1}^{10} n^2.$$

This time the thing that is being added up is the quantity n^2. But what is n?. Well, the machine starts with $n = 1$ and then does $n = 2$ and continues until $n = 10$, and then stops. Each time imagine the machine printing on the paper the value of n^2. So it start by printing 1^2 then it prints 2^2 and so on until it gets to 10^2. where it stops.

The important thing to realize here is that this is just shorthand. It is done to save writing. It does not involve any fancy calculation. It is simple. Suppose we only want to add up the numbers from 1980 to 1997 then

$$\sum_{n=1980}^{1997} n$$

means that the first value of n to use is 1980. We can also use 0 or even negative numbers.

Often we want to talk about adding up a set of data, perhaps the data in a table. To do this, one says something like: "let a_n be the n'th entry in the table." Then if there are, say, 30 entries in the table one can write the sum of the entries as

$$\sum_{n=1}^{30} a_n.$$

Or perhaps we want to first square every entry and then add up the results. In this case we write

$$\sum_{n=1}^{30} \left(a_n^2\right).$$

Perhaps, instead, we want to FIRST add up all these numbers and THEN square the result. Then this is written as

$$\left(\sum_{n=1}^{30} a_n\right)^2.$$

At other times, we might have a function $f(x)$ and want to add up $f(1) + f(2) + \cdots + f(42)$. This can be written as

$$\sum_{n=1}^{42} f(n).$$

The **average** of the numbers $a_1, a_2, \cdots, a_{100}$ can be written easily using summation notation as

$$average = \frac{1}{100} \left(\sum_{n=1}^{100} a_n\right).$$

Although this is just shorthand, it is very useful. There are a few things to be careful about. For example:

$$\sum_{n=1}^{5} n + 1.$$

What does this mean?. Does it mean

$$(1 + 2 + 3 + 4 + 5) + 1 \qquad or \qquad (1 + 1) + (2 + 1) + (3 + 1) + (4 + 1) + (5 + 1).$$

It is ambiguous. So you MUST be very careful when you use summation notation to use parentheses correctly. Thus,

$$\left(\sum_{n=1}^{5} n\right) + 1 = (1 + 2 + 3 + 4 + 5) + 1.$$

$$\sum_{n=1}^{5} (n + 1) = (1+1)+(2+1)+(3+1)+(4+1)+(5+1).$$

There are a few simple algebra rules. Here is a first example.

$$17\sum_{n=1}^{15} n^2 = \sum_{n=1}^{15} \left(17 \times n^2\right).$$

In other words if we multiply a sum of things by a constant, it is the same as multiplying each of the things we are adding up by the same constant (think: distributive property). Another property: if we have two sums of things, then we can combine into a single sum thus:

$$\left(\sum_{n=2}^{100} n\right) + \left(\sum_{n=2}^{100} n^3\right) = \sum_{n=2}^{100} \left(n + n^3\right).$$

BUT notice that this only works if both sums start at the same value of n (they both start at $n = 2$ in this example) and end at the same value of n ($n = 100$ in this example.) Also notice:

$$\left(\sum_{n=3}^{17} 1/n\right) + \left(\sum_{n=18}^{30} 1/n\right) = \left(\sum_{n=3}^{30} 1/n\right).$$

Do not try to memorize these formulae. Understand what they say, and use common sense; it will be a lot easier that way. You should now do some practise questions.

Some of the following problems involve **factorials.** The number

$$5! = 5 \times 4 \times 3 \times 2 \times 1.$$

In general the number $n!$ is all the whole numbers between 1 and n multiplied together. The first few factorials are shown in the table. Notice that they get large very quickly. In fact $100! \approx 10^{158}$. By convention $0! = 1$.

n	1	2	3	4	5	6	7	8	9	10
$n!$	1	2	6	24	120	720	5040	40320	362880	3628800

Problems

(5.3.1) Write out the following summations as a series of terms added together

$$(a) \sum_{n=1}^{5} n \qquad (b) \sum_{m=1}^{4} a_m \qquad (c) \sum_{k=2}^{6} (a_k + b_k) \qquad (d) \sum_{n=-2}^{2} f(n) \qquad (e)\, 3\left(\sum_{i=1}^{3} x_i\right)$$

$$(f) \left(\sum_{n=10}^{13} c_n\right) + \left(\sum_{n=14}^{15} c_n\right) \qquad (g) \sum_{n=2}^{5} (a_n + b_n)^2 \qquad (h) \sum_{n=1}^{5} 1/n! \qquad (i) \sum_{n=1}^{4} x^n/n$$

$$(j) \sum_{i=10}^{15} a_{i-9} \qquad (k) \sum_{n=1}^{5} 2^n \qquad (l) \sum_{n=1}^{5} (-1)^n \qquad (m) \sum_{n=1}^{4} (-x)^n$$

(5.3.2) Write out the following summations as a series of terms added together. For example

$$\sum_{n=1}^{100} n = 1 + 2 + 3 + \cdots + 99 + 100.$$

$$(a) \sum_{n=1}^{1000} \frac{a_n}{a_n^2 + 1} \qquad (b) \sum_{n=1}^{100} (a_{n+1} - a_n) \qquad (c)\, \frac{1}{N}\sum_{p=1}^{N} x_p \qquad (d) \sum_{n=1}^{200} \frac{x^n}{n!}.$$

(5.3.3) Combine into a single summation

$$(a) \left(\sum_{n=1}^{50} x_n\right) + \left(\sum_{m=51}^{100} x_m\right) \qquad (b) \left(\sum_{k=1}^{20} a_k\right) + 2\left(\sum_{k=1}^{20} b_k\right) \qquad (c) \left(\sum_{i=1}^{100} x_i\right) - \left(\sum_{i=71}^{100} x_i\right)$$

(5.3.4) Write the average of the numbers $a_1, a_2, \cdots, a_{1000}$ using summation notation.

(5.3.5) Show that the answer to 2(b) equals $a_{101} - a_1$. This kind of summation is called a **telescoping series** because all the terms except the first and last cancel out in pairs.

(5.3.6) (a) Show that if $a_k = k(k-1)/2$ that $a_{k+1} - a_k = k$. [hint: work out a_{k+1} by carefully replacing k by $k+1$ everywhere]
(b) Use the result of qu (5), and replace k by $a_{k+1} - a_k$ in the summation to show

$$\sum_{k=0}^{n} k = \frac{n(n+1)}{2}.$$

(c) use the formula in (b) to find the result of adding all the numbers from 1 to 100.

(5.3.7) Write out

$$\left(\sum_{n=1}^{2} a_n\right) \times \left(\sum_{m=1}^{2} b_m\right)$$

and show this is not the same as

$$\sum_{k=1}^{2} (a_k b_k).$$

(5.3.8) (a) How would you write the average of the numbers $x_1, x_2, \cdots x_{500}$ using summation notation.
(b) How would you write the average of the numbers $y_1, y_2, \cdots y_N$ using summation notation.
(c) How would you write the average of the **squares** of the numbers $a_{10}, a_{11}, \cdots a_{20}$ using summation notation.

(5.3.9) Calculate

$$(a) \sum_{i=1}^{100} 2 \qquad (b) \sum_{n=1}^{3} n^2 \qquad (c) \sum_{k=1}^{2} \left(\sum_{n=1}^{3} k \times n\right)$$

(5.3.10) Calculate

$$\left(\sum_{n=1}^{10} (n+1)\right) - \left(\left(\sum_{n=1}^{10} n\right) + 1\right).$$

Chapter 6

Straight Lines and Linearity.

Calculus is based on two fundamental ideas: slope and area. The concept of slope is VERY important. So you must be very good at slopes of lines! Fortunately, there really isn't much to know. Slope is just a measure of how steep a line is; this means how quickly it rises upwards as you move along it. A slope of 0 means horizontal,

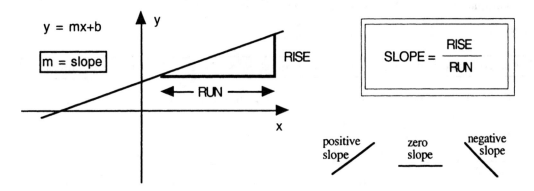

a vertical line has infinite slope, and a line at 45 degrees has slope of 1. Slope is "rise over run." This is easy to remember, but it's most important to understand its meaning. Look at the diagram. It should be clear that RUN stands for horizontal distance while RISE indicates vertical distance. The important point is that as we move along the line, the **ratio** of vertical distance traveled to horizontal distance traveled provides a good way to measure "steepness." However, there is a slight problem: we can move either left or right in the horizontal direction. Likewise, we can move either up or down in the vertical direction. This can be a bit confusing. But, we can deal with the ambiguity by fixing the convention that **RUN always stands for horizontal motion to the right.** We then decide that if a line moves up as we move to the right, then the line has positive RISE, and therefore positive slope. Conversely, if a line moves down as we move to the right, it has negative RISE, and hence negative slope. The summary of this is the following:

A straight line which moves UP as we move to the right has POSITIVE slope.

A straight line which moves DOWN as we move to the right has NEGATIVE slope.

Of course, there is a bit more to say. You might ask "what about a horizontal line?" Well, a horizontal line has no RISE at all, so we say it has slope 0. One the other hand, a vertical line has no RUN. This is tricky since the ratio for slope requires us to divide by the RUN. We can't do this since it would mean dividing by zero. But, if we use our idea that slope should measure steepness, it makes sense to say that a vertical line is "infinitely steep." Thus, we say that a vertical line has infinite slope.

The first key feature of a straight line is the following: as you move along it you move **steadily** up or down. For example consider a line with slope 2. If you move along this line one unit to the right in the x-direction, you move up in the y-direction exactly 2 units. If you move 3 units in the x-direction then you move 6 units in the y-direction. It does not matter where you start on this line: one unit change in x always produces two units change in y. This is because the **slope** of the line is **constant** at 2. Contrast this behavior with that of the parabola $y = x^2$ shown.

On a graph which is not a straight line the slope changes as you move along the graph.

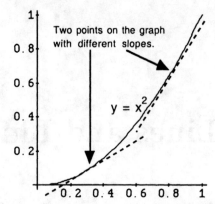

We might try to measure the "steepness" of this graph as well. However, different points on the parabola have different slopes. If you start at one point on the parabola and move along it one unit in the x-direction, then how much the y coordinate increases **depends on where you start.**

We can work out the different ways of writing the equation of a straight line from the key property:

> If a straight line has slope m and you move 1 unit in the x-direction, you move m units in the y-direction.

From this it follows that if you move along the line x units in the x-direction then you move mx units in the y-direction.

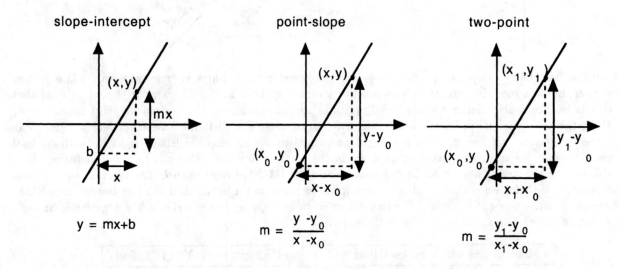

The point where a straight line crosses the x-axis is called the **x-intercept**. The point where it crosses the **y-axis is the y-intercept.** Here are three forms for the equation of a line:

(1) If you know that the slope is m and the y-intercept is b then the equation of the line is:

> y-intercept-slope form $y = mx + b.$

(2) If you know the slope is m and the line passes through the point (x_0, y_0) the equation is:

> point-slope form $y - y_0 = m(x - x_0).$

(3) If you know that the line passes through the two points (x_0, y_0) and (x_1, y_1) the equation is:

> | two point form | $y - y_0 = \left(\dfrac{y_1 - y_0}{x_1 - x_0}\right)(x - x_0).$ |

These equations and more will now be explained.

Suppose we know that a line crosses the y-axis at $y = b$ and has slope m. To write an equation for the line, we need to figure out what y is in terms of x. We know that the y coordinate changes by mx units if you move x units in the x-direction. If we start at $x = 0$ and $y = b$ and move to x-coordinate x, then the y coordinate must increase by mx units from b to $mx + b$. So we discover the y-intercept form:

$$y = mx + b.$$

Suppose instead we are told the line has slope m and that the point (x_0, y_0) is on the line. Then we want to know the y-coordinate when the x-coordinate is x. Well, if we move from x_0 to x the change in the x-coordinate is $x - x_0$. This means that the y-coordinate changes by $m \times (x - x_0)$. Now the change in the y-coordinate is $y - y_0$. Thus, this gives us the point-slope form:

$$y - y_0 = m(x - x_0) \qquad \text{which says} \qquad (\text{change in } y) = m \times (\text{change in } x).$$

Finally, if we are given that the points (x_0, y_0) and (x_1, y_1) are on the line then we can work out the slope of the line passing through these points as

$$SLOPE = \frac{RISE}{RUN} = \frac{y_1 - y_0}{x_1 - x_0}.$$

Now we know the slope; we also know that the line goes through the point (x_0, y_0). So we can use the point-slope formula we just worked out to get

$$y - y_0 = m(x - x_0)$$
$$= \left(\frac{y_1 - y_0}{x_1 - x_0}\right)(x - x_0).$$

As well as memorizing the 3 forms for the equation of a straight line, you need to be able to figure out the slope of a straight line when you are given certain information. For example if you are told that a line L passes through two points $(2, 5)$ and $(4, 11)$ you should be able to find out that the slope of L is 3 as follows:

Draw a neat little diagram with an x and y axis. Now mark on the two points $(2, 5)$ and $(4, 11)$ and draw the line through them. Remember that $SLOPE = RISE/RUN$. On your diagram, draw the triangle whose base

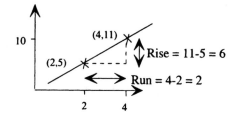

If you ALWAYS draw this diagram, you will avoid silly mistakes

slope = 6/2 = 3

has length $RUN = 4 - 2$ and height $RISE = 11 - 5$. Now calculate the slope.

There are other pieces of information you might be given from which you have to deduce the slope of the line. For example you might be told where the line crosses the x-axis and y-axis. Or, you may be asked something like this:

Example 6.0.1 *Find the slope of the line given by this equation:*

$$2y - 3x + 7 = 0.$$

Solution: you can always algebraically manipulate the equation into the form

$$y = m \cdot x + c.$$

Then the slope is just m. So, we solve for y in terms of x.

$$2y - 3x + 7 = 0$$
$$2y = 3x - 7$$
$$y = \frac{3}{2}x - \frac{7}{2}$$

Since this is now in slope-intercept form, we see that the line has slope $\frac{3}{2}$.

Example 6.0.2 *If the slope of the line L is 3 and if L goes through the point $(2,5)$ what is the equation of the line?*

Solution: we know that a point (x, y) is on the line L exactly if the slope of the line passing through $(2, 5)$ and (x, y) is 3. THINK ABOUT THIS!. ALWAYS draw a picture.

$$\text{slope} = 3 = \frac{\text{RISE}}{\text{RUN}} = \frac{y\text{-}5}{x\text{-}2}$$

This means that (x, y) is on the line provided they satisfy the equation

$$\frac{RISE}{RUN} = \frac{y-5}{x-2} = (slope\ of\ L) = 3.$$

Now do some algebra to get this equation into the form

$$y = 3x - 1.$$

Finally we can check this is the correct answer. It is a line with slope 3. Also, if we plug $x = 2$ in we find that $y = 5$, so the line does indeed go through $(2, 5)$.

We can summarize the meaning of all of this as follows: to find the equation of a line, we need either:

(1) The slope of the line along with a point on the line.

(2) Two points on the line.

It is far more important for you to understand these ideas than it is for you to just memorize the formulae. The basic LOGIC involved all revolves around the key idea that $SLOPE = RISE/RUN$. If you understand slope and you know that the equation of a line can be written as $y = m \cdot x + b$, then you should be able to work out the rest whenever you need it.

6.1 Parametric Equation of a Straight Line.

Imagine a plane flying, so that after t hours it has traveled $300t$ miles east and $400t$ miles north [by Pythagoras, its speed is $500mph$ because $3^2 + 4^2 = 5^2$.]. This can be expressed by:

$$x = 300t \qquad y = 400t.$$

Here

x is the number of miles east the plane has travelled after t hours

y is the number of miles north the plane has travelled after t hours

Since the speed in each direction is constant, the path of the plane is a straight line. We can find the equation of this line by eliminating t. This will give us an equation for y in terms of x:

$$t = \frac{x}{300} = \frac{y}{400}$$

Solving this gives $y = 4x/3$. This is indeed the equation of a straight line.

Another example: the equations

$$x = 3 + 5t \qquad y = 2 - 7t$$

define a line which again you can find as follows: use the two equations to get two expressions for t and then set them equal

$$\frac{x - 3}{5} = t = \frac{y - 2}{-7}.$$

This gives

$$-7(x - 3) = 5(y - 2) \qquad \Rightarrow \qquad y = (-7/5)x + (31/5)$$

which is the equation of a line.

The quantity t is called a **parameter.** The equations

$$x = a + b \cdot t \qquad y = c + d \cdot t$$

are the **parametric** equations of a line. Frequently t represents time. In this case, these equations simply tell you the x-coordinate and the y-coordinate of some object at time t. You need to be able to convert the parametric equation of a straight line into the form $y = mx + b$. This is done by eliminating t as above.

Problems

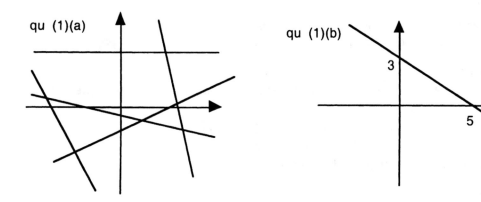

qu (1)(a)

qu (1)(b)

(6.1.1) (a) Copy the figure and mark each line as having positive, zero or negative slope.
(b) What is the equation of the line shown.
(c) number the lines in (a) in order of increasing (becoming more positive) slope.

(6.1.2) First sketch a diagram then find the equation of the line through $(2, -1)$ and $(-3, -5)$.

(6.1.3) First sketch a diagram then find the equation of the line through $(2, -1)$ and $(5, -1)$.

(6.1.4) Find the equation of the line through $(2, a)$ and $(4, b)$.

For problems 5-9, find the equation of the line indicated in the form $y = mx + b$.

(6.1.5) The line through $(3, 1)$ with slope 2

(6.1.6) The line which passes through the origin and the point $(3, 1)$

(6.1.7) The line with slope -1 which passes through the point with x coordinate 1 and y coordinate 2.

(6.1.8) The line through $(4, 1)$ which meets the x-axis at $x = 3$

(6.1.9) The line which meets the x-axis at $x = 5$ and the y axis at $y = -2$.

(6.1.10) Solve the equations $2x + 3y = 1$ and $x - y = 4$.

(6.1.11) Where do the lines $y = 3x + 1$ and $2y = x - 7$ cross?

(6.1.12) Find where the line which passes through the two points $(1, 2)$ and $(3, 5)$ intersects the line through the points $(2, 1)$ and $(-1, -1)$. Start by drawing a diagram.

(6.1.13) Two cars leave the origin at noon. One goes north at 60 mph. The other goes south at 40 mph. When are they 250 miles apart?.

(6.1.14) Car A leaves Santa Barbara at noon traveling south towards los Angeles at 50mph. At 1pm car B starts in Los Angeles traveling north to Santa Barbara at 100mph. The distance between the cars at noon was 100 miles. If the cops don't catch car B, what time do A and B meet?

(6.1.15) Two car companies ELVIS and HEARTS rent cars. The Elvis car costs \$30 plus 20 cents per mile. The Hearts car costs \$50 with 50 free miles and then costs 15 cents/mile. On a single diagram, graph the cost against distance driven for both car companies.
(a) When is it cheaper to get an Elvis?
(b) What is the significance of the **slopes** of the two graphs?.

(6.1.16) Water boils at 100° C (degrees celsius) which is 212° F (degrees Fahrenheit) and freezes at 0° C which is 32° F. Let x be the temperature of an object in ^{o} F and y the temperature of the same object in ^{o} C.
(a) express x in terms of y
(b) express y in terms of x
(c) use (b) to convert 85° F into celsius
(d) use (a) to convert your answer to (c) back into Fahrenheit.
(e) What temperature is the same in both celsius and Fahrenheit?
(f) Draw a graph showing Fahrenheit on the vertical axis and Celsius on the horizontal, for Celsius range -100^oC to $+100^oC$.

(6.1.17) A plane X starts at A and flies to B which is a point 300Km north and 400Km west of A. A plane Y starts at C (which is 200Km west of A) and flies north.
(a) Draw a diagram showing this information.
(b) where do the flight paths cross?
(c) If plane X travels at 500Km/hr, how fast must plane Y fly to crash into plane X. Assume the planes take off at the same time.

(6.1.18) (a) Find the equation of the line $x = 2t + 1$ and $y = 3 - 5t$ in the form $y = mx + b$.
(b) sketch this line in the xy plane.

(6.1.19) (a) What is the slope of the line $x = -3t + 1$ and $y = 4t + 2$.
(b) Where does this line cross the x-axis
(c) Where does this line cross the y-axis

(6.1.20) Find the point of intersection of the line $x = t - 2$ and $y = 3t + 1$ with the line $2x + 3y = 0$.

(6.1.21) If a collection of points all lie on the same straight line we say that the points are **collinear.** Given two points there is always a straight line passing through both of them. But given three points they may, or may not, be collinear.

(a) If P, Q, R are the three points that are collinear, what can you say about the **slope** of the line through P and Q, compared to the **slope** of the line through Q and R?. Sketch a diagram.

(b) If $P = (x_0, y_0)$ and $Q = (x_1, y_1)$ and $R = (x_2, y_2)$ write down the equation that says these slopes are equal in terms of this information.

(c) Use your answer to (b) to decide if the points $(2, 3)$ and $(5, 5)$ and $(-4, -1)$ are collinear.

(6.1.22) Suppose the points $(3, 5)$ and $(10, 17)$ are on line L.

(a) What point is midway between these points?

(b) What point is 1/3 of the way from $(3, 5)$ towards $(10, 17)$?

(c) If $(4, y)$ is on L, find y.

(d) if $(x, 8)$ is on L, find x.

6.2 Linear functions and Proportionality.

We are particularly interested equations like

$$y = 7 \times x.$$

They are very simple. If we plug in a value for x, the result is just 7 times this plugged in value. Here are some examples:

Suppose you earn \$7 per hour. How much money do you earn as a function of time worked? Well, if you work t hours, then you earn $7 \times t$ dollars.

If an object has a mass of x kilograms what does it weigh in pounds?. Since one kilogram weighs 2.2 pounds the object weighs $2.2 \times x$ pounds.

If 35% of the population has brown hair, out of x people how many should have brown hair? Answer: $0.35 \times x$.

If apples cost 15 cents each and you buy x apples, how much does it cost in dollars? Answer: $0.15 \times x$ dollars.

The point here is that multiplication is very common, so it is not a surprise that many equations just involve multiplying by some constant. In these situations we say that the two quantities are **proportional**. Thus, how much you earn is proportional to how many hours you work. And, how much an object weighs in pounds is proportional to its mass in kilograms. Now think about the first example: $y = 7 \times x$. We say that y is proportional to x and that 7 is the **constant of proportionality.** The symbol $\propto$ means "proportional to." Thus

$$y \propto x$$

means y is proportional to x.

There are some simple "tricks" when working with proportionality. If you were asked "how much money would you make if you worked twice as many hours per week as you do now" you can immediately say: "I would make twice as much money as I make now". This idea is by far the **most important** fact about proportionality:

> If y is proportional to x and you double x this causes y to double.

Here, double can be **replaced** by triple, quadruple, or even "multiplied by k."

It is often best to think about proportionality using the language of functions. A **linear function** is a function like $f(x) = 7x$. The most general form is

$$f(x) = C \times x.$$

The word "linear" means "straight". Not surprisingly, the graph of a linear function $f(x) = C \times x$ is a line which goes through the origin with **slope** C. The word "linear" reminds us that its graph is a straight line (LINE as in LINEar). Sometimes people say that a function like $f(x) = 3x + 7$ is linear. This is not strictly correct and we will refer to such a function as **linear plus constant.**

Example 6.2.1 *The cost of building a freeway is proportional to the length of the freeway. If it costs 2 billion dollars to build a freeway 100 miles long, how much would it cost to build a freeway 300 miles long?*

Solution: to build a freeway 3 times as long costs 3 times as much. This is because COST IS PROPORTIONAL TO LENGTH. So it would cost 6 billion dollars.

Example 6.2.2 *Suppose that the cost of building a road in is proportional to its length. Express cost as a function of road length.*

Solution: Since cost is proportional to length, we can immediately write

$$f(x) = C \times x.$$

Here, the input x represents length. The output $f(x)$ is the cost. Note that we are not told enough to find the constant of proportionality C. On the other hand, this function is a very good qualitative description of the relationship between length and cost since it tells us that cost CHANGES LINEARLY as length changes.

We now come to the idea of **inversely proportional**. Sometimes as one thing increases, another thing decreases. For example, if you have \$200 and cabbages cost \$$x$ each how many cabbages can you buy? The answer is $200/x$. This can be expressed by saying if

$$f(x) = (\text{number of cabbages I can buy with \$200 if they cost \$}x \text{ each})$$

then

$$f(x) = 200 \times x^{-1}.$$

Now looking at this we see that $f(x)$, the number of cabbages you can buy, is proportional to the quantity x^{-1}; that is, $f(x)$ is just 200 times this quantity. We express this by saying that the number of cabbages you can buy is **proportional to the inverse of x** or we say more simply the number of cabbages is **inversely proportional to x.** Thus

> *inversely proportional to x means proportional to the inverse of x.*

Here you should interpret the inverse of x to mean the reciprocal: x^{-1}. Sometimes the phrase **directly proportional** is used to mean "proportional" to help us distinguish "proportional" from "inversely proportional."

Example 6.2.3 *The time taken to build a freeway 100 miles long is inversely proportional to how many people work on it. If it takes 100 people 8 months to build, how long would it take 200 people?*

Solution: We are **doubling** the number of people so the time taken is multiplied by 1/2 because time is **INVERSELY PROPORTIONAL** TO NUMBER OF PEOPLE. Thus is will take 4 months.

Example 6.2.4 *The time taken to build a freeway is directly proportional to the length of the freeway and inversely proportional to the number of people who work. Express the time taken to build a freeway as a function of the length and the number of people working.*

Solution: We will use

$$x = (length\ of\ freeway\ in\ miles)$$
$$n = (number\ of\ people\ working\ on\ freeway)$$
$$t = (time\ taken\ to\ construct\ the\ freeway).$$

Then the answer is

$$t = K \times x \times n^{-1}$$

where K is the constant of proportionality. The time taken to build a freeway 1 mile long if 1 person works on it is given by this formula as $K \times 1 \times 1^{-1} = K$. So the constant of proportionality K is just how long it would take one person to build one mile of freeway (K is probably about 10 years). Looking at this formula, we see that increasing x causes t to increase. This is reasonable since increasing x means increasing the length of the freeway. On the other hand, increasing n makes t get smaller because we are DIVIDING by n. Again this is reasonable because increasing n means using more people to build the freeway. This means the freeway will take less time to build.

It often happens that a quantity x is proportional to two other quantities multiplied together. As an example, suppose x is proportional to the product of y and z. We express this by saying x is **jointly proportional** to y and z. This tends to show up in applications: the mass, M, of a rectangular slab of concrete is jointly proportional to the length, L and width W, of the slab. Thus $M \propto W \times L$.

Problems

(6.2.1) If y is proportional to x and x is 4 when y is 20 then what is y when (a) $x = 8$ (b) $x = 2$ (c) $x = a$ (d) $x = a^2$ (e) $x = 0$ (f) $x = (a + b)$ (g) $x = 1/w$.

(6.2.2) If x is proportional to y and y is proportional to z is x proportional to z? Explain!

(6.2.3) The amount of taxes a city collects is proportional to the population of the city. In 1980 the population was 2 million and it had increased to 3 million by 1992. If 6 billion dollars were taxes collected in 1980 how much was collected in 1992?

(6.2.4) The mass of a sphere made of gold is proportional to the diameter, d, cubed. Express the mass of the sphere as a function of the diameter.

(6.2.5) The rate that a certain disease spreads is proportional to the number of infected individuals and is also proportional to the number of uninfected individuals. The total population is P and the number of infected individuals is D. Express the rate that the disease spreads in terms of this information.

(6.2.6) The temperature, T, of a gas is jointly proportional to the pressure, P, of the gas and the volume, V, occupied by the gas.
(a) Express the temperature in terms of pressure and volume .
(b) Use you answer to (a) to express the pressure in terms of volume and temperature.

(6.2.7) The weight of a sphere is proportional to the radius cubed. If a sphere of diameter 1 cm has a mass of 2 grams what diameter sphere has a mass of 54 grams?

(6.2.8) The cost of moving rubble is proportional to the product of mass of the rubble with the distance the rubble is moved.
(a) Express the cost of moving rubble as a function of the mass and distance moved.
(b) If it costs \$50 to move 1 ton of rubble 1 mile, how much does it cost to move 30 tons 20 miles?

(6.2.9) (*) The cost of flying a passenger plane consists of fuel cost and hourly pay for the flight crew. The faster a plane flies, the more fuel it uses to fly each mile. Suppose that the cost of the fuel to fly one mile is proportional to the speed of the plane. The plane flies a distance of D miles.
(a) Express the time taken to fly D miles in terms of the speed of the plane
(b) Express the cost of the fuel and the cost of the flight crew in terms of the speed
(c) Express the total operating cost in terms of the speed.
(d) If the flight crew costs \$200 per hour and the fuel costs \$100 per hour when the speed is $300mph$ how much would it cost to fly 1200 miles at $600mph$?

(6.2.10) The **law of Gravitation** states that there is a force of attraction between any two objects. The strength of this force is proportional to the product of the masses of the two objects and inversely proportional to the square of the distance between them. Express the force in terms of this information. [the constant of proportionality is called **the universal gravitational constant** G.]

(6.2.11) U.S. hard disc drive companies attempt to double hard disc drive capacity every two years. Assuming they succeed, is hard disc drive capacity proportional to time?

(6.2.12) The time it takes to build a skyscraper is proportional to its height and inversely proportional to the number of construction workers. If it takes 10 workers 2.5 years to build a 20 story building, how long will it take 100 workers to build an 80 story building.

6.3   Linear Interpolation and Extrapolation.

To "extrapolate" mean to predict. Suppose that the Santa Barbara school district had 2100 second graders in 1992 and 2400 in 1997. Now many will there be 5 years from now?. We don't know, but we might guess based on this information. In the 5 years between 1992 and 1997 the number increased by 300 so we could guess that in the next 5 years the number will increase by another 300. Of course, all kinds of things could happen to make this guess wrong. However, we are more likely to be correct if we guess what will happen in 1998 since this is just one year later. We could do this as follows: in the previous 5 years the number of students increased on average by 60 students per year. If we assume that this will continue, there will be 2460 second graders in 1998. This method of guessing is called **linear extrapolation.** In this example, it means that we figure how quickly the

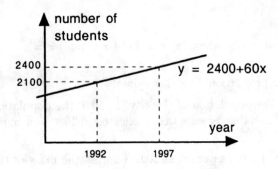

Extrapolating the number of students
based on the increase between 92 and 97

The two data points determine a straight
line. Using this line we can guess how many
students there will be in the future.

x = number of years **after** 1997

number of students is increasing and assume that this will continue. We can then give a formula: if you want to know how many students there are in a certain year, let x equal the number of years after 1997 that you want to know about. Since the number of students increases by 60 per year, the increase in x years will be $60x$ students. Note that there are already 2400 students in 1997. So we must add this to $60x$ to get the extrapolated number of students in the year $1997 + x$ is

$$2400 + 60x.$$

This formula is based on the **assumption** that the number of students will continue to increase at the rate of 60 per year. Therefore, it is a guess based on limited information. No one would expect this to give a very accurate prediction for the number of students 100 years from now. On the other hand, it might be quite good for the next few years. We can also use this formula to tells us how many students there were BEFORE 1997. Just make x negative! Thus, setting $x = -5$ tells us how many students there were 5 years before 1997 namely in 1992. The formula gives

$$2400 + 60(-5) = 2100.$$

This is exactly correct, but it should not be a surprise. We built the formula using the two pieces of information we were given. So the formula naturally gives us the correct answer for the dates 1992 and 1997. However if we try $x = -2$ we get

$$2400 + 60(-2) = 2280.$$

Now this is supposed to be how many students there were in 1995. As such, we are using the formula here to find out some information about a year BETWEEN the two years for which we already have information. This process is called **interpolation** ("inter" means between).

These kind of predictions are often very desirable: given a limited amount of information about some process, you want to interpolate or extrapolate. Now many concepts in life seem to obey the following semi-mystical principle: it is more efficient to think more abstractly about the ideas involed. To understand what this means in our context, suppose we are told that when $x = 4$ some quantity that depends on x, let us call it $f(x)$, has the value 1. Thus $f(4) = 1$. Suppose we also know that when $x = 6$ that $f(6) = 2$. Then, we can easily find a formula like the one above. One way to think about this is to draw a graph.

We mark the two data points $(4, 1)$ and $(6, 2)$. Now **draw a straight line through them.** This straight line describes a linear plus constant function. If you want to extrapolate to find $f(8)$ then look at the point on

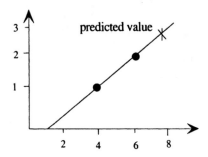

the line where $x = 8$. The y-coordinate of this point is the predicted value of $f(8)$. Using the ideas in the first section of this chapter, the equation of this line is

$$f(x) = x/2 - 1.$$

Thus, we can also use this to obtain $f(8) = 8/2 - 1 = 3$.

Example 6.3.1 *The world population in 1990 was 5.4 billion and in 1995 was 5.8 billion. What will the population be in 1996?*

Solution: We can use linear extrapolation to make this prediction. What this means is that we will assume the population is a linear plus constant function of time. We will measure time from 1990 so that x is the number of years **after** 1990. This means $x = 1$ is 1991 and $x = -1$ is 1989. We need only find the equation of the line passing through the points $(0, 5.4)$ and $(5, 5.8)$. This will have the form $f(x) = mx + b$.

$$m = \frac{5.8 - 5.4}{5 - 0}$$
$$= \frac{0.4}{5}$$
$$= .08$$

Now we have $f(x) = .08x + b$. Since we know that $f(0) = 5.4$, this tells us $f(0) = b = 5.4$. So our function is

$$f(x) = .08x + 5.4.$$

In 1998 we have $x = 8$ and the population is, $f(8) = .08(8) + 5.4 = 6.04$ billion people. It is important to notice that this formula only provides reasonable answers for a few years. In particular, we find our formula would give the world population in the year 1890 (ie $x = -100$) as $f(-100) = .08(-100) + 5.4 = -2.6$ So, according to the function, population was negative in 1890 ! Of course, this is completely silly. The world population is currently rising at about .08 billion (=80 million, 3 times the population of California) people per year and this has been going on for a number of years. Of course it cannot do so for ever.

This type of extrapolation is called **linear extrapolation** because we are using a **straight line** to make predictions. There are more sophisticated methods of extrapolation which use more complicated curves instead of straight lines. These are useful when we have more information available. However, the basic idea is the same: use a limited amount of information to make a function that we hope will give fairly accurate predictions.

6.4   The graphical method and interpolation.

We can use a graph to solve the equation $f(x) = 0$. To do this, we graph $y = f(x)$ and see where it crosses the x-axis. If we only want a rough idea about the answer, we can sketch the graph and then "eyeball it". But if we need a more accurate answer, we can use **linear interpolation.** The idea is this: once we have plotted the graph, and found roughly where it crosses the x-axis, we take the two closest points (x_1, y_1) and (x_2, y_2) that we plotted on either side of the crossing. Now we draw a straight line between these two points and calculate exactly where this straight line crosses the x-axis. This gives us a more accurate guess for the solution.

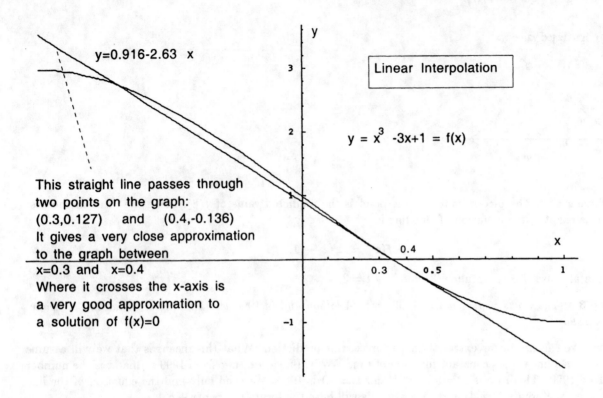

$y = 0.916 - 2.63\ x$

Linear Interpolation

$y = x^3 - 3x + 1 = f(x)$

This straight line passes through
two points on the graph:
(0.3,0.127) and (0.4,-0.136)
It gives a very close approximation
to the graph between
x=0.3 and x=0.4
Where it crosses the x-axis is
a very good approximation to
a solution of f(x)=0

In Chapter 2 we found an approximate solution for $f(x) = x^3 - 3x + 1 = 0$ of $x \approx 0.3$. We can use linear interpolation to improve this solution. By looking at the graph we see that the exact solution is somewhere in between 0.3 and 0.4. So we work out

$$f(0.3) = 0.127 \qquad f(0.4) = -0.136.$$

Now we find the equation of the straight line that goes through these two points on the graph

$$(0.3, 0.127) \qquad (0.4, -0.136)$$

Such a line is called a **secant** line. The slope of this line is

$$\frac{-0.136 - 0.127}{0.4 - 0.3} = -2.63$$

The point-slope equation for the line gives

$$y - 0.127 = -2.63(x - 0.3)$$

which simplifies to give

$$y = 0.916 - 2.63x.$$

This is the equation of the straight line that goes through these two points on the graph. It is very close to the graph between 0.3 and 0.4 as the diagram above shows. We work out where it crosses the x-axis and this gives a better approximation to a solution of $f(x) = 0$. This line crosses the x-axis when $0 = 0.916 - 2.63x$. Solving gives $x = 0.348289$. We can find out how good this approximation is by working out $f(0.348289) = -0.002617$. So this is pretty accurate.

If, however, we want even more accuracy then we should plot an extra point on our graph. That is we plot the point $(0.348289, -0.002617)$ The x-coordinate of this point is the value we just worked out above. The idea is the following: this point should be pretty close to where $f(x) = 0$. If we now repeat the linear interpolation using this new point, we will get an even more accurate approximation to the solution of $f(x) = 0$. We can repeat this process as many times as we want until the answer we have is accurate enough.

Here is the method to perform linear interpolation to improve the accuracy of a solution of the equation $f(x) = 0$.

(1) Plot the graph $y = f(x)$.
(2) Find where the graph crosses the x-axis.
(3) Take two x-values, close to this crossing point. Call them x_0, x_1.
(4) Work out $y_1 = f(x_1)$ and $y_0 = f(x_0)$.
(5) Find the equation of the straight line through the two points (x_0, y_0) and (x_1, y_1).
(6) Find the point where this line crosses the x-axis.
(7) The x-coordinate of this point is the answer.

Problems

(6.4.1) Refer to the above world population example a couple of pages back. Use linear extrapolation to find the population in 2010.

(6.4.2) For the function $f(x) = \sqrt{x}$ we know $f(4) = 2$ and $f(9) = 3$. (a) Use linear interpolation to find $\sqrt{5}$. Compare this to the actual value given in the tables at the end of the book. (b) What is the error?. (c) What is the percentage error?

(6.4.3) The average sea-level in 1900 at London-bridge was 33 feet. In 1990 is was 33.08 feet. Use linear interpolation and extrapolation to find (a) what the average sea-level was in 1950 (b) When the average sea level will be 34 feet.

Chapter 7

Logarithms and Exponentials.

An example of an **exponential function** is

$$f(x) = 10^x.$$

You should be very familiar with the table of powers of 10 shown. If x is positive then 10^x is a 1 followed by x

x	1	2	3	6	0	-1	-2	-6
10^x	10	100	1000	1000000	1	0.1	0.01	0.000001

zeroes since 10^x is 10 multiplied together x times. When the exponent is negative, the number of zeroes after the decimal place is **one smaller than x**. Thus, $10^{-3} = 0.001$. Here, there are 2 zeroes after the decimal place. It's easy to remember this if you note that $10^{-1} = 0.1$ has no zeroes after the decimal place; therefore, 10^{-2} has one zero after the decimal place, and so on.

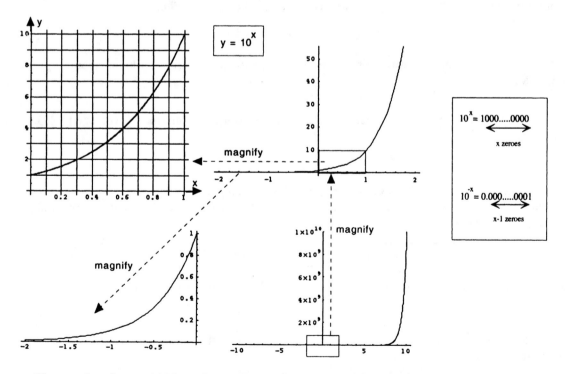

The graphs of $y = 10^x$ have been plotted for various x-values. The bottom right graph is for $-10 \leq x \leq 10$ and looks a bit strange. The reason is that the scale on the y-axis is ENORMOUS. This is to be expected since $10^{10} = 10,000,000,000.$ is a big number. The values of x less than 8, which cover most of the x-axis, yield comparatively small numbers no bigger than 10^8. So the curve is squashed so close to the x-axis that we cannot

91

even see it. The shaded box on this graph covers the range $-2 \leq x \leq 2$. This has been magnified and plotted in the top right graph. Since $10^2 = 100$ the y-axis for this graph should go up to 100, but it has been chopped off at about 50. Even with this graph, the portion for $x < 0$ is still squashed onto the x-axis. This is because $10^0 = 1$ and $10^{\text{negative number}} < 1$ so the graph is too close to the x-axis to distinguish. There are two more magnified views. The picture in the top left shows $0 \leq x \leq 1$. The bottom left graph details $-2 \leq x \leq 0$. The summary is

> If x is big then 10^x is GIGANTIC. If x is much less than 0 then 10^x is tiny.

Another important feature is that 10^x is never negative or zero. Yes, it gets close to zero, but it NEVER equals zero! Now you may be thinking "I know $10^2 = 100$, but I don't think anyone ever told me the meaning of $10^{2.517}$." This will be dealt with very shortly.

Exponential functions are VERY important for the following reason: exponentiation is just repeated multiplication. Let's discuss this in some more detail. The most basic mathematical operation is addition. It was invented to describe the most basic ideas. For example, if we have 3 eggs in a basket and we put 4 more eggs in the basket then adding 3 and 4 gives the number of eggs in the basket. The next idea is to think about REPEATED addition and this leads to multiplication. If we go to the store where eggs are sold in boxes of 12 and we put 4 boxes in the cart then the number of eggs we have is $12 + 12 + 12 + 12$. Of course, it is easier to get this by multiplying 4 **times** 12. To escape having to add 12 to itself 4 times, we invented a new operation called multiplication. It is much simpler to multiply 12 by 4 than to add 12 four times. The next step is to think about REPEATED multiplication. This idea leads to **exponentiation.** If we multiply four threes together we get $3 \times 3 \times 3 \times 3$. This is expressed as 3^4. The fourth power of 3 is the result of multiplying four 3's together. The number 4 is called an **exponent**.

If you understand this simple idea then you should realize that the formula

$$3^2 \times 3^4 = 3^{2+4}$$

says something which is just common sense. Namely, if we multiply 2 threes together and then multiply 4 threes together and multiply the results, then we get $2 + 4$ threes multiplied together. Thus,

$$(3 \times 3) \times (3 \times 3 \times 3 \times 3) = (3 \times 3 \times 3 \times 3 \times 3 \times 3).$$

This is called the **first law of exponents.** It is the most important thing there is to know about exponents. The general formula is

$$3^x \times 3^y = 3^{x+y}.$$

How would you explain this formula in simple English like the common sense explanation above? WRITE DOWN THIS SENTENCE HERE:

Does it REALLY make sense to you? Once you truly understand this idea, everything else about exponents and logarithms will make a lot more sense. Of course, the number 3 can be replaced by an arbitrary number. If we use a, for example, we have this:

> First Law of Exponents $a^x \times a^y = a^{x+y}$.

Now write down a simple sentence explaining this formula (it won't take more than a minute or two!!):

There is an analogy between the law of exponents and an important rule for multiplication. We know from a long time ago that

$$(2 + 4) \times 3 = (2 \times 3) + (4 \times 3).$$

This is expressed by saying we **distribute** multiplication over addition. Why is this formula true?. Really, it is just common sense. Remember that $(2 + 4)$ times 3 means $(2 + 4)$ threes added together. Well, if we add 2 threes to 4 threes, we get the same answer. We can think of this as follows: $(2 + 4) \times 3$ means "do the operation of addition repeatedly $(2 + 4)$ times using the number 3 to add each time." Now, for exponents, the expression 3^{2+4} just means to do the operation of multiplication $2 + 4$ times using the number 3 to multiply each time. This is the analogy between the distributive law for multiplication and the law of exponents.

This is a long explanation of the law of exponents. It is not enough to memorize this law. Of course, you must memorize it. But you must also UNDERSTAND it as well as you understand that $6 + 17$ is the same as $17 + 6$. Otherwise, you will not be able to understand logarithms.

Now we come to the tricky question: what is 10^0? We know that the correct answer is 1. But why? Here are two explanations. First, when we add numbers we start out with 0 and add the other numbers onto 0. Thus, 0 is the **starting point** for addition. So, we can think of 4×3 as **starting with** 0 and adding on 4 threes to it:

$$4 \times 3 = (((0 + 3) + 3) + 3) + 3.$$

Now with multiplication we do not want to start with 0 because multiplying by 0 just gives back 0. Instead, we want to start with 1. For example, 3^4 means to multiply 4 threes together as in

$$3^4 = (((1 \times 3) \times 3) \times 3) \times 3.$$

When multiplying 4 threes together, imagine starting with 1 and multiplying it by three 4 times. Now we can answer the question "what is 3^0?" Start with 1 and multiply it by three zero times. In other words leave the 1 unchanged; do not multiply it by anything at all. In this case, the answer is 1. The idea of this analogy is the following: if we don't add any numbers to 0, we remain at our starting point, namely, 0. The same is true with multiplication: if we do not have any numbers to multiply, we are left at the starting place of 1. This is the first explanation.

Here is the second view. Somebody once sat down (about 500 years ago) and thought "I wonder what 3^0 means." Nobody had ever thought about this before. There was no one to ask. So this person had to **invent** a meaning. This person thinks "whatever 3^0 turns out to be, it better be USEFUL, or else why bother." Now the most important thing about exponents is the first law. In fact, the first law is really the ONLY important thing about exponents. So whatever the person decides for 3^0, it better satisfy the law of exponents. Now the clever thing to realize is that this requirement FORCES 3^0 to equal 1. This is true because the first law of exponents says:

$$3^1 \times 3^0 = 3^{1+0}$$

This gives us

$$3 \times 3^0 = 3.$$

Thus, $3^0 = 1$. This ends the second explanation.

Second Law of Exponents $\qquad a^0 = 1.$

Now we come to negative exponents. What does 3^{-2} mean? After reading the above you say to yourself, "it means take -2 threes and multiply them together." But, what on earth could it possibly mean to take -2 threes?. It definitely does NOT mean take 2 negative threes! Our common sense approach seems to fail here. We just cannot imagine it. The approach then is to be practical . We agree that we do not know what value to give 3^{-2}. So, we will invent a meaning for it. The discussion for 3^0 suggests we should try to invent a meaning which works with the law of exponents. Once we have decided this point, the answer is forced on us:

$$3^2 \times 3^{-2} = 3^{2-2} = 3^0 = 1.$$

If we solve this to find the mysterious quantity 3^{-2}, our quest is over:

$$3^{-2} = \frac{1}{3^2}.$$

This suggests that 3^{-x} should be:

$$3^{-x} = \frac{1}{3^x}.$$

We have come up with this DEFINITION of 3^{-x} by demanding that negative exponents work with the law of exponents.

$$\text{Third Law of Exponents} \qquad a^{-x} = \frac{1}{a^x}.$$

Here is another explanation which may help you remember it. The number -2 is the **opposite** of the number 2. Division is the **opposite** of multiplication. Since 3^2 means **multiply** by 3 two times, then 3^{-2} should mean the opposite, namely, **divide** by 3 two times. Dividing by 3 two times is the same as multiplying by $1/(3 \times 3)$.

There is one more rule for exponents that you have seen before:

$$\text{Fourth Law of Exponents} \qquad (a^x)^y = a^{xy}.$$

For example,

$$\left(2^3\right)^5 = 2^{15}.$$

Again this is common sense. This equation says if we multiply a whole bunch of twos together one way, we get the same answer we would obtain by multiplying them another way. The left side is

$$(2 \times 2 \times 2) \times (2 \times 2 \times 2) \times (2 \times 2 \times 2) \times (2 \times 2 \times 2) \times (2 \times 2 \times 2).$$

In other words, 2^3 is 3 twos multiplied. We must multiply 5 lots of them together so the result is 15 twos multiplied together. You should memorize these rules:

$$a^{x+y} = a^x \times a^y \qquad a^0 = 1 \qquad a^{-x} = \frac{1}{a^x} \qquad (a^x)^y = a^{xy}.$$

One final point: the first law is the main one. We figured out all the other ones starting with it.

7.1 Fractional exponents: $7^{2.517}$

The summary of this section is: numbers raised to weird powers have a meaning which can be derived from the rules for exponents. We know what it means to raise the number 7 to the power of 3. But what does it mean to raise 7 to the power of 2.517? Common sense is at a loss here. How can we take 2.517 sevens and multiply them together? As in the cases we discussed in the last section, the idea is to make it mean something useful. You have seen many times

$$\sqrt{7} = 7^{1/2} = 7^{0.5}.$$

Why do people write this? Well, the first question to ask is "what is a square root?" The square root of 7 is the number x which when squared gives 7. Thus,

$$x^2 = 7.$$

Now the first law of exponents is the MOST USEFUL THING about exponents. If there was such a thing as $7^{0.5}$ we would want it to obey the rule of exponents. This means if we calculate

$$7^{0.5} \times 7^{0.5}$$

the rule of exponents would tell us the answer must be

$$7^{0.5} \times 7^{0.5} = 7^{0.5+0.5} = 7^1 = 7.$$

In other words, if the rule of exponents is going to be obeyed by this strange creature $7^{0.5}$ then $7^{0.5}$ is the number which when multiplied by itself gives the answer 7. But this can only mean that $7^{0.5}$ does in fact equal $\sqrt{7}$. People realized long ago that since common sense did not tell them $7^{0.5}$, they were free to invent a meaning. They decided that the rule of exponents is so useful that whatever meaning they invented must work with it. Once they fixed this requirement, they could figure out what $7^{0.5}$ ought to be.

The number 2 is the **cube root** of 8. This just means 2 cubed equals 8. The fourth root of 7 is (approximately) 1.62658 which just means that

$$(1.62658)^4 = 7.$$

In more generality, the number x is the **27'th root** of y just means that $x^{27} = y$. You might guess that the rule of exponents would demand that $7^{0.25}$ is the fourth root of 7. In other words, whatever the actual value of $7^{0.25}$, the law of exponents demands

$$\left(7^{0.25}\right)^4 = 7^{0.25 \times 4} = 7$$

so the guess is correct.

We are on our way to understand the question posed at the start of this section. Notice that $7^{0.1}$ is the tenth root of 7 meaning

$$\left(7^{0.1}\right)^{10} = 7^{0.1 \times 10} = 7.$$

Now this next point is crucial. Once we know what $7^{0.1}$ actually MEANS (namely the number with the property that ten of them multiplied all together gives the answer 7) we instantly understand something like $7^{0.3}$. It is just $7^{0.1}$ cubed because the rule of exponents says

$$\left(7^{0.1}\right)^3 = 7^{0.1+0.1+0.1} = 7^{0.3}.$$

This is progress. The number $7^{.01}$ is the 100'th root of 7. The number $7^{.001}$ is the 1000'th root of 7. Thus the number $7^{2.517}$ is the 1000'th root of 7 multiplied together 2517 times. Again, this follows from the rules:

$$\left(7^{0.001}\right)^{2517} = 7^{0.001 \times 2517} = 7^{2.517}.$$

Of course, this is not a very good way to calculate $7^{2.517}$. If you want to calculate it, use a calculator (we will also see a way to do it by hand with logarithms in the next section). But it does explain the number's meaning. The summary of this discussion is still: numbers raised to weird powers can be understood using the rules for exponents.

7.2 Logarithms.

The summary of this entire section is:

$$\boxed{x = log(y) \qquad \text{MEANS} \qquad y = 10^x.}$$

If you have heard of natural logarithms and logarithms to an arbitrary base, be patient. We shall deal with logarithms to base 10 for the time being. Once we firmly understand this case, the rest will be easy.

The opposite of addition is subtraction. The opposite of multiplication is division. And the opposite of exponentiation is logarithms. An answer to the question "what power of 10 gives the answer 1000" is 3; but another way to say it is $log(1000)$. In other words $log(1000) = 3$. The logarithm function is the **inverse function** to the function 10^x. What this means is the following:

$$\boxed{\text{10 to the power of } log(y) \text{ is } y. \qquad 10^{log(y)} = y}$$

This tells you what $log(y)$ IS by telling you what it DOES. The number $log(y)$ has a JOB to do: when you raise 10 to that power the result is y. Think REALLY HARD about this statement.

So $log(100) = 2$ because $10^2 = 100$, and $log(1000) = 3$ because $10^3 = 1000$. Also $log(0.1) = -1$ because $10^{-1} = 1/10 = 0.1$. The table on the next page shows some powers of 10 and the equivalent facts about logarithms.

x	1	2	3	6	0	-1	-2	-6
$y = 10^x$	10	100	1000	1000000	1	0.1	0.01	0.000001

y	10	100	1000	1000000	1	0.1	0.01	0.000001
$x = log(y)$	1	2	3	6	0	-1	-2	-6

Table 7.1: Table of values for $log(y)$

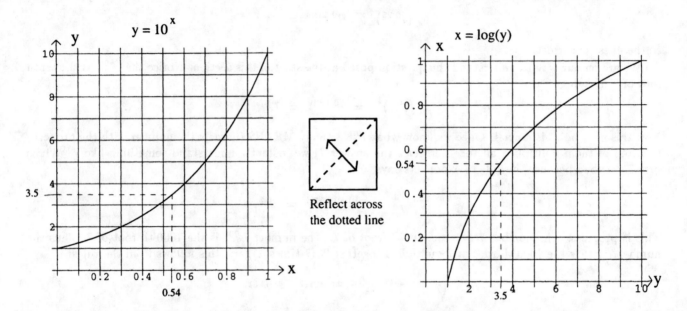

The graph of $y = 10^x$ is shown alongside the graph $x = log(y)$. The graph I have drawn of $log(y)$ is a bit unusual but only because I have labeled the horizontal axis y and the vertical axis x. I will explain why I did this. First, you should look at the graphs. Notice that they seem very similar. In fact they are "mirror images" of each other. Imagine putting a mirror along the line through the origin at 45°. Then each graph is the mirror image of the other one reflected across this imaginary mirror. This is indicated in the middle of the diagram.

To explain this, look at the point $(0.54, 3.5)$ on the left graph indicated by the dotted lines. This point tells us that

$$10^{0.54} = 3.5.$$

But (this is the key point) it also gives us that $log(3.5) = 0.54$. So on the graph on the right we see the point $(3.5, 0.54)$ which is again shown by dotted lines. Every point on the left graph gives a point on the right graph. **But the x and y coordinates get switched.** This happens whenever we graph the inverse of a function. The graph of the inverse function is the graph of the original function reflected. This is true simply because the relationship between x and y is reversed for the inverse function: if $y = f(x)$ then $x = f^{-1}(y)$.

Looking at the graphs below we see that if $x < 1$ then $log(x)$ is **negative**. This makes sense because $10^{negative} < 1$. For example, $log(0.1) = -1$ and $log(0.0001) = -4$. If x is very close to 0, then $log(x)$ is a very BIG negative number. It's easy to see that this is expected by looking at the graph of 10^x and THINKING about the meaning of $log(x)$. Another thing to remember is that as we make x bigger, $log(x)$ also gets bigger. The graph of $log(x)$ rises as we move to the right. BUT it rises VERY slowly. For example, $log(1,000,000) = 6$. Imagine drawing a graph with an x-axis one MILLION units long. When you plot the graph the y value will only have increased to 6. And this is only the start! Since

$$log(10^n) = n$$

to find a number whose log is 20 you have to take the log of 1 **followed by 20 zeroes.** The summary is that $log(x)$ gets big VERY VERY slowly as we increase x.

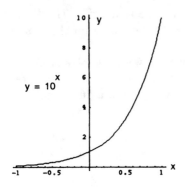

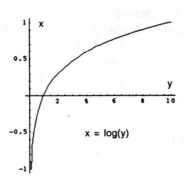

Here is the reason we are interested in logarithms. We saw that 10^x means to multiply x lots of 10 together. We also saw that if you multiply x lots of 10 by y lots of 10, then the result is $x + y$ lots of 10 multiplied together. This is just an English statement of the first law of exponents. In symbols, we have

$$10^x \times 10^y = 10^{x+y}.$$

Now this is pretty neat. In order to **multiply** the two numbers 10^x and 10^y we end up **adding** x and y together. Addition is a lot easier than multiplication; it takes some effort to multiply 25.12 by 79.43 without a calculator. But once we know that $10^{1.4} = 25.12$ and that $10^{1.9} = 79.43$ it is very easy to add the exponents $1.4 + 1.9 = 3.3$ and then say the answer is just $10^{3.3}$. We can look this up in a table and find that $10^{3.3} \approx 1995$.

What we have here is a marvelous method that converts **multiplication** into **addition**. Now, if we want to multiply 25.12 and 79.43 using this method, the first thing we have to do is find x so that $10^x = 25.12$ and y so that $10^y = 79.43$. We do this by looking up the **logarithm** of 25.12 and of 79.43. Once we find these values, we just **add** them. This is $x + y$ [$= 3.3$ in the example above] but we want to find 10^{x+y}. So we look up in the log table which number has logarithm equal to $x + y$. This gives us the answer to 25.12×79.43. Thus, with the help of a log table, we can multiply very complicated numbers using only addition. The key fact is:

> Logs Convert **Multiplication** into **Addition**.

Here is the summary of how to multiply x and y using logs:
(1) Find the logs of x and y
(2) add these logs together
(3) find the antilog of the result of step (2)

The next marvelous fact we will discuss is that

> Logs Convert **Exponentiation** into **Multiplication**.

We saw that

$$(10^x)^y = 10^{x \times y}.$$

Remember that this is because 10^x is x lots of 10 multiplied together. Now to raise **this** number to the power of y, we take y lots of (x lots of 10 multiplied together) and multiply all these together. So the result is just to multiply $y \times x$ lots of 10 together. This means that if we want to work out what we get by taking the y'th power of 10^x all we need do is compute $x \times y$. Of course, this still applies if x and y are weird numbers as in the last section. In this case, multiplication is far easier than working out the y'th power of a number. For example, to find the square root of a number we raise it to the power $1/2$. In particular, if we are finding the square root of 10^x, all we need to do is compute $(1/2) \times x$ and look up

$$10^{(1/2)x}$$

in a table. Let's clarify all of this with an example.

Example 7.2.1 *Find* $\sqrt{71}$ *using the log table.*

Solution: we want to know $\sqrt{71}$. The first thing to do is to find out what x should be so that $10^x = 71$. We do this by looking up the **logarithm** of 71 in a table. We find $10^{1.8513} = 71$, so $x = 1.8513$. Now we want the square root. This means we multiply by $1/2$ to obtain $(1/2)x = 0.9256$. So $\sqrt{71} = 10^{0.9256} = 8.426$.

Now it is true that calculators can do these things more quickly. However, very often we must solve algebraic equations using logarithm techniques. It is IMPOSSIBLE to do so without understanding these basic features of logarithms. Besides, it is pretty neat to be able to calculate $3.765^{5.147}$ without having to ask a dumb machine for the answer!

There are laws for logarithms just like there are laws for exponents. Logarithms are the **opposite** of exponentials. Every formula with exponents has a counterpart for logarithms. It is a bit like saying the same sentence in two different languages; the words sound different but the meaning is the same.

The first law of exponents:

$$10^a \times 10^b = 10^{a+b}$$

has a counterpart with logarithms:

First Law of Logarithms	$log(x \times y) = log(x) + log(y).$

> These formulae might look like they have nothing in common, but they actually say the same thing. Here is the explanation. The number $log(x \times y)$ is the power of 10 that gives $x \times y$, because that is the $log's$ JOB! In other words
>
> $$10^{log(x \times y)} = x \times y.$$
>
> But if we compute using the **first law of exponents** we get:
>
> $$10^{log(x)+log(y)} = 10^{log(x)} \times 10^{log(y)} = x \times y.$$
>
> These are the same so
>
> $$10^{log(x)+log(y)} = 10^{log(x \times y)}.$$
>
> This means that $log(x) + log(y)$ must equal $log(x \times y)$.

The first law is the most important fact about logarithms. It is as important as the first law of exponents. It even says the same thing in a disguised way. The other three laws of exponents can also be translated to logarithm language.

Second law of Logs	$log(1) = 0.$

The reason here is that the job of $log(1)$ is to be the number such that

$$10^{log(1)} = 1.$$

We know from the **second law of exponents** that

$$10^0 = 1$$

and so

$$10^{log(1)} = 10^0$$

so $log(1) = 0$.

Third law of Logs	$log(1/x) = -log(x).$

The **third law of exponents** tells us that

$$10^{-log(x)} = \frac{1}{10^{log(x)}}.$$

But we can simplify this because $10^{log(x)} = x$ so that

$$10^{-log(x)} = \frac{1}{x}.$$

This means that $-log(x)$ is the power of 10 we have to take to end up with the answer $1/x$ and so $log(1/x)$ MUST be $-log(x)$. This might seem confusing at first: it is a bit like looking in a mirror where everything has been switched around. However, if you think about it for a while you will get it all straight.

The first and third laws combined tell us

$$\begin{aligned} log(x/y) &= log(x \times y^{-1}) \\ &= log(x) + log(y^{-1}) \quad \text{using the first law of logs.} \\ &= log(x) + (-1)log(y) \quad \text{using the third law of logs.} \\ &= log(x) - log(y). \end{aligned}$$

Thus, **logs convert division into subtraction**, just as they convert multiplication into addition.

We know that $\quad log(a^2) = log(a \times a) = log(a) + log(a) = 2log(a)$.
Also $\qquad\qquad log(a^3) = log(a \times a \times a) = log(a) + log(a) + log(a) = 3log(a)$.

This pattern continues. Since a^p is a multiplied together p times, and taking logs converts multiplication into additions we find:

$$log(a^p) = log(a \times \cdots a) = log(a) + \cdots + log(a) = p \cdot log(a).$$

$$\boxed{\text{Fourth law of Logs} \qquad log(a^p) = p \times log(a).}$$

First the brief explanation. The number a^p is p lots of a multiplied together. Each a is $log(a)$ lots of 10 multiplied together. So a^p is p lots of ($log(a)$ lots of 10) all multiplied together. In other words $plog(a)$ lots of 10 multiplied together. Now a more algebraic explanation. If we work out

$$10^{p \times log(a)}$$

using the **fourth law for exponents** we get

$$10^{log(a) \times p} = \left(10^{log(a)}\right)^p.$$

But $10^{log(a)} = a$, so this simplifies to be a^p Thus,

$$10^{p \times log(a)} = a^p.$$

Since $p \times log(a)$ is the power of 10 which gives a^p we know that $log(a^p)$ equals $p \times log(a)$.

A common example of this law involves the log of a square root:

$$log(\sqrt{x}) = log(x^{1/2}) = \frac{1}{2}log(x).$$

We have seen that every rule for exponents can be translated into one for logarithms. This just reflects that $log(x)$ is the **inverse function** of 10^x.

In summary we have the four laws of logarithms which correspond to the four laws of exponents, and you should memorize them.

$$\boxed{log(x \times y) = log(x) + log(y) \qquad log(1) = 0 \qquad log(1/x) = -log(x) \qquad log(a^p) = p \times log(a).}$$

$$\boxed{\text{Also} \quad log(10^x) = x \qquad 10^{log(x)} = x \qquad log(x/y) = log(x) - log(y) \quad log(10) = 1.}$$

WARNING the following is a common mistake!

$$log(x + y) \neq log(x) + log(y).$$

For example $log(1) = 0$ and $log(2) \approx 0.3010$ but

$$0.3010 \approx log(2) = log(1 + 1) \neq log(1) + log(1) = 0 + 0.$$

There is NO WAY to simplify $log(x + y)$. Logs convert multiplication into addition, but they do not convert addition into anything at all!

7.3   Using Log tables.

The summary of this section is: to find the logarithm of a number between 1 and 9.999 use the log table at the end of the book. To find the log of any other number, count how many places you would need to move the decimal point to get a number between 1 and 9.999. Add this to the log of the number you get between 1 and 9.999.

Log(2) = 0.3	Log(0.2) = 0.3 - 1 = -0.7	**multiplying by 10 adds 1 to the Log**
Log(20) = 1.3	Log(0.02) = 0.3 - 2 = -1.7	**dividing by 10 subtracts 1 from the Log**
Log(200) = 2.3	Log(0.002) = 0.3 - 3 = -2.7	
Log(2000) = 3.3	Log(0.0002) = 0.3 - 4 = -3.7	

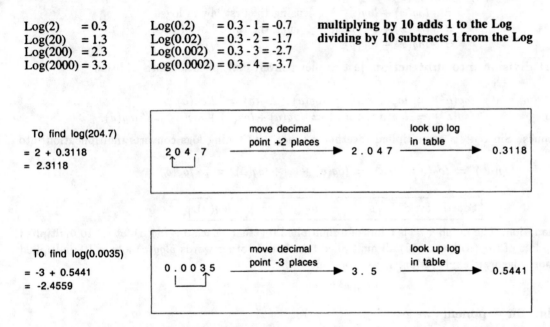

To find log(204.7)

= 2 + 0.3118

= 2.3118

move decimal point +2 places   2.047   look up log in table   0.3118

To find log(0.0035)

= -3 + 0.5441

= -2.4559

move decimal point -3 places   3.5   look up log in table   0.5441

It is amazing just how important logarithms and exponentials are. A calculator will tell you $log(71)$ or $10^{2.7}$ very quickly. But a calculator does not help you to understand logarithms and exponentials. This is a bit like music. You can turn on the radio and hear music, but that is not the same as being able to play a musical instrument. Why should anyone want to play a musical instrument?. The music the person plays has probably been recorded by a better musician and can be heard with much less effort by playing a CD. Why should anyone actually calculate with logarithms? Calculators do it much more quickly and accurately. However, you will not understand logarithms and exponentials if you just use a calculator. And understanding is very important! Without understanding, the instant answers provided by the calculator really have no value to you at all. So in this class you will get practise calculating by hand. In the process, the mystery of logarithms will disappear.

There are two ways you will be asked to calculate logarithms and exponentials. The first method is graphical. The second way uses a table of logarithms. These two methods are designed to teach you several skills. In particular, you will gain experience at using the laws of logarithms and exponents in many cunning ways. You will also come to understand the meaning of the phrase "the logarithm is just the inverse function of the exponential function." At the back of these notes is a table of logarithms. Of course, the table does not give the logarithms of all numbers since that is impossible. However, it does give the logarithms of numbers between 1 and 10. The nice thing is that we can use the table to find the logarithm of ANY number.

Example 7.3.1 *Find log(20) using the table.*

Solution: The first thing to realize is that the rules of logarithms are very helpful:

$$log(20) = log(2 \times 10) = log(2) + log(10).$$

Now we can look up $log(2)$ in the table: it is 0.3010. We do not need to look up $log(10)$ because we know $log(10) = 1$. Think about this! Thus, $log(20) \approx 1.3010$.

Scientific Notation. The number 3914 is written in scientific notation as 3.914×10^3. To write any number in scientific notation, write it as a number between 1 and 9.999 multiplied by a power of 10. Thus $0.00456 = 4.56 \times 10^{-3}$.

Similarly $log(200) = log(2) + log(100) \approx 2.3010$. We can summarize this little procedure: to find the logarithm of a number x write x in scientific notation as $y \times 10^n$ where n is a whole number and y is between 1 and 9.999. Then $log(x) = n + log(y)$. We can look up $log(y)$ in the table. In practise, what you should do to find n is this: **count how many places you need to move the decimal point until you get a number between 1 and 10. This count is** n.

Example 7.3.2 *Find* $log(0.0002)$ *using the table.*

Solution: To find $log(0.0002)$ we need to move the decimal point 4 places. Therefore,

$$log(0.0002) = -4 + log(2) \approx -4 + .3010 \approx -3.6990.$$

Common mistake this is **not** the same as -4.3010. When you work out $log(x)$ if x is bigger than 10 you can just put the number of places you moved the decimal point in front of the log. But when x is less than 1 you can not. **Understand** this example. Likewise, if we want to find $log(7000.)$ we have to move the decimal point 3 places so

$$log(7000) = 3 + log(7) \approx 3.8451$$

The next step is to learn how to use the log tables to convert the logarithm of a number back into the original number. Suppose you are told that the logarithm of a certain number a is 0.4771. If you are asked to find the number a, then you must look through the table to find the number you take the log of to get 0.4771. This turns out to be 3. Hence, $a = 3$ is the answer. This is expressed by saying that 3 is the **antilog** of 0.4771. Thus,

> 0.4771 is the log of 3   so   3 is the antilog of 0.4771.

Now **antilog** is just another name for the function $f(x) = 10^x$. In other words:

> $$antilog(x) = 10^x$$

It is more convenient to talk about the "antilog" function instead of the "10-to-the-power function". It also helps us remember how to find the antilog from a table of logarithms: we just use the table backwards.

But what if we want to find the antilog of 1.4771. We look in the table and it is not there! The only logarithms we see are all between 0 and 1. Think! We are trying to find $antilog(1.4771) = 10^{1.4771}$. This equals $10^1 \times 10^{0.4771}$. We can find $10^{0.4771} = 3$ in the tables and we know $10^1 = 10$. So the answer is $10^{1.4771} = 30$. We summarize this by giving the answer to a slightly different problem which can be solved using the same technique (and which you should solve now):

> $$antilog(6.4771) = 10^6 \times antilog(0.4771)$$

Now we come to a **tricky** point: find the antilog of a **negative** number. To find $antilog(-2.8)$ do the following. First write $-2.8 = -3 + 0.2$. The reason we do this is because we can look up the antilog of 0.2 in the table. What we have done is to express -2.8 as a **negative whole number** plus a number between 0 and 1 for which we can look up the antilog. Having done this:

> $$antilog(-2.8) = 10^{-2.8} = 10^{-3+0.2} = 10^{-3} \times antilog(0.2) \approx 10^{-3} \times 1.58 \approx .00158.$$

7.4 Using the graph of 10^x.

At the back of these notes is a graph of the function 10^x for x between 0 and 1. We can use this graph to find $10^{0.3}$. Do it!. But the graph allows us to find any power of 10 using the laws of exponents.

Example 7.4.1 *Find* $10^{2.3}$ *using the graph.*

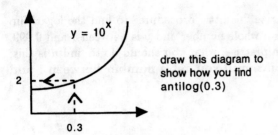

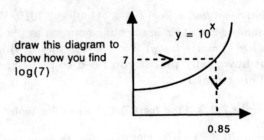

Solution: We know this equals $10^2 \times 10^{0.3}$. We also know that $10^2 = 100$ and we can use the graph **as shown in the diagram** to find $10^{0.3} \approx 2$. So the answer is $10^{2.3} \approx 200$.

As another example,
$$10^{-3.7} = 10^{0.3} \times 10^{-4} \approx 2 \times 0.0001 = 0.0002.$$

So the graph is good for finding powers of 10.

On the other hand, it is also good for finding logarithms. We just work backwards.

Example 7.4.2 *Find log(7) using the graph.*

Solution: We must look for a number along the x-axis so that the corresponding point on the graph has y-value 7. We discover that $x = 0.85$ **as shown in the diagram**: so we find that $log(7) \approx 0.85$.

Again we can use the graph to find the log of any number. For example, to find $log(2000)$, we just notice that $log(2000) = log(2) + log(1000)$. We then use the graph to find $log(2)$ and we know $log(1000) = 3$. Work this out to be sure you understand.

7.5 Using Logs to Multiply and Divide

The next thing is to learn how to use logarithms to do multiplication. Logarithms were invented to make multiplication easier. Long multiplication is a slow business. To calculate 27.35×4.762 by hand takes a long time. But addition is much faster. The trick then is to use logarithms to convert a multiplication problem (which takes a long time) into an addition problem (which can be done quickly). This is done as follows: set $x = 27.35 \times 4.762$. Our job is to find x. The clever bit is to take logs of both sides:

$$log(x) = log(27.35 \times 4.762) = log(27.35) + log(4.762).$$

So now we find using our log tables that $log(4.762) \approx 0.6778$ and $log(27.35) \approx 1.4370$. We then add these to get 2.115. Now, this is NOT the answer we want. This is $log(x)$. So, we must again use our tables to find $x = antilog(2.115) \approx 130.2$

Now this is indeed a much slower way to multiply than to use the calculator. That is not the point. Doing this will help you to understand logs. In summary, here is how to multiply x and y using logs:

(1) Find $log(x)$ and $log(y)$.
(2) Add these: $log(x) + log(y)$. This is the log of the answer.
(3) Find the antilog of the result of step (2).
(4) Ask yourself "is the answer reasonable?"

Why does this work? Well we know
$$log(x \times y) = log(x) + log(y).$$

We want to work out $x \times y$. So instead we work out $log(x \times y)$ and find the antilog of the result.

Now that you have mastered multiplication, let's see how to use logs to do division. Essentially, it is the same method as multiplication except that we SUBTRACT instead of ADDING the logs. Here is how to divide x by y using logs:

(1) Find $log(x)$ and $log(y)$.
(2) Subtract: $log(x) - log(y)$. This is the log of the answer.
(3) Find the antilog of the result of step (2).
(4) Ask yourself "is the answer reasonable?"

Why does this work? Again, from the rules

$$log(x/y) = log(x) - log(y).$$

We want to calculate x/y. So instead we work out $log(x/y)$ and then find its antilog.

7.6 Using Logs to find powers of numbers.

How do we calculate $\sqrt{3} = 3^{1/2}$ without a calculator?. Use logs!. The problem is to find $x = 3^{0.5}$. This is tough. But, if we take logs of both sides we find

$$log(x) = log(3^{0.5}) = 0.5 \times log(3).$$

Now we need only work out $0.5 \times log(3) = 0.2385$. We then use the table to find the antilog of this result to get $x \approx 1.73$. It's important to see how to use the table to find antilogs properly. In this case, we want to find x so that $log(x) = 0.2385$. To do this, we search the table for the number 0.2385. In the table, look at the horizontal row corresponding to 1.7. On this row we find that $log(1.73) = 0.2380$ and $log(1.74) = 0.2405$. But our desired value 0.2385 is not there; it lives IN BETWEEN 0.2380 and 0.2405. Therefore, we approximate by using $x = 1.73$. Make sure you see how to carry out this procedure. In summary: if the number you want is not in the table, use the **closest** one.

We can use the same trick to find 2^{30}. It would take a long time to multiply 30 twos together, but logs give us the approximate answer quickly. Thus, $log(2^{30}) = 30 log(2) \approx 9$. This means $2^{30} \approx 10^9$. In fact, it is easy to work out a power of 2 this way. We only need to remember that $log(2) \approx 0.3$. Then to work out any power of 2, for example 2^n, just compute $0.3 \times n$. This is the log of the answer. To get an approximate answer, we forget about the fractional part. In particular, to find 2^{64}, we compute $0.3 \times 64 = 19.2 \approx 19$. Thus

$$2^{64} \approx 10^{19}.$$

Powers of 2 are very important. They show up in many contexts such as computer science and population growth.

Logs can be used to calculate any number raised to any power.

$$\boxed{a^x = 10^{x log(a)}}$$

Explanation: the fourth law tells us this works because

$$a = 10^{log(a)}.$$

Raising both sides to the power of x gives

$$a^x = \left(10^{log(a)}\right)^x = 10^{x \cdot log(a)}.$$

Here is how to calculate a^x using log tables.
(1) Find $log(a)$.
(2) Multiply by x giving $x \times log(a)$. This is the log of the answer.
(3) Find the antilog (same as finding 10 to the power) of the result from step (2).
(4) Ask yourself if your answer is reasonable.

We can convert powers of any number a into powers of any other number b. For example:

$$2^x = 10^{x \times 0.3010} \qquad 10^x = 2^{3.322 \times x}$$

We shall work through one now.

Example 7.6.1 *Convert 5^y into a power of 3.*

Solution: first we find (explained below)

$$3^{1.465} \approx 5.$$

Once we know this, we just raise both sides to the power of y. This gives the answer:

$$3^{1.465 \times y} = 5^y.$$

In order to find the power of 3 that gives 5, we have to solve the equation

$$3^x = 5.$$

We do this by taking logs of both sides to get

$$log(5) = log(3^x) = x \times log(3)$$

We discover the following:

$$x = \frac{log(5)}{log(3)} \approx \frac{.6990}{.4771} \approx 1.465$$

7.7 Solving Algebraic Problems Using Logs.

Logs are used to solve equations where the unknown, x, appears as an exponent. This is very common in equations which come up in biology and other sciences. As a very simple example, suppose we want to solve

$$3^x = 7.$$

To do this problem, we need to get to a situation where we have $x = something$. Alas, x is an exponent and this causes serious difficulty. No amount of algebraic trickery will help us here. But there is hope. We can start by **taking logs of both sides.** Thus, we have that

$$log(3^x) = log(7).$$

The point of this trick is that

$$log(3^x) = x log(3).$$

This means that we get

$$x \times log(3) = log(7)$$
$$\Rightarrow x = \frac{log(7)}{log(3)} \qquad \text{dividing both sides by } log(3).$$

We can now find x easily using our log tables. The idea to remember is this:

| If we have a^x and we need to find x, we can "bring x down to ground level" by taking logs. |

Example 7.7.1 *Solve the equation $M = M_0 2^{-t}$ for t.*

Solution: this equation describes, for example, the mass, M, at time t of something which cuts its mass in half every 1 unit of time. This happens for a radio-active substance which is decaying with time. Using this example, the problem really asks "if we start with a mass of M_0 how long until the mass decays to M?" The unknown is t; the quantities M and M_0 are considered known constants in this problem. So our answer will express t in terms of M and M_0. To get started, take logs of both sides:

$$log(M) = log(M_0 2^{-t})$$

Now use $log(M_0 2^{-t}) = log(M_0) + (-t) \times log(2)$ and solve to get

$$t = \frac{log(M_0) - log(M)}{log(2)}.$$

Example 7.7.2 *Solve* $A = BK^{t-t_0}$ *for* t.

Solution: this time, the quantities A, B, K and t_0 are regarded as "given." We must find t in terms of these constants. The same idea applies: t is an exponent, so we "undo" the exponentiation by taking logs of both sides. The only difference here is that the exponent is not just t but $t - t_0$. This means that we will obtain $t - t_0 = something$. We then just add t_0 to both sides. Thus,

$$
\begin{array}{rll}
& A = BK^{t-t_0} & \\
\Rightarrow \quad log(A) = & log(B \times K^{t-t_0}) & \text{taking logs of both sides.} \\
= & log(B) + log(K^{t-t_0}) & \text{using } log(xy) = log(x) + log(y). \\
= & log(B) + (t - t_0) \times log(K) & \text{using } log(x^a) = a \times log(x). \\
\Rightarrow \quad (t - t_0) \times log(K) = & log(A) - log(B) & \text{subtracting } log(B). \\
\Rightarrow \quad (t - t_0) = & \frac{log(A) - log(B)}{log(K)} & \text{dividing by } log(K). \\
\Rightarrow \quad t = & t_0 + \left(\frac{log(A) - log(B)}{log(K)} \right) & \text{adding } t_0.
\end{array}
$$

Notice how we proceed **one step at a time**. Each step is on a new line. Each step is also clearly explained. Writing out this explanation helps the person reading it understand what is going on. More importantly, it helps the person doing the problem organize their thoughts logically. The benefits of this approach are best realized if you take the time to read aloud each line in ENGLISH. So, I might read the first two lines of the above solution in the following way: "A equals B times K to the t minus t zero implies that the *log* of A equals the *log* of B times K to the t minus t zero. This follows by taking *logs* of both sides of the first equation." It might seems pedantic, but it's actually very helpful. It reminds us that mathematics is really a language that SAYS something. If you want an "A" in this course, you will take great care to write your answers in this manner.

The types of algebraic problems above will frequently arise in textbooks and lectures in some of the other classes you will take. You only need to remember two things:
(1) Use logs when the unknown is an exponent.
(2) Do one step at a time.

7.8 Graphing with Logarithmic Coordinates.

In many situations, it is very helpful to use logarithms to plot graphical data. Suppose we have some data in the form of 25 pairs of values

$$(x_1, y_1), (x_2, y_2), \cdots, (x_{25}, y_{25}).$$

Then we might plot this data on a graph by making a point on the graph for each of these pairs. Often when we do this we get something which suggests that the points might lie on some kind of curve. The question then is, "which curve?" As we shall explain in a moment, it often helps to plot a different graph by plotting the points

$$(x_1, log(y_1)), (x_2, log(y_2)), \cdots, (x_n, log(y_{25})).$$

Of course, this will give a different graph. Mathematically this can be expressed by saying that instead of graphing the function $y = f(x)$ you graph the function $Y = log(f(x))$. This means $Y = log(y)$. So in a sense we are converting one graph into another. The advantage is that in many situations this new graph may be a straight line. Naturally, this does not always happen. However, it does work in many contexts which arise in economic or biological systems.

Let's see what happens when we do this with the function $f(x) = 2 \times 3^x$. If we draw the graph of this function, it is a not very revealing curve. But suppose instead that we plot the graph $Y = log(2 \times 3^x)$. What should we see? Well, we can find out by simplifying

$$Y = log(2 \times 3^x) = log(2) + x \times log(3).$$

Now, $log(3) = 0.4771$, and $log(2) = 0.3010$. So the graph of $Y = 0.3010 + 0.4771 \times x$ is just a straight line. This is actually very powerful. In an application, we will not usually have an explicit formula for $f(x)$. Instead, we will only have a few data points. But if we use this procedure, we can figure out an explicit formula for $f(x)$

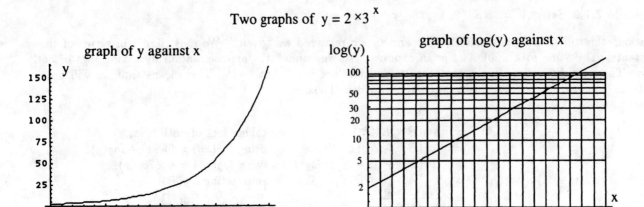

Two graphs of $y = 2 \times 3^x$

using the straight line. This is truly remarkable. The graph of $y = f(x)$ initially is just some curve we don't understand; but, once we convert it to a straight line, we can draw many useful conclusions.

For example, suppose that we are graphing the number of people with flu in New York on a day by day basis. We plot the data for 25 days and see a curve that does not mean too much to us. So, we try plotting the logarithm of the data instead. Now we see an almost straight line starting on day 1 and ending on day 25. If we need to predict how many people will have flu in 3 days time, we might continue the straight line graph and guess that it would give the correct answer. The point here is that we have actually brought linear extrapolation to bear on a graph that is not a straight line. We do this via trickery; that is, we use *logs* to convert our graph into a straight line. We extrapolate using the line and then convert back using antilogs. The summary is this: if we plot the graph of $Y = log(f(x))$ and we get the straight line $Y = m \cdot x + b$, then the function $f(x)$ is given by

$$f(x) = K \cdot a^x \qquad \text{where } K = 10^b \text{ and } a = 10^m.$$

The explanation here is just the following. If $f(x) = K \cdot a^x$ then

$$Y = log(K \cdot a^x) = log(K) + log(a^x) = log(K) + x \times log(a)$$

is a straight line with slope $m = log(a)$ and y-intercept $b = log(K)$.

Let's try to understand the significance of K and a in the flu example. On day 0 the number of infected people is $f(0) = K \cdot a^0 = K$. Also

$$\frac{\text{number of sick people on day n}}{\text{number of sick people on the previous day, (n-1)}} = \frac{K \cdot a^n}{K \cdot a^{n-1}} = a.$$

Thus, K is the number of people with flu on day 0 and each day there are a times as many infected people as there were the day before.

7.9 Applications of Exponential functions.

The phrase "exponential growth" is used often by people to mean something is getting big very quickly. For example, saying "the number of people using the internet is growing exponentially" does not just mean that every day more people are using it. It means that the number of people using it is increasing more and more rapidly. To make this a bit more concrete, suppose there are two million users in 1995, four million users in 1996, eight million users in 1997, and so on. Each year the number of users doubles. So, the number of users after n years is 2^n million people where $n = 1$ means the year 1995.

Things that reproduce tend to grow exponentially. For example, suppose a pair of rabbits produce 6 children and then go to bunny heaven. The first generation has 2 rabbits. The next generation has 6 rabbits. These 6 form 3 pairs, so that the third generation has 3×6 bunnies. Quite naturally, these form 3×3 pairs, so the fourth generation has $3 \times 3 \times 6$ bunnies. If we think about it, every parent bunny gets replaced by 3 children. So each generation has 3 times as many bunnies as the previous generation. We start with 2 bunnies in generation 1 so after n generations there are

$$2 \times 3^{n-1}$$

bunnies. Think very hard about this: why 3^{n-1} and not 3^n? We can also write the equation as

$$\text{number of bunnies after n generations } = (2/3) \times 3^n.$$

Remember that exponentials are about REPEATED multiplication. That "rabbits multiply" and that each generation multiplies what went before should lead us to have guessed that bunny populations grow exponentially (unless, of course, there is something that eats the bunnies). If we replace bunnies by viruses, or people, the story above is unchanged.

Any situation in which the same thing happens over and over again is likely to lead to exponential functions. For example, suppose you dive down into the ocean. As you go deeper it gets darker, because the light gets absorbed by water more completely as you dive deeper. Why does this happen? Well, if you shine light into water, for every meter of water that the light travels through, some portion of the light is lost. If the water is very clear, then you only lose a little of the light. On the other hand, if the water is cloudy you lose a much bigger portion of light for every meter of depth. Now suppose the water is really cloudy and that half the light is lost as it travels through one meter of water. How much light is lost in traveling through 2 meters of water? If we start with 1 unit of light, then 1/2 a unit gets through the first meter of water. Now this light must travel through the next meter of water. So only half of it gets through, so that $(1/2) \times (1/2)$ units of light get through 2 meters of water. At 3 meters depth only $(1/2) \times (1/2) \times (1/2)$ units survive. For every meter of depth, you lose half the light

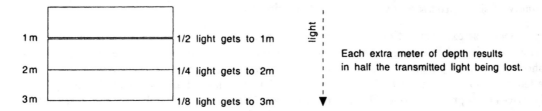

you had before. So at a depth of d meters the amount of light that gets through from one unit of light is $(1/2)^d$. For example, if $d = 10$ then $(1/2)^{10} \approx .001$ of a unit gets through. The brief explanation is that each meter of depth multiplies how much light can pass through by 1/2. We have here an example of repeated multiplication; in other words, an exponential function. The same thing applies to light shining through fog or the leaves of trees.

In the financial world, money earns interest. There are various schemes which decide exactly how much interest a given clump of money will earn. Money deposited in a bank earns **compound interest**. This provides another example of exponential functions. What does it mean have to compound interest? Suppose we start with a clump of 300.00 dollars (this is called the principal). We invest this money into an account which compounds interest annually at 3 percent. At the end of the first year, the account will have $300.00 + .03(300.00) = 309.00$. Now suppose we leave this money in the account for a second year. This time the account manager will compute interest using 309.00 to give us $309.00 + .03(309.00) = 318.27$. The point is that in the first year we earned 9.00 dollars interest whereas in the second we earned 9.27 dollars; we earn more interest with each passing year. This is actually an example of an exponential function. The general formula for compound interest is:

$$\boxed{A = P(1+i)^n}$$

where: $P = $ the principal, or initial amount of money.
$i = \dfrac{\text{Annual Interest Rate}}{\text{Number of Compoundings Per Year}}$.
$n = $ the number of compounding periods.
$A = $ total amount after n compounding periods.

The explanation for this formula is as follows. Suppose the bank breaks the year up into twelve periods for compounding interest. The annual rate of interest is 5%. Then in one 1/12 of a year the interest is 1/12 of 5% of the money in the bank. This means that at the end of the month you have $1 + (0.05/12)$ times as much money as you had at the start. If we set $i = 0.05/12$ then each month that goes by your money is multiplied by $1 + i$. So after n months you have $(1 + i)^n$ times what you started with.

Credit cards use this method by turning their customers into banks! They normally compound interest on debt EVERY MONTH.

Example 7.9.1 *A college student has a credit card. He has purchased a car stereo for 300.00 dollars using this card. The student is very pleased because the credit card offers a special deal: no payments for the first six months. Of course, the card compounds monthly and interest on the debt will continue to accrue. The credit card has a typical annual percentage rate (APR) of 17.9 percent. How much money does the student owe at the end of the six months?*

Solution: we can use the formula above. Here $P = 300.00$ dollars and $n = 6$ months. Since interest is compounded monthly, there are 12 compoundings per year. So we have

$$i = \frac{.179}{12} = .0149.$$

Now we write

$$A = 300 \times (1 + .0149)^6 = 327.87.$$

This is amazing. The student is paying 27.87 dollars to take advantage of the "special offer !"

Exponential functions show up in many other contexts. Examples include radioactive decay, electronics, etc. The basic reason is that many of these processes involve repeated multiplication.

There is a useful way to visualize exponentiation which has various applications. A **binary tree** is shown at the top of the page. It is a collection of lines called **branches** and blobs called **nodes** or **vertices.** They are like family trees which show children of parents and so on. Trees are useful in computer science and mathematics. The top vertex is called the **root.** The next layer has 2 vertices. The 3'rd layer has 4 vertices. Each layer has twice the number of vertices than preceding one. Often computers solve problems by searching many possibilities. To do this they construct a "search tree." The computer that recently defeated the human world chess champion (and sent him wailing from the room) used search trees.   If a problem involves searching a tree, then the number

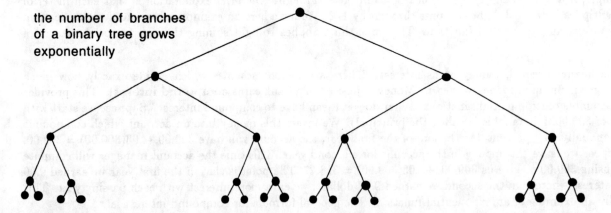

the number of branches
of a binary tree grows
exponentially

of possibilities grows exponentially. In practise, this means that even the most powerful computers soon run into difficulties.

7.10   Logarithms in Science.

Many mathematical situations involve numbers that are too large to easily contemplate. One way to overcome this is to use the logarithm since it makes size more manageable. We shall describe several examples in this section.

The **Richter Scale** is used to measure the strength of an earthquake. This is actually a logarithmic scale. An earthquake of magnitude 6 is 10 times as strong as one of magnitude 5. The Richter magnitude of an earthquake is

$$log(\text{distance ground moves up and down}).$$

Remember that the graph of the function $log(x)$ increases VERY slowly as x increases. So, the part of the Richter scale from 0 to 9 covers a range of 10^9 or one billion.

An Englishman called Norman Robert Pogson created the system we use to describe the brightness of a star in the night sky. This is called **stellar magnitude.** A large magnitude means NOT very bright. So, if one star is 5 magnitudes more than another star, this means first star is 100 times dimmer than the second. In other words,

$$\text{stellar magnitude} = -2.5 log(\text{brightness}).$$

The minus sign is to ensure dimmer stars get a bigger magnitude. The Sun's magnitude is -26.7, the full Moon is -11. Thus, the difference in magnitude between Sun and Moon is 15.7. Since a difference of 5 means 100 times dimmer, this means the Sun is about a million times brighter than the moon. The brightest star, Sirius, is -1.6. The faintest stars visible in large telescopes are magnitude 20. Technically, what is described here is called **apparent magnitude** since it indicates the brightness we see.

Think about chemicals. Some chemicals are acidic, some are alkaline, and others are neutral. Acidity is the opposite of alkalinity. The scale used to measure this is called th pH scale. 7 is neutral (for example water) , below 7 is acidic, above 7 is alkaline. The scale is also **logarithmic.** A ph of 5 is 10 times as acidic as a pH of 6. A pH of 4 is 10 times as acidic as a pH of 5. The acidity of a chemical depends on how many hydrogen ions H^+ are in it. (A hydrogen atom has one electron, remove the electron and we get a hydrogen ion which therefore has a positive charge.) To measure this, we need some units. A **mole** or **gram molecule** is 6.02×10^{23} molecules (Avogadro's number.) A molecule of water, H_2O has a **molecular weight** of 18 (1 for each of the two hydrogen atoms and 16 for the oxygen atom). A mole of molecules has a weight in grams equal to the molecular weight.

The pH value for acids is

$$-log(\text{number of moles of } H^+ \text{ ions in one liter}).$$

A neutral chemical has only 10^{-7} moles of H^+ ion per liter. So the pH for this is

$$-log(10^{-7}) = 7.$$

An extremely strong acid has 1 *mole/liter* and so a pH of $-log(1) = 0$. So we are (usually) taking the log of a number less than 1 and this is negative. The minus sign makes the answer positive.

There is a similar definition for alkalinity. How alkaline something is depends on how many hydroxide OH^- ions it has. (A hydroxide ion is one atom of oxygen plus one atom of hydrogen bonded plus one extra electron giving the ion a negative charge.) If we combine H^+ with OH^- we get one molecule of water H_2O which is neutral. Thus, the acidity of one H^+ ion exactly cancels the alkalinity of one hydroxide ion OH^-. The pH of an alkaline substance is

$$14 + log(\text{number of moles of } OH^- \text{ ions in one liter}).$$

One mole of OH^- ions and one mole of H^+ ions have the same number of ions each. So, if we mix equal quantities of an acid with pH $(7 - 2)$ and an alkali with pH $(7 + 2)$ they exactly cancel to leave us with a neutral pH of 7.

An important consequence of pH involving logs is this: if we have 1 gram of acid and we want to dilute it to increase the pH by 1 then we must add 9 grams of water. This will divide the concentration of H^+ ions by 10. Instead of the original number of H^+ ions being in 1 gram of liquid they are now spread out among 10 grams of liquid. So

$$\begin{aligned}
\text{pH of diluted acid} &= -log(\text{conc. in original acid}/10)\\
&= -[log(\text{conc. in original acid}) - log(10)]\\
&= -[log(\text{conc. in original acid}) - 1]\\
&= 1 - log(\text{conc. in original acid})\\
&= 1 + (\text{pH of original acid}).
\end{aligned}$$

There is a story of a biologist who wanted to find out the average pH of the lemons from three different kinds of lemon tree. So, the biologist took one lemon from each tree, combined the juices, and then measured the resulting pH. Did the biologist get the right answer? Does the idea of average pH make any sense at all?

Have you ever been to a rock concert? Everyone knows that Eddie Van Halen plays REALLY loud. If fact, he plays so loud that before you can leave for the show, even your mother is screaming "You'll come back deaf from that concert. He plays at 200 decibels!" But just what does 200 decibels really mean? Here is a list of sound sources together with their approximate decibel output.

Sound Source	Total Loudness Power	Decibels
Airplane Engine Nearby	1,000,000,000,000	120dB
Inside Airplane Cabin	10,000,000,000	100dB
Pneumatic Drill	1,000,000,000	90dB
Noisy Office	10,000,000	70dB
Ordinary Conversation	10,000	40dB
Quiet Conversation	1000	30dB
Soft Rustle of Leaves	10	10db

This system of units was invented by the Bell Telephone Company to measure signal loudness in telephone equipment. The basic unit is the "Bel" which is named after Alexander Graham Bell. These units are now used in virtually every sound application. Now one Bel is a large unit, so the standard practice is to divide by 10 to get decibels. This is a logarithmic scale. The formula is

$$\text{number of dB} = 10 log(\text{Total Loudness Power}).$$

As an interesting exercise, let's see how many decibels change occurs when we double a given loudness power. Suppose we have guitar amp which is set to produce sound with P total loudness power. Then our formula gives us $10 log(P)$ dB. Now suppose we crank the amp to double the loudness to $2P$. We have

$$10 log(2P) = 10(log(P) + log(2)) = 10 log(P) + 3.01$$

dB. So doubling the loudness power produces only 3dB change! To cement this in your mind, you might compute the loudness power corresponding to 200dB. If your mother is right, how will Eddie compare to a nearby airplane engine?

7.11   Half Life and Doubling Time.

Some chemical elements are radioactive. To be more precise, certain **isotopes** of certain elements are radioactive. This means that over a period of time such an isotope spontaneously changes into something else. It does not do this all at once. Instead, over a given period of time, a certain proportion will decay. The time taken for half an initial amount to decay is called the **half-life** of the isotope.

Suppose that an isotope called W has a half-life of 10 years. This means that if you have any amount, (say A grams) of it, then after 10 years only half that amount ($A/2$ grams) remains. After 20 years only 1/4 of that amount remains. This is because the amount is multiplied by 1/2 for every 10 years that passes. After 30 years only $(1/2)^3 = 1/8$ remains. After some number, n of half-lives the amount that remains is

$$\left(\frac{1}{2}\right)^n$$

times what you started with. Thus if you start with A grams of isotope then

$$\boxed{(\text{amount remaining after } n \text{ half lives }) = A \left(\frac{1}{2}\right)^n.}$$

Thus with a problem involving half-life, the idea is to **measure time in units of half-lives.** This is done by dividing the time span t [in years] by the half-life K. Thus

$$(t \text{ years}) = (t/K \text{ half lives.})$$

Once you have converted time measurements into units of half-lives, since each half-life multiplies what you had by 1/2, after t years you end up with

$$\left(\frac{1}{2}\right)^{t/K}$$

times what you started with.

The isotope carbon-14 has a half-life of about 5730 years. Carbon-14 decays into (non-radioactive) Nitrogen. This is very useful for dating things. The method of using Carbon-14 to date things is called **radiocarbon dating.** It works best over the period 50,000 B.P. (=before present) to 500 B.P. This is very useful for dating objects from recent human history. At any moment the amount of Carbon-14 in the entire world is only 70 tons and this is continually decaying. However, Nitrogen atoms in the atmosphere are continually converted into Carbon-14 by impact with cosmic rays. So the amount of Carbon-14 in the atmosphere is fairly constant. Plants absorb carbon-dioxide from the atmosphere, and with it they get some carbon-14. Animals that eat plants then get the carbon-14. Animals that eat those animals then get it. The ratio of carbon-14 to ordinary carbon in a living organism is the same as the proportion that appears in the atmosphere (actually there is a small correction). When the organism dies, it stops absorbing new Carbon-14. Thus, the Carbon-14 in its body begins to decay. So the ratio of Carbon-14 to regular carbon decreases once the organism has died. Therefore, to date a dead organism, one measures the proportion of Carbon-14 to regular carbon in the organism.

We will now describe a good method to do problems involving half-lives. Think about the function

$$f(t) = A\left(\frac{1}{2}\right)^{t/K}.$$

Because the exponent is t/K instead of t we find that it is necessary to **increase t by K in order to halve** $f(t)$. In other words:

$$
\begin{aligned}
f(t + K) &= A\left(\tfrac{1}{2}\right)^{(t+K)/K} \\
&= A\left(\tfrac{1}{2}\right)^{1+(t/K)} \\
&= \left(\tfrac{1}{2}\right) \times A\left(\tfrac{1}{2}\right)^{t/K} \\
&= \left(\tfrac{1}{2}\right) \times f(t).
\end{aligned}
$$

If you have an isotope with a half life of K years, and if you start with an amount A at $t = 0$ then the amount after t years is

$$A\left(\frac{1}{2}\right)^{t/K}.$$

The number t/K is the time in years divided by the half life. Thus it is the **number of half lives** that occur during t years. For example if the half life is 10 years, then a time period of 30 years is a period of 3 half-lives. The amount of isotope decreases by a factor of 1/2 for each half-life that passes by. Thus during 3 half-lives the proportion you end up with is $(1/2)^3$ what you started with. This is what the formula gives.

Example 7.11.1 *The half-life of a certain element is 40 years. If there was 17 grams in 1900 how much is there in 1950.*

Solution: The period from 1900 to 1950 is 50 years which is $50/40 = 1.25$ half lives. The initial amount is multiplied by $(0.5)^{1.25}$ in this time. So in 1950 there is

$$17 \times \left(\frac{1}{2}\right)^{1.25} \approx 7.15 \text{ grams.}$$

Some quantities double in a given period. For example bacteria growing with an unlimited food supply and plenty of space will double their number in a certain period of time called the **doubling time.** The situation is very similar to radio-active decay and half-life. The only mathematical difference is that we set K to be the doubling time. Then a period of t years is t/K doubling times. After this number of doubling periods, the amount (of bacteria, or whatever is doubling) has been multiplied by

$$2^{t/K}.$$

Thus if the initial amount is A at time $t = 0$ and the doubling time is K

$$\boxed{\text{(amount after } t \text{ years)} = A \times 2^{t/K}.}$$

Example 7.11.2 *Bacteria are doubling in mass every 7 hours. Initially there are 20 grams. When will there be 900 grams?*

Solution: Let t be the time in hours. Then t hours is $t/7$ doubling periods. Thus after t hours there is $2^{t/7}$ times what we started with. We start with 20 grams so after t hours there is

$$20 \times 2^{t/7}.$$

We want to know when this equals 900 so we need to solve the equation

$$20 \times 2^{t/7} = 900.$$

Thus

$$2^{t/7} = 900/20 = 45.$$

This equation can be solved using logs. The answer is

$$t = 7\log(45)/\log(2) \approx 38.4 \text{ hours}.$$

This answer is reasonable, since 38.4 hours is about 5 doubling periods (of 7 hours). During 5 doubling periods, the amount increases by $2^5 = 32$. We started with 20 grams, and 32 times this is 640 grams, which is the right ball-park.

7.12   Exponential Functions.

The functions

$$f(t) = A2^{t/K}$$

played a key role in the previous section. Using the number 2 is very convenient for problems involving half-life or doubling time. However with slightly more effort any other number can be used. The number $e = 2.71828..$ is often used. Thus the formula

$$f(t) = Ae^{rt}$$

will often be seen in books when discussing half-life for example. The point is that all exponential functions share a fundamental property: they grow "exponentially." This means:

$$\boxed{\text{The time taken to double is always the same } \textbf{provided f(t) is an exponential function.}}$$

For example, consider the exponential function

$$f(t) = K \times a^t.$$

The amount of time that must elapse for $f(t)$ to double is always the same. This time depends on the number a but not on K. Thus, if $a = 4$ then the time it takes for $f(t)$ to double is $1/2$ because

$$f(t + 1/2) = K \times 4^{t+.5} = K \times 4^t 4^{0.5} = K \times 4^t \times 2 = 2f(t).$$

If $a = 10$ then the doubling time is 0.3010 (because $10^{0.3010} \approx 2$) since

$$f(t + 0.3010) = K \times 10^{t+.3010} = K \times 10^{0.3010} 10^t = K \times 2 \times 10^t = 2f(t).$$

The doubling time (call it T) for the function $f(t) = K \times a^t$ is

$$\boxed{T = \frac{log(2)}{log(a)}}.$$

This comes from solving the equation

$$a^T = 2.$$

The reason this gives the doubling time is because after T units of time we want $f(t + T)$ so:

$$2f(t) = 2 \times (K \times a^t) = f(T + t) = K \times a^{T+t} = K \times a^T a^t = a^T \times (K \times a^t) = 2f(t).$$

Dividing by $K \times a^t$ we find we must solve $a^T = 2$. Take logs to get

$$log(a^T) = log(2)$$
$$\Rightarrow \quad T \times log(a) = log(2)$$

$$\boxed{T = \frac{log(2)}{log(a)}}.$$

If a quantity doubles in 10 years, then go backwards 10 years and it is half what it is now. Thus doubling time is **negative** of half life. Radio-active isotopes have positive half-life and negative doubling time. For bacteria it is the reverse. So the formula above can also be used to find the half life. This is simply:

$$\boxed{T = -\frac{log(2)}{log(a)}}.$$

We can also derive this by solving:

$$a^T = \frac{1}{2}.$$

7.13 Logarithms to other bases

The logs and antilogs we have talked about so far involve base 10. Thus,

$$antilog(x) = 10^x \qquad log(10^x) = x.$$

We can actually use any other positive number b in place of 10 and consider **logarithms to base b**. Thus

$$\boxed{x = log_b(y) \qquad \text{MEANS} \qquad y = b^x.}$$

Read $log_b(x)$ as "logarithm to base bee of exx". For example $log_2(8) = 3$ because $2^3 = 8$. Also $log_5(25) = 2$ because $5^2 = 25$. The most important bases for logarithms are 10, 2 and 2.718281.... Logarithms to base 2 are very useful for computers, cybernetics, combinatorics and information theory. The strange number 2.718281.... is called e and we will have more to say about it later. Logarithms to base e are very useful in science, engineering and math. Instead of writing $log_e(x)$ the notation $ln(x)$ is used. This refers to **natural logarithms**. The word "natural" is code for e. You will discover later that

$$\frac{d}{dx}(e^x) = e^x \qquad \frac{d}{dx}(ln(x)) = \frac{1}{x} \qquad \int x^{-1} \, dx = ln(x).$$

These nice formulae are some of the main attractions of e.

$$\boxed{ln(x) = log_e(x) = log_{2.71828..}(x).}$$

There is a simple method to convert between natural logs and logs to base 10.

$$\boxed{ln(x) = ln(10)log(x) \approx 2.302 log(x).}$$

This works because

$$e^{ln(10)log(x)} = \left(e^{ln(10)}\right)^{log(x)} = (10)^{log(x)} = x.$$

This means that $ln(10)log(x)$ equals the natural log of x because raising e to the power of this number does indeed give x. All the properties of log also work for natural log (except that $log(10) = 1$ does not work for natural log.)

Problems

(7.13.1) Use the log table to find the logarithms of the following. [two are impossible] (a) 6.842
(b) 6000   (c) 6842   (d) 100   (e) 0.5189   (f) -3 (g) 890.0006   (h) 0   (i) 0.0006842

(7.13.2) Use log tables to find the logarithms of the following WITHOUT DOING ANY MULTIPLICATIONS. BRIEFLY INDICATE what you are doing.
(a) 376×413 (b) 8.79×43.21 (c) 345×4.6412 (d) 89×64.3 (e) 0.0045×3.03

(7.13.3) Use log tables to find the logarithms of the following WITHOUT DOING ANY DIVISIONS. BRIEFLY INDICATE what you are doing.
(a) $1/4$ (b) $2/100$ (c) $345/4.6412$ (d) $89/64.3$ (e) $0.0045/3.03$

(7.13.4) Use log tables to find the logarithms of following. BRIEFLY INDICATE what you are doing. You may use a calculator in this problem to MULTIPLY only in (c),(d) and (g).
(a) 2^{100} (b) 100^2 (c) $(3.456)^{2.091}$ (d) $23^{-3.7}$ (e) $\sqrt{17}$ (f) $1/\sqrt{17}$ (g) $1/(45^{7.5})$

(7.13.5) Use log tables to find the anti-logarithms of following. BRIEFLY INDICATE what you are doing.
(a) 0.5293   (b) 1.6200   (c) 5.5   (d) 2   (e) 20   (f) 3.6123   (g) -2.5 (h) -3.8714
(i) $(-3) + 0.3010$ (j) $(-2) + 0.4771$

(7.13.6) Use logs and antilogs to perform the following calculations. BRIEFLY INDICATE what you are doing.
(a) 376×413 (b) 8.79×43.21 (c) 345×4.6412 (d) 89×64.3 (e) 0.0045×3.03

(7.13.7) Use log and antilogs to perform the following calculations. BRIEFLY INDICATE what you are doing.
(a) $1/4$ (b) $2/100$ (c) $345/4.6412$ (d) $89/64.3$ (e) $0.0045/3.03$

(7.13.8) Use log and antilogs to perform the following calculations. BRIEFLY INDICATE what you are doing. You may use a calculator in this problem to MULTIPLY only.
(a) 2^{100} (b) 100^2 (c) $(3.456)^{2.091}$ (d) $23^{-3.7}$ (e) $\sqrt{17}$ (f) $1/\sqrt{17}$ (g) $1/(45^{7.5})$

(7.13.9) Combine the following into a single logarithm
(a) $log(20) + log(30)$ (b) $log(20) - log(30)$ (c) $5log(2)$ (d) $log(x) + log(x)$
(e) $2log(x) + 3log(x)$ (f) $log(x^3) - log(x^5)$ (g) $log(x) + log(y)$ (h) $x \times log(y)$ (i) $(x + y)log(a)$
(j) $(a + 3)log(1)$ (k) $a \times log(2) - b \times log(2)$

(7.13.10) Split the following up into logs of single variables combined with $+, -, \times$. (One of them is impossible)
(a) $log(x \times y)$ (b) $log(x^3 \times y^2)$ (c) $log(\sqrt{x})$ (d) $log(x + y)$ (e) $log(x \times log(y))$

In problems 11-16 use logarithms to solve the equations for x [hint: take logs of both sides and simplify]

(7.13.11) $10^x = 0.3010$

(7.13.12) $10^{x+3} = 0.6$

(7.13.13) $2^x = 0.6020$

(7.13.14) $10^x = y$

(7.13.15) $2^x = 5 \times 3^x$

(7.13.16) $1/(5^x) = 3^y$

(7.13.17) A population of rabbits doubles every year, and there were 1 million rabbits at the start of 1990.
(a) How many rabbits are there in 1995 [hint: how many in 1991?, 1992?,...]
(b) When will there be 10 million rabbits? [hint: use logs]

(7.13.18) A bank pays 7% interest compounded annually. There is $1000 in the account in January 1990. You may use a calculator.
(a) How much money will be in the account in January 1991?
(b) How much money will be in the account in January 1992?
(c) How much money will be in the account in January 1993?
(d) How much money will be in the account in January 2000?
(e) How much money will be in the account x years after 1990?
(f) In what year will there be one million dollars in the account? [hint: use your answer from (e) and logs]

(7.13.19) The half-life of element X is 50 years. If there are 80g initially
(a) How much is there after 200 years? [there is an easy way]
(b) How much is there after 17 years? [use logs]
(c) When will 40g remain? [there is an easy way]
(d) When will 50g remain? [use logs]

(7.13.20) Bacteria are growing exponentially in an environment of unlimited space and food. The doubling time is 1 hour.
(a) If there is initially x milli-grams of bacteria, express the mass of bacteria as a function of time t.
(b) Use your answer to (a) to write down an equation whose solution is the time at which there are $3x$ milligrams of bacteria.
(c) Solve your equation from (b)
(d) Your answer to (c) should be between 1 and 2 hours. Why?. Is it?

(7.13.21) The population of rabbits on an island is growing exponentially. In 1970 there were 500 rabbits and by 1980 this population had grown to 3000 rabbits. Thus the population of rabbits t years after 1970 is given by $R(t) = R \times 2^{t/K}$.
(a) Find R and K . (b) Explain why the doubling time for the population is K
(c) Find the doubling time. (d) Explain why your answer to (c) is reasonable in terms of the data.

(7.13.22) The level of radio-activity on the site of a nuclear explosion is decaying exponentially. The level measured in 1990 was found to be 0.7 times the level measured in 1980. What is the half life?. [hint: set up an equation similar to the one in (21)]

(7.13.23) On regular graph paper, draw the graph of $y = log(x)$ for $0 < x \leq 10$. Use your graph to find $log(6.7)$ How accurate is it? What is the **slope** of the graph at $x = 6.7$. [to find this: draw a tangent line and work out the slope]

Do the problems 24-29 using the graph of $y = 10^x$ in the back of these notes. Do not use a calculator or log tables (except for checking your answers). In each case BRIEFLY DESCRIBE what you are doing. Also **draw a sketch diagram** for each problem showing the numbers on the x-axis and y-axis **as was done in the text** that are involved for that problem.

(7.13.24) Find (a) $log(6.5)$ (b) $log(1.5)$ (c) $log(22)$ (d) $log(0.0005)$ (e) $log(100000)$

(7.13.25) Find (a) $10^{0.5}$ (b) $antilog(0.35)$ (c) $antilog(2.68)$ (d) $10^{-2.3}$ (e) $antilog(1000)$

(7.13.26) Find 312×85.3.

(7.13.27) Find $429/0.583$.

(7.13.28) Find $\sqrt{79}$ [hint: what is $log(\sqrt{x})$ in terms of $log(x)$?]

(7.13.29) In this problem you may use a calculator to multiply only. Find $(3.78)^{2.71}$. [same hint as (28)]

(7.13.30) Solve the equation $2^x = 10^{C \cdot x}$ for C.

(7.13.31) Solve $P = Q \times 3^{t+a}$ for t.

(7.13.32) Solve $a \times b^t = c \times d^t$ for t.

(7.13.33) Express 4^y as a power of 10.

(7.13.34) Express 10^a as a power of 2.

(7.13.35) Sketch a sketch graph of 10^{-x} for $0 \leq x \leq 3$. What does this graph tell you about the limit of 10^{-x} as $x \to \infty$?

(7.13.36) Sketch a sketch graph of $1 - 2^{-x}$ for $0 \le x \le 10$. What does this graph tell you about $1 - 2^{-x}$ when x is large?[hint: first sketch 2^{-x}.]

(7.13.37) Solve $10^{x+4} = 100^x$

(7.13.38) Solve $2^{3x+1} = 3^{2x-1}$

(7.13.39) An area is contaminated with an element which has a half-life of 20 years. How long does it take until only 1% of the element remains?

(7.13.40) The half-life of carbon-14 is 5730 years. A bone is discovered which has 20% of the carbon-14 found in the bones of living animals. How old is the bone?

(7.13.41) Solve $log(log(x)) = 8$. [the answer is rather BIG!] The function $log(log(x))$ sometimes arises. You can make $log(log(x))$ as big as you want, by choosing the right value for x but x has to be huge.

(7.13.42) Use the method described in these notes to estimate the following powers of 2 without a calculator or doing lots of multiplications. (a) 2^{10} (b) 2^{30} (c) 2^{100}.

(7.13.43) Some biologists at UCSB have carefully recorded the number of elephant seal births in the Channel Islands from aerial photographs since the 1950's. The table records this data. Draw a graph showing the **logarithm** of the data plotted against time. Draw the best straight line you can through the data. How many seal births would you expect in 1997?

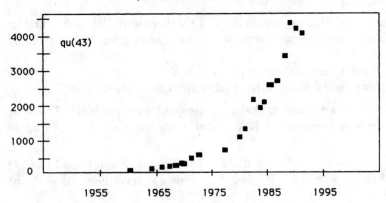

Elephant Seal Births in Channel Islands

year	59	64	66	68	69	70	71	72	77	80	81
seal births	49	74	134	181	221	245	330	446	665	1339	1547

year	82	83	84	85	86	87	88	89	90	91
seal births	1885	2341	2180	2554	2613	2820	3352	4417	4011	3987

(7.13.44) Reread section 7.8. The table gives the number of cells in a dish over a 5 hour period. Plot a graph of the **logarithm** of the number of cells against time. The graph you draw should be the best straight line you can draw for these points. The horizontal axis should be labelled 0 to 10. The Vertical axis should be labelled 1 to 8.
(a) What do you expect the population to be after 6 hours?
(b) What do you expect the population to be after 10 hours?

Hours	0	1	2	3	4	5
Cells	10,000	18,220	33,200	60,490	110,200	200,850

(7.13.45) On the planet Maximillian live **Sprogs** and **Graks**. Initially there were 800 Sprogs and 100 Graks. Sprogs double every 10 years and Graks double every 5 years.

(a) How many Graks were there after $2\frac{1}{2}$ years?

(b) When are there as many Sprogs as Graks?

(7.13.46) (**) In the year 1900 in the country Acirema, there were 100 Lawyers and 4 million people. Every 10 years the number of Lawyers double, and the population increases by 2 million. Write down an equation whose solution tells you in which year 20% of the population are lawyers. DO NOT SOLVE this equation. Carefully explain how you got it, and what all the variables you use are.

(7.13.47) Use the graph of 10^x to find an approximate solution to the equation

$$10^x = 20(x - 0.3).$$

(7.13.48) E. Coli bacteria are growing in a hamburger exponentially. Initially there are 100,000 bacteria. After 20 minutes there are 150,000. How many are there after an hour?

(7.13.49) solve $log(w + 4) - log(w - 2) = 1$.

(7.13.50) use **natural logs** to solve the following

(a) $3^x = 7$ (b) $2^x = 1/100$ (c) $4 \times 3^x = 108$ (d) $5/7^x = 1/6$ (e) $4^{x/2} = 64$.

(7.13.51) use **natural logs** to solve the following

(a) $e^x = 10$ (b) $e^{2x} = e^8$ (c) $4^x = 7$ (d) $6 \times 2^x = 3^x$

7.14   Summary

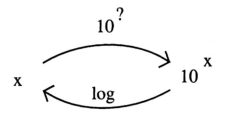

Log is the inverse of "10 to the power of.."

$log(x \times y) = log(x) + log(y)$ $log(1) = 0$ $log(1/x) = -log(x)$ $log(a^p) = p \times log(a)$.
$log(10^x) = x$ $10^{log(x)} = x$ $log(x/y) = log(x) - log(y)$ $log(10) = 1$.
$a^{x+y} = a^x \times a^y$ $a^0 = 1$ $a^{-x} = \frac{1}{a^x}$ $(a^x)^y = a^{xy}$.

Chapter 8

Derivatives.

We use mathematics to describe things with numbers. One of the most important things to describe is how quickly a certain quantity is changing. The speed of a car along the road is a measure of how quickly it is covering distance. The acceleration of a car is how quickly the speed is changing. The inflation rate is how quickly prices are increasing. Population growth is how quickly the population is increasing. If rain falls at a rate of 2 inches per hour then this is how quickly the water level in a bucket rises as rain falls. The extinction rate is how quickly species go extinct. The infection rate is how quickly a disease spreads. Calculus was invented partly to help calculate how quickly things change. One of the main reasons for studying how quickly things change is that **the scientific description of many situations is given by equations which express in mathematical terms "how quickly one quantity changes."** Many of the most fundamental and important relationships people have discovered, from nuclear physics to evolution, from economics to psychology are expressed in these terms. This should come as no surprise. After all, most of the things which interest us are not constant but continually changing.

A single number like 50 does not change. When we talk about quantities changing we are talking about many numbers. If you only know that the temperature was 50° F at 6am then you can't say anything about how quickly the temperature is changing. To discuss how temperature changes, you need to know what the temperature was at other times.

8.1 Average Rate of Change and Secants.

Suppose that the temperature at 6am was 50° F and it has risen to 68° F by 8am. The **average rate of change** of temperature between 6am and 8am is found by taking the **change** in temperature between 6am and 8am and dividing by the length of time taken for the change (the length of time between 6am and 8am.) The change in the temperature between 6am and 8am is how much the temperature increases by in that time. The temperature increases from 50° F to 68^0 F so the increase is $68 - 50 = 18^o$ F. The time taken for this increase in temperature is 2 hours. Thus

$$\frac{\text{change in temperature}}{\text{time taken}} = \frac{68 - 50}{8 - 6} = 9^o \ F/hr.$$

On the average, the rate of change of temperature was 9 degrees per hour. So that in the two hours between 6am and 8am the total increase in temperature was twice this amount (because the time span from 6am to 8am is 2 hours and each hour the temperature increased by 9°.)

The graph shows the temperature plotted against time. The average rate of change can be seen on this graph. It is the slope of the **secant line** shown. The secant line (nothing to do with trigonometry!) is the diagonal line that starts at $(6, 50)$ and ends at $(8, 68)$. It connects the two points on the graph corresponding to the times we are looking at. The average rate of change is the change in temperature (that is height of the triangle shown) divided by the time taken (which is the width of the triangle shown). Thus it is the slope of the diagonal side of the triangle.

Let us consider a more general situation. There is some quantity, $f(t)$, which is a function of the time t. In

119

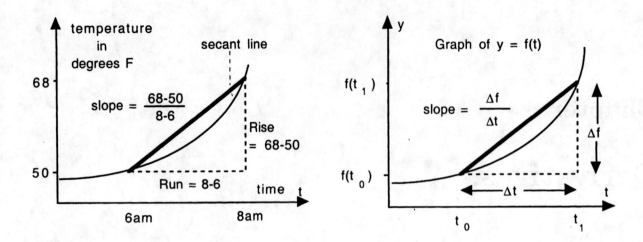

the above example $f(t)$ is the temperature at time t. The **change** in $f(t)$ between two times t_0 and t_1 is simply

$$(\text{change in f(t) between } t_0 \text{ and } t_1) = f(t_1) - f(t_0).$$

In the example $t_0 = 6$ and $t_1 = 8$ and the change in $f(t)$ between these times is the how much the temperature increases between 6am and 8am. The **average rate of change** of $f(t)$ is the how much $f(t)$ changes by during this time interval divided by the time taken for the change. More simply stated it is the change in $f(t)$ divided by the change in t. We are talking about what $f(t)$ is at time t_0 and what it is at time t_1 so the time interval from t_0 to t_1 is the time taken for the change to happen. Thus the time taken for the change is $t_1 - t_0$ so

$$(\text{average rate of change in f(t) between } t_0 \text{ and } t_1) = \frac{f(t_1) - f(t_0)}{t_1 - t_0}.$$

In the example above this is 9° F **per hour.** The units for the rate of change are obtained by taking the units for $f(t)$ and dividing by the units for t. Can you see why? The expression

$$\frac{f(t_1) - f(t_0)}{t_1 - t_0}$$

is called a **difference quotient.** Quotient means divide. The numerator is the difference of two numbers $f(t_1)$ and $f(t_0)$. Also the denominator is the difference of two numbers. This is why the ratio is called a difference quotient.

It is convenient to denote the change in $f(t)$ by Δf which simply means

$$\Delta f = (\text{change in f(t) between } t_0 \text{ and } t_1) = f(t_1) - f(t_0)$$

and the time interval by Δt thus

$$\Delta t = \text{time taken} = t_1 - t_0.$$

Using this notation we can write the average rate of change in a very simple way

$$\text{average rate of change} = \frac{\text{change in f}}{\text{change in t}} = \frac{\Delta f}{\Delta t}.$$

This is all shown on the figure. The Greek letter Δ is pronounced "delta" and is like an English "D." Δ is used to remind us of the word "difference". Thus $\Delta f/\Delta t$ is the difference in f divided by the difference in t. We see from the diagrams that the slope of the secant line connecting two points on the graph equals the average rate of change between the two times represented by these points:

> Average Rate of Change equals Slope of a Secant Line.

Problems

(8.1.1) The distance in miles of a car from it's starting point after t hours is given by the formula $f(t) = 50t$.
(a) What is the average rate of change of $f(t)$ between $t = 2$ and $t = 4$. Include units
(b) What is the average rate of change of $f(t)$ between $t = 2$ and $t = 3$.
(c) What is the average rate of change of $f(t)$ between $t = 2$ and $t = 2.1$.
(d) Explain the meaning of your answer to (b).
(e) Sketch the graph $y = f(t)$ for $0 \le t \le 4$.
(f) What is the slope of this graph? What are the units of this slope?

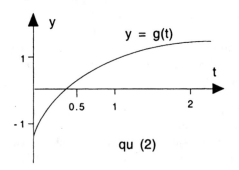

qu (2)

time hrs	0	0.5	1	1.5	2	2.5	3	3.5	4
alcohol mg	65	55	47	40	34	29	25	22	20

qu (3)

(8.1.2) Using the graph $y = g(t)$ shown find
(a) The average rate of change of $g(t)$ between $t = 0$ and $t = 2$.
(b) The average rate of change of $g(t)$ between $t = 0$ and $t = 1$.
(c) The average rate of change of $g(t)$ between $t = 0$ and $t = 0.5$.

(8.1.3) The table of data shows how much alcohol is in the blood of a rat during a period of 4 hours. On average, how quickly is the amount of alcohol decreasing during the time period between
(a) $t = 1$ and $t = 3$.
(b) $t = 1$ and $t = 2$.
(c) $t = 1$ and $t = 1.5$.
(d) When was the rate of decrease greatest?

(8.1.4) A tanker ship is off-loading its oil tank at a terminal. It takes 20 minutes to fill a 500 gallon truck. What is the average flow rate of oil into the truck in gallons per hour?

8.2 Speed, Velocity and Instantaneous rate of change.

$$speed = \frac{distance\ travelled}{time\ taken}$$

To measure the speed of an object, we measure the distance it moves in a certain amount of time and then divide by this amount of time. The units of speed are the units you use to measure distance divided by the units you use to measure time. Now, this formula works if the object moves with constant speed. On the other hand, if the speed of the object changes then this formula gives the **average speed** during the time interval in question. For example if you drive from Santa Barbara to Los Angeles airport, a distance of 100 miles and it takes 2 hours (because of bad traffic in Los Angeles) then your average speed is $100/2 = 50$ *mph*. Even though you went through Ventura at 95 *mph*, your average speed is much lower.

In everyday speech the words **speed** and **velocity** are used interchangeably. In science there is a subtle difference: velocity can be negative but speed is always positive. A negative velocity simply means that the object is moving BACKWARDS. Of course, we must first distinguish between forwards and backwards. Once we choose these directions, a velocity of -20 *m/s* will mean the object is moving backwards with a speed of 20 *m/s*.

This is sometimes expressed by saying that velocity has a **magnitude** and a **direction** whereas speed has only a magnitude. The reason for making this distinction in science is that calculations are usually easier with velocity.

Here is an example which illustrates the advantage of using velocity. Imagine two cars moving along a road which runs north-south. We decide traveling north is a positive velocity and traveling south is a negative velocity. Car A has velocity 50 *mph* and Car B has velocity 60 *mph*. How fast does Car B pass Car A? The answer is that car B passes car A at a relative velocity of 10 *mph*. If we use speed to compute this, we reason as follows. Since the cars are going in the same direction we can find out how much faster Car B is going than car A by subtracting one speed from the other. Suppose now that car A has velocity u *mph* and car B has velocity v *mph*. Then car B passes car A at a relative velocity of $v - u$ *mph*. (The relative velocity is how fast you would think car B is going if you are sitting in Car A and imagine you are not moving. It is how fast car B would hit you.) In the first example $v = 60$, and $u = 50$ and $v - u = 10$. The nice thing about velocity is that the formula $v - u$ works even if the cars are going in OPPOSITE directions. To see this, suppose car A is going in the opposite direction to car B. Since the cars are going in opposite directions, the relative speed is now obtained by **adding** the speeds of each car. On the other hand, the velocity of car A is $u = -50$ *mph* so that the velocity of car B relative to car A is $60 - (-50) = 110$ *mph*. Evidently, the formula to work out relative speed is confusing: sometimes we add, sometimes we subtract. However the formula for relative velocity is simpler. We always subtract one velocity from the other. The point here is that speed does not say anything about direction. Somehow, we must account for direction in calculations. Velocity provides the simplest method to accomplish this.

Suppose we want to know the speed of the car minute by minute, not just the average speed during the whole 2 hours. This could be done by recording how far the car travels every minute and then working out the average speed during that minute. During one minute the speed of the car usually does not change very much (unless you slam the brakes on). If we wanted to know the speed of the car even more accurately, we might record how far it travels each second. This would let us calculate the average speed over each second. This would be a very accurate result.

The summary is: if we want to know the speed very accurately, we measure the distance traveled during a very short time. The shorter the time we use, the more accurate is our result.

Example 8.2.1 *Suppose that an object is moving along the ground and that the distance (in cm) it has traveled after t seconds is given by the formula $f(t) = t^2$ cm. How fast is the object moving after 1 second?.*

Solution: It is easy to see that the object's speed is increasing. That is, using the formula we can calculate that in the first second it travels only $1cm$. But in the next second it travels 3 *cm* (this is $f(2) = 4$ minus $f(1) = 1$.) The first table shows the distances traveled for certain selected times. The second table shows the average speed during various time intervals. We see that during the time interval from $t = 1$ until $t = 10$ the average speed is $11cm/sec$.

time t in seconds	0	1	1.001	1.01	1.1	2	10
distance in cm, $f(t) = t^2$	0	1	1.002001	1.0201	1.21	4	100

Table 8.1: Distances traveled

time interval	1 to 10	1 to 2	1 to 1.1	1 to 1.01	1 to 1.001	1 to 1.000001
average speed	11	3	2.1	2.01	2.001	2.000001

Table 8.2: Average speeds

We will try to find the speed at $t = 1$ second by approximating with average speeds which are computed using short time intervals. Here is how to work out the average speed between $t = 1$ and $t = 1.1$ seconds. The average speed between these two times is the distance traveled between these two times divided by the time taken. To find out how far the object moves between these two times we figure out how far it has gone at $t = 1$ and subtract this from how far it has gone at $t = 1.1$ That is, the distance traveled during this time is

$$f(1.1) - f(1) = (1.1)^2 - 1^2 = 1.21 - 1 = 0.21 \ cm.$$

The time taken to cover this distance is the time from $t = 1$ until $t = 1.1$ which is 0.1 seconds. Thus,

$$\begin{aligned}
(\textit{average speed between } t = 1 \textit{ and } t = 1.1) &= \frac{\textit{distance travelled}}{\textit{time taken}} \\
&= \frac{f(1.1) - f(1)}{0.1} \\
&= \frac{0.21}{0.1} \\
&= 2.1.
\end{aligned}$$

Repeating this for the other time intervals we can calculate the average speeds shown in the table. Looking at the table we see that the numbers $11, 3, 2.1, 2.01, 2.001$ are getting closer and closer to 2. In other words, as we work out the average speed over shorter and shorter time intervals the answer gets closer and closer to 2. If we continue these calculations, we discover that the average speed during 1 millionth of a second after $t = 1$ is 2.000001 cm/s. So if we ask what was the speed at time 1, it looks like the answer ought to be 2 cm/s. This is correct.

This example illustrates a very important idea: as we take successively shorter time intervals and compute the average speeds during these short times intervals we obtain average speeds which get closer and closer to the number 2. We say that **the limiting value of the average speeds is 2 cm/s.** The phrase "limiting value" refers to "the average speed over shorter and shorter time intervals."

Now this may all seem rather complicated. Why bother with all this average speed stuff. We want to know the speed at $t = 1$. Surely there is a simple way to get the answer $2cm/s$ without all this bother. Yes and No!. You see there is an interesting PHILOSOPHICAL point here. What EXACTLY do you mean when you talk about the speed of the object at precisely $t = 1$? After all during the single instant $t = 1$ THE OBJECT DOES NOT MOVE AT ALL. To understand this, we can draw an analogy to a strip of movie film which captures the motion of our object. When we consider a single instant, it's as though we have chosen to view one particular picture frame in this movie strip. In this case, you might ask "how can you talk about the speed of the object at a single instant. At that instant, the object is at one definite place. It does not move!" On the other hand, it is "obvious" that the object does indeed have some definite speed at $t = 1$. But it is not so obvious how we find that speed. A practical person would say "look, just work out the average speed during one tenth of a second. That is such a short time interval that whatever answer you get is good enough." Of course, if we insist on getting an even more accurate answer we could use an even shorter time interval. Perhaps we might try one hundredth of a second. In other words our "practical person" is really saying "take the limiting value of the average speed during shorter and shorter time intervals." So we answer our philosophical question as follows: this limiting value is what we really MEAN when we talk about speed. It is important to realize that this discussion about the limiting value of an average rate of change during a very short time interval applies not just to speed but to every rate of change. So we can use the same ideas for the rate of change in temperature, weight, money, etc. For any of these quantities, the summary is "compute the average rate of change during a REALLY short time interval."

The function $f(t)$ in the above example gives the distance the object travels in t seconds. We used it to work out the speed of the object. We saw that **speed is the rate of change of distance traveled.** To be more precise, the **exact** speed at $t = 1$ is the **instantaneous rate of change of distance**. This means the limiting value of the average speed over shorter and shorter time intervals. At $t = 1$ the object has traveled $f(1)$ cm, and at time $1 + \Delta t$ the object has traveled $f(1 + \Delta t)$ cm. So in the Δt time interval from $t = 1$ until $t = 1 + \Delta t$ the object travels a distance $\Delta f = f(1 + \Delta t) - f(1)$. The average speed during this time span is $\Delta f / \Delta t$. The exact speed at $t = 1$ is

$$(\textit{exact speed at } t = 1) = \lim_{\Delta t \to 0} \frac{\Delta f}{\Delta t}.$$

This formula says to take the limiting value of the average speed over the time interval as the length of the time interval Δt gets shorter and shorter.

This is what we mean by **instantaneous rate of change.** It is the limiting value of the rate of change during shorter and shorter time intervals. This is our solution to the philosophical question "what do you mean by the **instantaneous** rate of change?" It does seem a bit complicated. Fortunately the practical upshot is simple: the instantaneous rate of change is "the average rate of change during a **REALLY** short time interval." But it is important to remember that "really short time interval" is a quick way to say "take the limiting value for shorter and shorter time intervals."

The instantaneous rate of change of $f(t)$ when $t = 1$ is written mathematically as $f'(1)$. This is pronounced "eff prime of one." So, the statement that "the instantaneous rate of change of $f(t)$ when $t = 1$ is 2" is written

$$f'(1) = 2.$$

In our example above this says that the velocity at $t = 1$ is 2. We also say that the **derivative** of $f(t)$ at $t = 1$ is 2. The word "derivative" comes from "derive" which simply means "obtain." Thus, the derivative $f'(t)$ is something we obtain from the function $f(t)$. It is the instantaneous rate of change of $f(t)$ at time t.

Problems

(8.2.1) A car drives at $30mph$ for 1 hour then at $60mph$ for the next 30 minutes, and then is stationary for the next 30 minutes. Sketch a graph showing distance traveled against time taken. What was the average speed over the entire two hours?

(8.2.2) Everyday I ride my bicycle to work. There is a short steep hill I have to climb at the start. Then it is fairly flat, but half way along I usually have to wait at lights for a minute. I usually take 23 minutes to cover the 5 miles.
(a) Sketch a graph showing distance against time.
(b) Sketch a graph showing my speed against time.

(8.2.3) The table shows the position of a point on the x-axis during the time interval $0 \le t \le 1$ seconds.
(a) Estimate the speed at $t = 0.5$. Explain what you did.
(b) When was the speed greatest?
(c) What was the average speed during the 1 second?

t	0	0.2	0.4	0.6	0.8	1
x	3	3.8	4.8	6.0	7.4	9.0

(8.2.4) In **example** 8.2.1 2 pages ago, calculate the average speed during the 1 millionth of a second after $t = 1$.

(8.2.5) The graph shows the height of a ball thrown into the air on the planet Brontitaur. The vertical units are meters and the horizontal units seconds. Make a table showing the average speed during the 1/2 second intervals, starting at $0, 0.5, 1, 1.5$. What was the speed of the ball at $t = 1$.

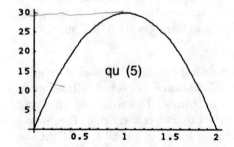

(8.2.6) The temperature at t hours after noon is given by the function $f(t)$ in degrees Fahrenheit.
(a) What are the units of $f'(t)$
(b) What is the meaning of the statement that $f'(7) = -3$.
(c) What is the significance of $f'(t)$ being positive? versus being negative?

(8.2.7) The total profit that a company has made measured in millions of dollars is $p(t)$ where t is the time measured in years with $t = 0$ corresponding to 1990.
(a) What is the meaning of $p'(5) = 0$
(b) What are the units of $p'(t)$

(c) What does the sign of $p'(t)$ tell you about the company?

(8.2.8) The population (in millions) in a certain country t years after 1900 is given by the function $p(t)$. There were 450 million in 1900. If $p'(t) = 5$ throughout the time span 1900 to 1910 what was the population of the country in 1907? [hint: what is the practical significance of $p'(t) = 5$]

(8.2.9) After t years I have $f(t)$ thousand dollars in the bank.(a) What are the units of $f'(t)$? (b) What is the practical meaning of $f'(7) = 0.3$? Write a complete sentence!

8.3 The slope of a graph: secants and tangents.

The next idea to understand is

> Instantaneous Rate of Change equals Slope of Graph equals Slope of Tangent line.

Here is an interesting question: what is the slope of a graph? In previous chapters, we said that slope was a measure of "steepness." It was easy to find the slope of a straight line by using the ratio *Rise/Run*. This worked because the steepness of a straight line is the same everywhere on the line. But the steepness of a curve changes from one point to another. What can we do? A good idea is to use what we know about lines to help us understand more complicated curves. To measure the steepness of a graph at a point p, we draw a straight line through p which has the same steepness as the graph. We then say that the slope of this line is the slope of the graph at the point p. To draw such a line, take your ruler and move it so that p is on the edge. Move the ruler so that it points in the same direction as the curve at the point p. Now draw the line. This line is called a **tangent line** to the curve at p. It is now easy to find the slope of the graph at the point p. We simply find the slope of the tangent line. Of course, if we draw a tangent line at another point q we will very likely obtain a different slope.

Now here is the key idea. Both figures show the graph of a function $f(t)$. The instantaneous rate of change of $f(t)$ when $t = 3$ equals the slope of the graph $y = f(t)$ at $t = 3$ shown on the right. Think about it: look at the figure and ask yourself what significance the slope of the graph might have. The graph is sloping upwards: as you move to the right the graph rises. The slope measures how quickly the graph rises, which in turn tells us how quickly $f(t)$ increases. In the figure the slope at p is 2. This means that if we start at p and move along the graph a small distance to the right (for example from $t = 3$ to $t = 3.1$, a distance of 0.1) then the graph rises (approximately) a distance 2 times as much as this small distance ($2 \times 0.1 = 0.2$ in the example).

We can see this a different way. We know that the average rate of change of $f(t)$ is given by a difference quotient. But a difference quotient also tells you the slope of a secant line to the graph. The exact rate of change of the

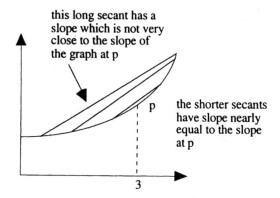

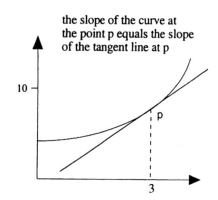

function at $t = 3$ is the limit of the average rates of changes computed for smaller and smaller time intervals. Each one of these average rates of change is the slope of a secant line. As the time intervals get shorter, the secant lines get shorter. Again, look at the figure. We see that when the secant lines are very short they have almost exactly the same slope as the curve at $t = 3$.

> The limit of the slopes of shorter and shorter secant lines equals the slope of the tangent line.

Here are a few things to realize about the derivative. Since the slope of the graph $y = f(x)$ is given by the derivative $f'(x)$, we can say the following: the graph has positive slope at, for example, $x = 3$ if $f'(3)$ is positive. So, we say that the graph and the function are **increasing** at $x = 3$ if $f'(x) > 0$. Likewise, we say they are **decreasing** if $f'(3) < 0$. Of course, there is nothing special about 3. That is, for any number t the graph and function is increasing at t if $f'(t) > 0$. It is very important to know whether a function is increasing or decreasing. For example, if you are sick, and your temperature is increasing then you are getting sicker. If it is decreasing, then (hopefully) you are getting better.

8.4   Notations for Derivatives.

The derivative of $f(t)$ is often written as

$$\frac{df}{dt}$$

(pronounced "dee eff dee tee"). This means exactly the same thing as $f'(t)$. The notation comes from $\Delta f/\Delta t$. In fact,

$$\frac{df}{dt} = \lim_{\Delta t \to 0} \frac{\Delta f}{\Delta t}$$

appears quite often. This is just another way to indicate that the derivative is the limit of the change in $f(t)$ divided by the change in t.

For example

$$f'(3) = \lim_{\Delta t \to 0} \frac{\Delta f}{\Delta t}$$
$$= \lim_{\Delta t \to 0} \left(\frac{f(3 + \Delta t) - f(3)}{\Delta t} \right)$$

Each notation has advantages and disadvantages. A nice thing about the $f'(x)$ notation is that if we want to talk about the rate of change of $f(x) = x^2$ when $x = 3$ then (as we shall see later) we can write

$$f'(x) = 2x \qquad \text{and so} \qquad f'(3) = 6$$

Thus $f'(3) = 6$ is the rate of change of $f(x) = x^2$ when $x = 3$. This is very convenient.

An advantage of the d/dx notation is that we can write out derivatives in one line. Thus

$$\frac{d}{dx}\left(3x^4 + 7x^2 + 2\right) = 12x^3 + 14x.$$

On the other hand, if we use the $f'(x)$ notation we must write more:

$$\text{Set } f(x) = 3x^4 + 7x^2 + 2. \qquad \text{Then} \qquad f'(x) = 12x^3 + 14x.$$

That is, it's possible to use the d/dx notation without actually giving the function $3x^4 + 7x^2 + 2$ a **name** before we write down its derivative. But, with the $f'(x)$ notation we first must **name** $f(x) = 3x^4 + 7x^2 + 2$.

A disadvantage of the d/dx notation is that it is much more cumbersome to express the numerical value of the derivative when $x = 3$. To do this we have to write

$$\left.\frac{d}{dx}\left(x^2 + 7\right)\right|_{x=3} = 6.$$

So both notations have their advantages. You should be familiar with both notations! Actually, there is one more notation you will sometimes see, particularly in physics books. The derivative of $f(t)$ is often written $\dot{f}(t)$. This notation has always seemed a bit **dotty** to me.

Warning. Do not fall into the trap of thinking

$$\frac{df}{dt}$$

means $d \times f$ divided by $d \times t$. The df is not two things multiplied. It is short-hand to remind us of a small change Δf and that the difference quotient $\Delta f/\Delta t$ is very nearly equal to df/dt provided Δt is small enough.

Problems

(8.4.1) An oven is heating up. You tell a philosopher "at 1pm the oven was heating up at exactly 3° C per minute." The philosopher replies "what do you mean **precisely**?" What would you tell her if she does not know what a limit is?

(8.4.2) The volume (in m^3) of water in my (large) bathtub when I pull out the plug is given by $f(t) = 4 - t^2$. This formula is only valid for the 2 minutes it takes my bath to drain. Use a calculator if you wish.
(a) Find the average rate the water leaves my tub between $t = 1$ and $t = 2$
(b) Find the average rate the water leaves my tub between $t = 1$ and $t = 1.01$
(c) What would you guess is the exact rate water leaves my tub at $t = 1$?
(d) In this bit h is a very small number. Find the average rate the water leaves my tub between $t = 1$ and $t = 1 + h$ [just work out the difference quotient]
(e) What do you get if you put $h = 0$ in the answer to (d) ?
(f) Why are all you answers negative?

(8.4.3) An ice cube is melting. The mass of the ice cube after t minutes is $m(t)$ grams. You are told that the rate of change of $m(t)$ is -3 $grams/min$.
(a) How much mass does the ice cube lose in 5 minutes?
(b) If the ice cube starts out with a mass of $90gms$ how long until it has all melted?
(c) What would it mean if the rate of change was $+3$ $grams/min$?

(8.4.4) **Acceleration** is the rate of change of velocity. The velocity (positive means DOWN) of a pumpkin thrown off the top of Cheadle hall t seconds after launch is $32t$ ft/sec (until it hits the ground).
(a) What is the average rate of change of velocity between $t = 1$ and $t = 2$.
(b) What is the average rate of change of velocity between $t = 1$ and $t = 1.1$.
(c) If the pumpkin lands after 2.5 seconds, what is its **speed** when it hits ?

(8.4.5) Use a calculator for this. The function $f(t) = t^2$
(a) What is the average rate of change of $f(t)$ between $t = 1$ and $t = 1.001$.
(b) What is the average rate of change of $f(t)$ between $t = 2$ and $t = 2.001$.
(c) What is the average rate of change of $f(t)$ between $t = 3$ and $t = 3.001$.
(d) What is the average rate of change of $f(t)$ between $t = 4$ and $t = 4.001$.
(e) Now plot the answers you got for (a)-(d) to get a graph of velocity against time. Label the horizontal axis t and label it for $0 \le t \le 4$. For example if you got a slope of 17 for (b) put a point at $(2, 17)$.
(f) What do you notice about these four points?.
(g) Using you answer to (f) what would you guess is the average rate of change of $f(t)$ between $t = 1.5$ and $t = 1.501$.
(h) Use a calculator and check if your guess is right.

(8.4.6) Use the graph of $y = x^2$ on the next page.
(a) Try to carefully draw a tangent line at $x = 1.5$. Now carefully measure the rise and run of the tangent and figure out the slope.
(b) Repeat this at the point $x = -1$. [watch out you get the signs right!]
(c) Draw a secant line going through the points $t = 0.5$ and $t = 1$ and measure its slope.

(8.4.7) Copy the graph on the next page and clearly indicate
(a) Where the derivative is positive
(b) Where the derivative is zero
(c) Where the derivative is largest (and positive)
(d) Where the derivative is most negative

(8.4.8) Draw the graph $y = f(x)$ of a function with the following properties

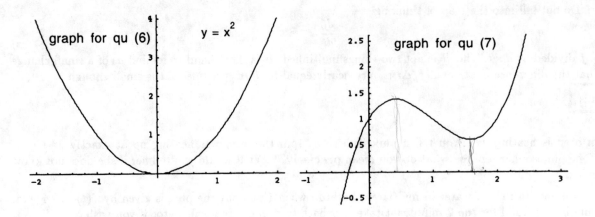

(a) $f(2) = 3$.   (b) The derivative is negative only for $2 \le x \le 3$.   (c) $f'(x) = 1$ when $x \le 0$.

(8.4.9) The number of hours of daylight in Santa Barbara varies throughout the year. This is shown on the graph.

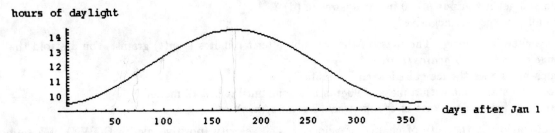

(a) Which is the longest day ?(most hours of daylight)
(b) Which is the shortest day ?
(c) On which day is the number of hours of daylight increasing most rapidly
(d) On which day is the number of hours of daylight decreasing most rapidly
(c) Approximately how much more daylight is there on March 21'st compared to March 20'th? [use the slope]

(8.4.10) Use the graph of $f(x) = 10^x$ at the back to find
(a) The average rate of change of $f(x)$ between $x = 0.3$ and $x = 0.4$
(b) Measure the slope of the tangent line to the graph at $x = 0.35$
(c) Use your answer to (b) to say what the slope of the tangent line to $g(x) = 2 \times 10^x$ is at $x = 0.35$
(d) Same as (c) but for $h(x) = C \times 10^x$. Where C is a constant.

(8.4.11) Use the table for e^x at the back to find (approximately) the derivative of e^x at each of the points $x = 1, 2, 4$. Do this by finding the average rate of change over a suitably short interval. What do you notice about your answers?

(8.4.12) Use the table of square roots at the back to find the average rate of change of $\sqrt{x}$
(a) between $x = 4$ and $x = 9$
(b) between $x = 4$ and $x = 5$
(c) between $x = 4$ and $x = 4.1$
(d) between $x = 4$ and $x = 4.01$
(e) what would you guess the instantaneous rate of change at $x = 4$ is?

(8.4.13) repeat the previous question but with $ln(x)$ in place of $\sqrt{x}$.

8.5   Rate of change without time.

The examples we have looked at so far are for quantities that change with time. There are, however, many situations where some quantity, y, depends on another quantity, x. This means y changes if we change x. For example, if you put adrenaline into your blood then your heart speeds up. The speed of your heart, y, depends on the amount of adrenaline, x, in your blood. In particular, increasing x causes y to increase. We might ask the question how much does the heart speed up with the addition of one milligram of adrenaline? This question is asking "what is the rate of change of heart beat as the quantity of adrenaline increases?". This has nothing to do with time. The "rate of change" here concerns the amount of adrenaline. Now if a medical book were to say that the rate of change is 5 bpm/mg, this would tell us that the heart rate goes up by 5 beats per minute if we add one extra milligram of adrenaline.

If we have any function $f(x)$ that depends on a quantity x then we can speak about the rate of change of $f(x)$ as x varies. How is this possible? Suppose we know that the rate of change of $f(x)$ at $x = 3$ is $f'(3) = 7$. This means that if x is increased by some small number, say Δx, then the corresponding increase in $f(x)$ is $7\Delta x$. For example if x is increased by 0.01 then $f(x)$ increases by 0.07. Expressing this in math:

$$f(3 + \Delta x) - f(3) \approx 7\Delta x \qquad \text{so} \qquad f(3.01) - f(3) \approx 0.07$$

It is important to realize two things:
(a) $f'(3) = 7$ tells us how much $f(x)$ increases between $x = 3$ and $x = 3.01$ but NOT between other values like $x = 5$ and $x = 5.01$
(b) The increase in $f(x)$ is very nearly but NOT EXACTLY 7 times the increase in x.

Example: Consider federal tax. Suppose one year you earn \$40000 and you pay \$8000 in tax. Now, the amount of tax you pay is not a fixed proportion like 20% of your income (that is, federal tax is not a "flat" tax). This means that people who do not earn a lot of money pay a small percentage of their income as tax while those who earn more pay a bigger percentage. You might ask the question "how quickly does the tax I pay increase as I

Income in \$	5000	10000	15000	20000	30000	40000	40001
Tax in \$	500	1100	1900	2900	5200	8000	8000.24

Table 8.3: Tax Table

earn more money?" In other words, if you were to earn one extra dollar, how much extra tax would you pay? In the table above, if you earn \$40001 then you pay \$8000.24. So, you pay an extra \$0.24 in tax when you earn one extra dollar. This means that 0.24 is the rate of change of tax as the amount you earn increases. So if you earn an extra \$100, then you pay an extra 0.24 × \$100 in tax. Economists call this the **marginal rate of taxation.** You imagine the extra one dollar above \$40000 that you earn as at the side of those \$40000. The "marginal rate" is the extra tax, \$0.24 at the "side" of the \$8000 tax you paid. Again, the rate of change here has nothing to do with time. The quantity, y, that is changing is how much tax you pay. The quantity, x, that is causing it to change is how much money you earn.

Example: You grow wheat. Putting more fertilizer on the field results in more wheat. You are interested in the number of tons, y, of wheat that you harvest. You can vary the number of tons, x, of fertilizer that you put on the field. An important question is "how much extra wheat do you get if you use an extra ton of fertilizer?". It depends. A bit of fertilizer will help. If you use too much you will poison the crop. Suppose you increase the amount of fertilizer from 5 tons to 6 tons and that y increase from 200 tons to 220 tons. Then 1 extra ton of fertilizer produces 20 extra tons of wheat. Thus, the average rate of change of y is 20 when you change x from 5 to 6. Again, the rate of change here has nothing to do with time; rather, it indicates how the quantity of wheat obtained changes with the quantity of fertilizer used. This example shows that it's very important to realize that the rate of change may vary. If too much fertilizer is used, the soil will not support growth. At that point, if you were to continue adding fertilizer, your yield would decrease. This means that the rate of change of y is negative for large enough values of x.

Example: A road goes through some mountains. The height of the road above sea-level changes along the road. The distance, x, along the road is measured from the town called Origin. The height of the road x meters from Origin is a function $h(x)$. Some stretches of road are flat. For example, between $x = 10,000$ and $x = 15,000$ is a 5 kilometer long flat stretch 1000 meters above sea-level. So $h(x) = 1000$ when $10,000 \leq x \leq 15,000$. Along this part of the road the rate of change of height is zero: $h'(x) = 0$. But, there are stretches where the road goes steeply upwards. Here $h'(x)$ is positive. At $x = 20,000$ (20 kilometers from Origin) there is a sign by the road which says "gradient 14%." This is steep. It means that for every 1000 meters we move along the road, we move 140 meters upwards. It might not sound like a lot, but it IS a very steep road. Here $h'(20,000) = 0.14$ (since 0.14 is 14%). In other words, the derivative of the height measures how steeply the road goes up (or down.)

Example: The temperature of ocean water drops as depth underwater increases. It cools down to about 3° C at the lower depths. This is the temperature at which water is densest. Water that is colder than 3^0 C is lighter and so rises to the surface. The temperature, T, of water depends on the depth, x, of the water. Thus, temperature is a function of depth. This is expressed by writing $T(x)$. The derivative $T'(x)$ is the rate of change of temperature with depth. This tells us how quickly the temperature changes as depth increases. For example, $T'(100) = -0.1$ means that at a depth of 100 meters the temperature drops by 0.1 of a degree for each meter deeper we dive. If $T'(300) = 0$ then at a depth of 300 meters, if we dive 1 meter deeper, the temperature does not get any colder.

Problems

You will find that making a sketch graph, and labeling the axes will help you think about the following problems. Look at your sketch and think what the slope of the graph represents in each case. Confirm this by thinking about the units of the derivative.

(8.5.1) Let $A(r)$ be the area enclosed by a circle with radius r.
(a) What are the units of $A'(r)$
(b) What is the meaning of $A'(3) = 6\pi$.

(8.5.2) The height of an airplane above the ground when it has flown x miles is $h(x)$ feet.
(a) What are the units of $h'(x)$
(b) What is the meaning of $h'(10) = 500$.
(c) During which part of the flight would you expect $h'(x)$ to be negative?
(d) What would it mean if $h(x)$ is negative?

(8.5.3) The volume of water in a reservoir depends on the height of the water measured on a marker on a dam. If the volume is $V(x)$ gallons when the height is x meters
(a) What are the units of $V'(x)$
(b) What is the meaning of $V'(30) = 5 \times 10^7$.

(8.5.4) If a commodity is priced at $\$p$ the number of items that sell is $Q(p)$.
(a) What does $Q(50) = 20,000$ mean ?
(b) What are the units of $Q'(p)$?
(c) What does $Q'(50) = -200$ mean ?

(8.5.5) The number of people who will develop an infectious disease depends on the percentage of people inoculated. If x is the percentage of people inoculated then $f(x)$ people get the disease.
(a) What does $f(0) = 10^6$ mean ?
(b) What does $f'(50) = -10^4$ mean ?
(c) What does $f^{-1}(100) = 85$ mean?

8.6   The Tangent Line Approximation.

Suppose that the temperature at 7 am is 55°F and that it is increasing at a rate of 20° per hour. What is the temperature going to be at 6 minutes past 7am ? We can work this out as follows. 6 minutes is 1/10 of an hour. Since the temperature rises 20° in 1 hour, it rises by 1/10 of this in 6 minutes. Thus the temperature rises 2°. The temperature starts out at 55° so 6 minutes later it has risen to $55 + 2 = 57°$. The method we have used to do this is called the **tangent line approximation.**

Let's see what this has to do with tangent lines. On a graph which shows the temperature plotted against time the slope of the tangent line is how quickly the temperature is rising: 20° per hour. If we move along the tangent line from 7am to 7:06am the increase in y-coordinate represents 2°F. Thus the y-coordinate of the point at time 7:06am is 57°F. The reason for the word **approximation** is that the temperature at 7:06am is not exactly 57°F because although the temperature is rising at a rate of 20°F per hour **at 7am** it is not rising **steadily**. The graph of temperature is not exactly a straight line. So our answer of 57°F is not exactly correct, only an approximation.

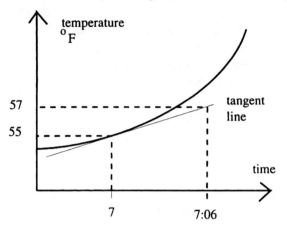

The slope of the tanget line is how quickly the temperature is rising: 20°F per hour.

The height of the point on the tangent line at 7:06am is 57°F.

We will now give a more detailed discussion. For the moment, suppose $f(x)$ is some given function. Suppose also that we are told that $f'(3) = 7$. Then we have this important fact:

$$\boxed{f'(3) = 7 \qquad \text{means that} \qquad f(3 + \Delta x) \approx f(3) + 7\Delta x.}$$

This works best if Δx is small. What "small" means depends on the situation and on how much accuracy is desired. But it helps us to understand the meaning of the derivative if we pretend this is reasonably accurate when Δx is as big as 1. Then

$$f'(3) = 7 \text{ means that} \qquad f(3 + 1) \approx f(3) + 7.$$

The point is that $f'(3)$ is telling us how rapidly $f(x)$ increases as we increase x.

$$\boxed{f'(3) = 7 \text{ means that each unit increase in } x \text{ produces approximately 7 units increase in } f(x).}$$

Using this we can **approximate** the function $f(x)$ by the **tangent line**. Remember that the tangent line to the graph $y = f(x)$ at $x = 3$ is a straight line which has the same slope as the graph at $x = 3$. The equation of the tangent line (using that the slope is $f'(3) = 7$ and that $(3, f(3))$ is a point on the tangent line) is

$$y = f(3) + 7(x - 3).$$

If we do not move very far from the point $x = 3$ then the tangent line gives a **good approximation** to the function $f(x)$. We call the function

$$\boxed{f(3) + 7(x - 3)}$$

the **tangent line approximation.** It is a linear plus constant function and its graph is the tangent line. Incidentally, since $\Delta x = x - 3$ this is the same as the first formula of this section. The general formula for the tangent line approximation to a function $f(x)$ at (for example) $x = 3$ is

$$\boxed{\text{Tangent Line Approximation at } x = 3 \qquad f(3) + f'(3)(x - 3)}$$

We can be a bit more precise by what we mean by "a good approximation."

We say that one number, y is **small in comparison to another number** x if the **ratio**

$$\frac{y}{x}$$

is small. This does not necessarily mean that y itself is small. It just means that compared to x it is insignificant. For example, 1 is small *compared* to one million but 1 is not small compared to 0.5. Similarly, 0.001 is small compared to 1 but it is not small compared to 0.00001. What we have here is a notion of **relative smallness.**

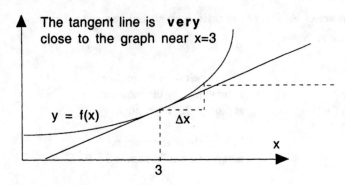

The tangent line is **very** close to the graph near x=3

y = f(x)

Δx

x

3

The **error** in the tangent line approximation is the height of the gap between the graph and the tangent line. It is small compared to Δx

It's easy to see what this means. As an example, a cat is large compared to a fly, but small compared to an elephant. Now, one of the ways of thinking about the derivative is to realize that it says something about one quantity being small *compared* to another quantity. This idea will now be explained.

When we change x by a small amount, Δx, then $f(x)$ will change by some small amount. The small number Δx is the **cause** of the change in $f(x)$. It seems sensible to measure things in comparison to this small change. Suppose we use the tangent line approximation to calculate the approximate change in $f(x)$ and we find that the error is only 0.0001 when Δx is 0.01. Then this approximate answer is good in comparison to the magnitude of the change Δx. But, if the error was 0.1 this is an error which is 10 times bigger than our scale of smallness $\Delta x = 0.01$. So it would not be a good approximation in comparison to Δx.

The **error** involved in the tangent line approximation is the difference between the tangent line approximation and the exact value given by the function. When we say that the tangent line approximation is a **good approximation** we mean this:

> The error is small in comparison to Δx.

The value of $f(x)$ when $x = 3.01$ is $f(3.01)$. The point on the tangent line with $x = 3.01$ has

$$y = f(3) + 7(3.01 - 3) = f(3) + 7(.01).$$

The error between the real value, $f(3.01)$, and what the tangent line predicts is

$$\text{error} = (\text{real value}) \text{ - } (\text{predicted value}) \ = \ f(3.01) \ - \ (f(3) + 7(.01)).$$

The relative error **compared to the change in x** is

$$\frac{f(3.01) \ - \ (f(3) + 7(.01))}{0.01}.$$

The point is that this **relative error** is very small.

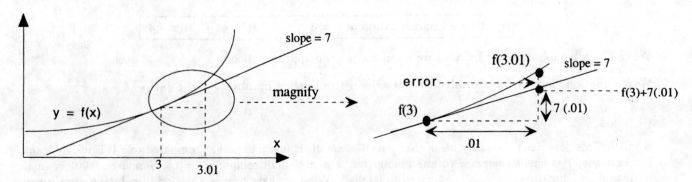

Perhaps the simplest explanation for that the relative error is small is obtained just by **looking** at the graph of $f(x)$ and noting that the tangent line is **really** close to this graph near to $x = 3$. This means that the error is small compared to Δx.

Here is another explanation of why the error is small compared to Δx.

$$f'(3) \;=\; \lim_{\Delta x \to 0} \frac{f(3 + \Delta x) - f(3)}{\Delta x}.$$

If we subtract $f'(3)$ from both sides we get

$$\lim_{\Delta x \to 0} \frac{f(3 + \Delta x) - f(3)}{\Delta x} - f'(3) \;=\; 0.$$

Now put this over a common denominator to get

$$\lim_{\Delta x \to 0} \frac{[f(3 + \Delta x) - f(3)] - [\, \Delta x \times f'(3)\,]}{\Delta x} \;=\; 0.$$

Re-arranging this we see

$$\lim_{\Delta x \to 0} \frac{f(3 + \Delta x) - [\, f(3) + \Delta x \times f'(3)\,]}{\Delta x} \;=\; 0.$$

The numerator is the difference between the real value of $f(x)$ at $x = 3 + \Delta x$ and the tangent line approximation to this. Thus the numerator is the error, so:

$$\lim_{\Delta x \to 0} \left(\frac{\text{error}}{\Delta x} \right) \;=\; 0.$$

This means that if we make Δx very small, then not only is the error small, but **the error is small compared to Δx.**

Problems

(8.6.1) If $f(3) = 7$ and $f'(3) = 2$ what does the tangent line approximation give as an approximate value for
(a) $f(3.1)$ (b) $f(3.01)$ (c) Which of these answers would you expect to be more accurate and why?

(8.6.2) Use the graph of $y = 10^x$ at the back of the book. Find the tangent line approximation to $f(x) = 10^x$ at $x = 0.5$. Use this approximation to calculate $10^{0.6}$ What is the percentage error?

(8.6.3) (a) You will see later that if $f(x) = \sqrt{x}$ then $f'(4) = 1/4$. Use this to:
(b) Use the tangent line approximation at $x = 4$ to $f(x)$ to find $\sqrt{4.4}$
(c) What is the percentage error in your answer to (b)?
(d) Use the tangent line approximation at 4 to $f(x)$ to find $\sqrt{9}$
(e) What is the percentage error in your answer to (d)?
(f) Why is the percentage error bigger in (e) than in (c)?

(8.6.4) The following are tangent lines for a function $f(x)$:
At $x = 1$, the tangent line is $y = 4x + 5$.
At $x = 3$, the tangent line is $y = x + 12$.
At $x = 4$, the tangent line is $y = 17$.
At $x = 6$, the tangent line is $y = -3x + 25$.
Sketch the graph of a function $f(x)$ in the range $1 \leq x \leq 6$ which has these properties.[hint: draw the tangent lines first]

(8.6.5) Suppose $f(x)$ has derivative $f'(x)$. In this problem, you will discover the derivative of the function $C \times f(x)$ where C is a fixed constant.
(a) First, suppose that $f(x) = ax + b$. Find $f'(x)$. (Hint: derivative is slope; what is slope of a straight line?)
(b) Using your answer to (a) , find the derivative of $C \times f(x)$ if $f(x)$ is a linear plus constant function.
(c) Now suppose $f(x)$ is an arbitrary function. Based on your results from (a) and (b), what do you expect is

the derivative of $C \times f(x)$? (Hint: explain this by using that the tangent line is a good approximation to $f(x)$).

(8.6.6) (a) If $f(x) = ax + b$ and $g(x) = cx + d$, what are $f'(x)$ and $g'(x)$?

(b) Write down the formula for $f(x) + g(x)$ and find its derivative.

(c) Suppose $f(x)$ and $g(x)$ are arbitrary functions. What is the derivative of $f(x) + g(x)$? Explain as in part (c) of the above problem.

8.7   Differentiation.

There is an easy algebraic way to find the rate of change of simple functions. This is called **differentiation.** In this section, we shall see how to differentiate polynomials. There are just a few basic rules to remember. The **power rule** says

$$\boxed{\text{The derivative of } x^n \text{ is } n \times x^{n-1}.}$$

Thus the derivative of x^3 is $3x^2$. Next, to differentiate a polynomial, just apply this rule to each one of the terms.

Example 8.7.1 *Find the derivative of $f(x) = 2x^5 - 4x^3 + x^2$.*

Solution: use the boxed rule above on each term to obtain

$$f'(x) = 2(5x^4) - 4(3x^2) + (2x) = 10x^4 - 12x^2 + 2x.$$

Another point to remember is this:

$$\boxed{\text{The derivative of a constant is 0.}}$$

Thus, if $f(x) = x^2 + 13$, then these rules tell us that $f'(x) = 2x$. The point is that the derivative of 13 (a constant) is **zero** because derivative is rate of change, and a constant does not change, so has rate of change 0.

This allows us to easily calculate many rates of change. For example, if we want to know the rate of change of $f(x) = x^2$ when $x = 3$ we first differentiate $f(x)$ by writing

$$f'(x) \; = \; 2x.$$

Now the rate of change of $f(x)$ when $x = 3$ is $f'(3)$ which we get by plugging $x = 3$ into the formula $f'(x) = 2x$. Thus, $f'(3) = 6$. It's also important to realize that this value is the slope of the tangent line to $f(x) = x^2$ at $x = 3$. This means the equation of the tangent line is just

$$f'(3)(x - 3) + f(3) \; = \; 6(x - 3) + 9 \; = \; 6x - 9.$$

We can immediately put this to good use.

Example 8.7.2 *A circle of radius r encloses an area given by $A(r) = \pi r^2$. Find the rate of change of area with*

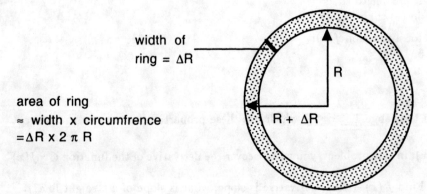

width of
ring = ΔR

R

R + ΔR

area of ring
≈ width x circumfrence
= ΔR x 2 π R

respect to radius.

Solution: using the rules above, the derivative is given by

$$A'(r) = \pi(2r^1) = 2\pi r.$$

Notice that the derivative of the area function is the function which gives the circumference of the circle. This will be explained. The diagram shows a circle a radius R and one of slightly larger radius $R + \Delta R$. The shaded area is the area of the big circle minus the area of the small circle. The shaded area is approximately equal to the width of the shaded area times the circumference of the small circle. So the shaded area is approximately $\Delta A \approx \Delta R \times 2\pi R$. Thus $\Delta A / \Delta R \approx 2\pi R$. The approximation gets better and better in the limit as $\Delta R \to 0$. "In the limit" the approximation becomes perfect. This is "common sense." How quickly the area of the circle grows must just depend on the perimeter of the circle because this is where new area appears as the circle grows.

The rule that

$$\frac{df}{dx} = nx^{n-1}$$

works even if n is negative, or not a whole number. For example

$$\frac{d}{dx}\left(\sqrt{x}\right) = \frac{1}{2}x^{-1/2} \qquad \frac{d}{dx}\left(x^{-1}\right) = -x^{-2}.$$

Why do these rules work? First if C is a constant and you multiply a function $f(x)$ by this constant to give a new function $C \times f(x)$ then the new function is C times as big as the old one. So it changes C times as quickly. If a certain change in x causes $f(x)$ to increase by 3 then $C \times f(x)$ will increase by $C \times 3$. For example if you work twice as many hours as your friend and the hourly rate of pay in increased by \$1 per hour then your pay increase will be twice that of your friend. The function that describes what you earn is 2 times the function for your friend. Thus

$$\frac{d}{dx}(C \times f(x)) = C \times f'(x).$$

The formula for the derivative of x^n can be obtained using limits, and this will be done in section (8.9). But here is a simpler explanation. The derivative of x is just 1. This is easy to see if you think of the derivative as the slope of the graph. The graph $y = x$ is a line with slope 1, so the derivative is 1. Another way to see it is that derivative is rate of change. The rate of change of x as x changes is just 1 : increase x by some amount and that is exactly how much x changes by!

To understand why the derivative of x^2 is $2x$ think of x^2 as $x \times x$. Two x's multiplied together. If x is increased by some amount Δx then **both** these x's are increased. The first x multiplies the second x. So any **increase** in the **second** x is multiplied by the **first** x. If the second x is increased by Δx then this is multiplied by the first x to give an increase of $x\Delta x$. BUT the **first** x is also increased by Δx and this increase is multiplied by the **second** x to give an increase of $\Delta x \times x$. The **combined** effect of these two increases is an increase in x^2 of $x \times \Delta x + \Delta x \times x$ in other words . This is how much x^2 increases by when x is increased by Δx. So the **rate** of increase of x^2 is $2x\Delta x$ divided by Δx which gives $2x$. A similar explanation works for x^3. This time there are three x's multiplied together. And so on.

Problems

Find the derivatives of the functions in the first five problems below:

(8.7.1) (a) $f(x) = x^3$ (b) $f(x) = 2x^4$ (c) $f(x) = 7x^4 - 3x^2 + 1$ (d) $f(x) = x^{-1}$.

(8.7.2) (a) $g(x) = 1 - 7x$ (b) $g(x) = (2 + 3x)^2$ [multiply out first!] (c) $g(x) = 17 + 2^3$.

(8.7.3) (a) $(x^2 + 1)^2$ (b) $(-3)x^{-2} + 3x$ (c) x^3/x^7

(8.7.4) (a) $(x + 1)(x + 2)(x + 3)$ (b) $f(x) = ax^2 + bx + c$ (c) $f(x) = \frac{x^2 + 2x + 1}{x + 1}$ (hint:factor the numerator).

(8.7.5)

$$1 + x + \frac{x^2}{2!} + \frac{x^3}{3!} + \frac{x^4}{4!} + \frac{x^5}{5!}$$

simplify your answer [remember how factorials work: $5! = 5 \times 4 \times 3 \times 2 \times 1$]

(8.7.6) If C is any constant and $f(x) = Cx$ show that $f'(x) = C$.

(8.7.7) The height of a missile at time t is given by $f(t) = -at^2 + bt$ where a, b are constants. Find the function which gives the rate of change of the missile's height.

(8.7.8) Find the equation of the tangent line to $f(x) = x^3 + x^2 + x + 1$ at $x = 4$. [hint: use the derivative to find the slope of the tangent line]

(8.7.9) Suppose $f(x) = 3x^2 + 4$
(a) Sketch the graph of $f(x)$ for $-2 \le x \le 2$.
(b) Write down the formula for $f'(x)$. Draw its graph for $-2 \le x \le 2$.
(c) Now find the formula for $f''(x)$ by differentiating your formula for $f'(x)$ above (this is the second derivative of $f(x)$). Draw the graph of $f''(x)$ for $-2 \le x \le 2$.

(8.7.10) A sphere of radius r has volume $V(r) = \frac{4}{3}\pi r^3$. Find the rate of change of volume with respect to radius. What quantity for a sphere does the formula for $V'(r)$ give?

(8.7.11) A piece of PVC pipe which is open at both ends encloses a volume $V(r, l) = \pi r^2 l$ where r is the radius of the pipe and l is its length. Suppose for a fixed length l we vary the radius r of the pipe. What is the rate of change of volume with respect to radius? What quantity does the formula for this rate give?

(8.7.12) A square with side length x has area $A(x) = x^2$. Find the rate of change of area with respect to side length.

(8.7.13) Find the derivatives with respect to x of
(a) $3x^{-3}$ (b) $x^{-2/3}$ (c) $\sqrt{x}$ (d) $3x^{-4.2}$ (e) $(\sqrt{x} + x^{-1/2})^2$.

(8.7.14) A grain storage silo is in the shape of a giant beer can $100m$ high and the base is a circle of radius $20m$. The height of grain remaining in the silo after t days is $h(t) = 100 - (t^2/4)$.
(a) when is the silo empty?
(b) how quickly is height changing after 10 days.
(c) use your answer to (b) to say approximately how many cubic meters of grain were removed on day 10. Briefly explain.

(8.7.15) Find the equation of the tangent line to $y = x^2 + x$ at $x = 1$.

(8.7.16) Find the equation of the tangent line with slope 5 to $y = 2x^2 - 3x$

(8.7.17) Where does the tangent line to $y = e^x$ at $x = 1$ cross the y-axis?

8.8   The exponential function e^x

We are going to need a function $f(x)$ with the amazing property that when we differentiate $f(x)$, the result is the same function that we started with, namely $f(x)$. In other words, we want to find a function so that

$$f'(x) = f(x).$$

Now there is a very disappointing solution to this: the function $f(x) = 0$ has derivative 0. But there are other possibilities. Here is one

$$f(x) = 1 + x + \frac{x^2}{2!} + \frac{x^3}{3!} + \frac{x^4}{4!} + \frac{x^5}{5!} + \cdots$$

If we differentiate this function, then we get

$$f'(x) = 0 + 1 + \frac{2x}{2!} + \frac{3x^2}{3!} + \frac{4x^3}{4!} + \frac{5x^4}{5!} + \cdots$$

Now remember that $5! = 5 \times 4 \times 3 \times 2 \times 1$ and so we can cancel the 5 on top with the 5 in $5!$ on the bottom to get

$$\frac{5x^4}{5!} = \frac{x^4}{4!}.$$

Doing this for each term we see that indeed $f'(x) = f(x)$. The $\cdots$ on the end means that we keep on adding more terms for ever. Actually, in practise one does not need to do this. By the time we get to, say, the 100'th term

$$\frac{x^{100}}{100!}$$

is so small that we can safely ignore it. The point is that 100! is enormous. Just think about multiplying together all the numbers from 1 to 100. Since one divides by this number, the result is insignificant. So for most practical purposes we can stop with the first 100 terms (unless a really accurate answer is needed: then we might use the first 1000 terms).

If we work out what $f(1)$ is we get

$$1 + 1 + \frac{1}{2!} + \frac{1}{3!} + \frac{1}{4!} + \frac{1}{5!} + \cdots \approx 2.71828...$$

This number goes by the name of e. It is a number like $\pi \approx 3.14159...$ that we can't write down exactly. But it appears in many places. Now it turns out that this mystery function $f(x)$ is actually e^x. Thus

$$e^x \ = \ 1 + x + \frac{x^2}{2!} + \frac{x^3}{3!} + \frac{x^4}{4!} + \frac{x^5}{5!} + \cdots$$

The most important thing about this function is that differentiating it gives back the exact same function again. There is a table showing the values of e^x for x between -9 and 9 in the back of this book. Look at it for a minute or two. You will immediately see that e^9 is pretty big and e^{-9} is pretty small. This should be no surprise. Since e is bigger than 2 then e^9 must be bigger than $2^9 = 512$. Similarly, since $1/e < 1/2$ if follows that $e^{-9} < 2^{-9} < 1/512$.

We started by asking for a function which equals its own derivative. We found e^x has this property. Are there others ? Yes, but they are all just multiples of this one. In other words each constant K gives a function

$$f(x) \ = \ K \times e^x$$

and these are the ONLY functions which equal their own derivative. Memorize this!

Example: the function $f(x) = e^{17x}$ has derivative

$$f'(x) = 17e^{17x}.$$

The rule for finding the derivative of such a function is to bring the constant multiplying x out in front BUT DO NOT SUBTRACT 1 AS YOU WOULD WHEN DIFFERENTIATING x^{17} to get $17x^{16}$. The general rule is that

$$\frac{d}{dx}\left(e^{Kx}\right) \ = \ Ke^{Kx}.$$

Memorize this!

$\boxed{\text{Problems}}$

(8.8.1) If $f(x) = e^{5x}$ find
(a) $f'(x)$ (b) $f'(3)$ (c) $f'(0)$ (d) $f''(x)$

(8.8.2) Find $\frac{d^2}{dx^2}\left(2e^{-3x}\right)$.

(8.8.3) Find a non-zero function such that $f'(x) = 4f(x)$.

(8.8.4) Find a non-zero function such that $f''(x) = 4f(x)$.

(8.8.5) The two functions $s(x)$ and $c(x)$ are given by

$$s(x) \ = \ x - \frac{x^3}{3!} + \frac{x^5}{5!} - \frac{x^7}{7!} + \frac{x^9}{9!} - \cdots$$

$$c(x) \ = \ 1 - \frac{x^2}{2!} + \frac{x^4}{4!} - \frac{x^6}{6!} + \frac{x^8}{8!} - \cdots$$

Show that $s'(x) = c(x)$ and $c'(x) = -s(x)$. Deduce that $s''(x) = -s(x)$ and $c''(x) = -c(x)$. [hint: look at how we differentiated e^x and how things canceled with factorials.] In fact $s(x) = \sin(x)$ and $c(x) = \cos(x)$.

(8.8.6) Use the first 3 terms of the formula for e^x to work out $e^{0.1}, e^{0.3}, e^{0.7}$. Compare your answer to the values given in the table of e^x at the end of these notes. Work out the percentage error in each case. You should notice that the percentage error gets bigger as x gets bigger.

(8.8.7) If we replace x by $5x$ in the formula for e^x we get

$$e^{5x} = 1 + (5x) + \frac{(5x)^2}{2!} + \frac{(5x)^3}{3!} + \frac{(5x)^4}{4!} + \frac{(5x)^5}{5!} + \cdots$$

Differentiate this and show that the result is indeed $5e^{5x}$.

(8.8.8) A bomb falls with velocity $v(t) = C(1 - e^{-kt})$ where C and k are constants.
(a) What is the terminal velocity of the bomb; that is, what is the maximum velocity the bomb can reach, regardless of the height of the bomber?
(b) How long must the bomb fall to reach this velocity?

(8.8.9) What is the equation of the tangent line to the graph of $y = 3e^{2x}$ at $x = 0$

(8.8.10) The mass in mg of a colony of bacteria after t hours is $20e^{t/2}$. How quickly was the colony growing after 4 hours?

(8.8.11) The temperature in $^\circ$F of a corpse t hours after death is $60 + 38e^{-t/24}$.
(a) How quickly is the temperature decreasing after 3 hours.
(b) What is the temperature of the surroundings of the corpse.
(c) What was the temperature at the point of death?

8.9   Calculating the Derivative using Limits.

We have discovered that if we start with a particular function $f(x)$, we get a new function $f'(x)$ called the derivative of $f(x)$. This is a wonderful thing. If we want to know the rate of change of $f(x)$ when $x = 7$, the answer is just $f'(7)$. The JOB of the derivative is to tell us the rate of change of the original function $f(x)$ at any number x we like.

In the last section we listed some rules for differentiating polynomials. The rules we gave for finding the derivative are not so mysterious. For example, suppose we start with a CONSTANT function such as $f(x) = 42$. Well, this is a dull function. Plug in anything for x and the answer is always 42. So what is the rate of change of this function? Since the constant 42 never changes it makes sense that the rate of change of $f(x)$ is always 0.

What about a function like $f(x) = 7x$? This is a linear function. The derivative is $f'(x) = 7$. This is also reasonable: For example, if you increase x by 0.01 then $f(x)$ increases by 7×0.01. In other words,

$$\Delta f = f(x + 0.01) - f(x) = 7(x + 0.01) - 7x = 7x + 7(0.01) - 7x = 7 \times 0.01.$$

So the rate of change of $f(x)$, given by Δf [=the change in $f(x)$] divided by Δx [=the change in x, which is 0.01] is 7.

From the power rule we know that the derivative of $f(x) = x^2$ is $f'(x) = 2x$. This means that the graph of $y = x^2$ has slope $2x$ at x. Look at the graph and check out if this is right. We see that the graph has slope zero at $x = 0$ [which is good because the derivative is $2x$ is 0 when $x = 0$]. Looking at the graph, we see that the slope is negative when x is negative [again the derivative $2x$ is negative when x is negative]. The derivative tells us that when $x = 1$ the slope is $2(1) = 2$. Look at the graph to confirm this.

We can also work out the rate of change of $f(x) = x^2$ algebraically. We imagine that Δx is a small number (like 0.01). We then work out how much $f(x)$ increases when x is increased to $x + \Delta x$.

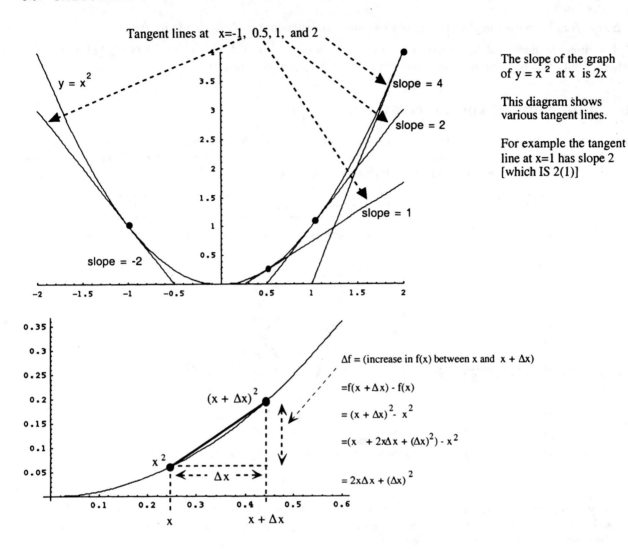

Tangent lines at x=-1, 0.5, 1, and 2

The slope of the graph of $y = x^2$ at x is 2x

This diagram shows various tangent lines.

For example the tangent line at x=1 has slope 2 [which IS 2(1)]

Δf = (increase in f(x) between x and x + Δx)

= f(x + Δx) - f(x)

= (x + Δx)^2 - x^2

= (x^2 + 2xΔx + (Δx)^2) - x^2

= 2xΔx + (Δx)^2

To find the average rate of change between x and $x + \Delta x$ we divide Δf by Δx. Thus

$$\frac{\Delta f}{\Delta x} = \text{(average rate of change of } f(x) \text{ between } x \text{ and } x + \Delta x)$$

$$= \frac{2x\Delta x + \Delta x^2}{\Delta x}$$

$$= 2x + \Delta x \qquad \text{canceling an } \Delta x \text{ on top and bottom}$$

Now we take the limit of $2x + \Delta x$ as Δx gets smaller and smaller. This limit is just $2x$. So this is the rate of change of $f(x)$ at x.

Problems

(8.9.1) You might enjoy doing a similar calculation to find the rate of change for x^3. You need to work out the change in $f(x) = x^3$ when x is increased by a small number Δx to $x + \Delta x$. So you will work out $f(x + \Delta x) - f(x)$. Then do some algebra to simplify this. You should get $3x^2\Delta x + 3x\Delta x^2 + \Delta x^3$ [be careful when you work out $(x + \Delta x)^3$ to multiply out the parentheses one step at a time.] Then divide this by Δx to get the average rate of change. Now think about what the limit of this is as Δx goes to 0. You should finally end up with $3x^2$ which is the derivative of x^3.

(8.9.2) Repeat the algebraic procedure using limits to find the derivative of $f(x) = 1/x$. To do this, simplify

$f(x + \Delta x) - f(x)$ by converting the fractions to a common denominator. Now divide by Δx.

(8.9.3) Use limits to show that the derivative of the function $g(x) = C \times f(x)$ is $g'(x) = C \times f'(x)$. Here, C is a constant. [hint: this problem is easier than you think: just pull out a factor of C]

8.10   The Sign of the Derivative.

Again, consider the function $f(x) = x^2$. If $x > 0$ then the graph of $f(x)$ slopes upward. So we expect the tangent lines to $f(x)$ in this range to have positive slope. In fact, this is exactly what the derivative $f'(x) = 2x$ tells us: if $x > 0$ then $2x > 0$. This idea works in general; that is, given any function $f(x)$,

> the derivative $f'(x)$ is **positive** wherever the graph of $f(x)$ slopes upwards.

Notice that the derivative of x^{-1} is $-x^{-2}$. The minus sign in front tells us that the derivative is negative. We

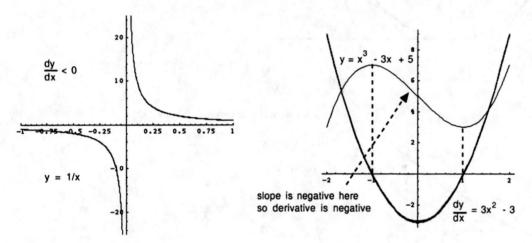

can see why this is true by looking at the graph of $y = 1/x$. Notice that it has negative slope everywhere. In fact, if we increase x then $1/x$ decreases because we are dividing by a larger number. Thus as x gets bigger $1/x$ gets smaller. So, again we see that the rate of change is negative.

The derivative of $f(x) = x^3 - 3x + 5$ is $f'(x) = 3x^2 - 3$. Both of these functions are shown on the same graph. The graph of $f'(x)$ is the darker curve. Whenever the slope of $f(x)$ is negative the derivative $f'(x)$ is negative. You can see this in the diagram. The part of the graph of $f(x)$ between the dotted lines is where the slope is negative. We see that in this region that $f'(x)$ is indeed negative.

Conversely, the sign of the derivative $f'(x)$ is negative wherever the graph of $f(x)$ slopes downwards. At places where the derivative $f'(x) = 0$, the tangent line is horizontal. At these places the graph is level so that it slopes neither up nor down.

> Problems

(8.10.1) The height in meters of a rocket above the ground t seconds after launch is $h(t) = 20t + 5t^2$.
(a) What is the formula for the speed of the rocket at time t?
(b) What is the speed of the rocket at $t = 2$?
(c) How long did it take for the speed of the rocket to become 120 m/s?
(d) When was the height of the rocket 100 m?

(8.10.2) The amount of money a company has in the bank after t years is $180 + 3t - t^2$ thousand dollars
(a) How much did the bank account increase during the second year?
(b) What was the average rate of increase of money in the account during the second year?
(c) What was the instantaneous rate of increase of money in the account at the start of each of the four quarters

of the second year?

(d) When did the bank account run out of money?

(8.10.3) (a) What is the slope of the graph $y = x^2$ at $x = 3$?.

(b) Use your answer to (a) to write down the equation of the tangent line to this graph at $x = 3$.

(c) Make a sketch of the graph and the tangent line you just found

(d) A plane flies along this graph so that it is moving towards the origin. On the front of the plane is a very bright light. The light shines straight ahead. When the plane is at the point on the graph where $x = 3$ what point on the y-axis does the light shine on? [look at your diagram]

(8.10.4) The price of a certain computer stock t days after it is issued for sale is $p(t) = 100 + 20t - 5t^2$ dollars. The price of the stock initially rises, but eventually begins to fall. During what period of time does the stock price rise?. If you owned the stock, when would you sell it?

(8.10.5) A circular cloud of poison gas from a factory explosion is expanding so that t hours after the explosion the radius of the cloud is $R(t) = 50 + 20t$ meters. How fast is the area of the cloud increasing 3 hours after the explosion?

(8.10.6) Air is pumped into a spherical balloon, so the balloon expands. The volume of a sphere of radius R is $4\pi R^3/3$. If the radius of the sphere after t seconds is $2t$ *cm* at what rate is air being pumped in when $t = 3$?. [hint: the rate air is pumped in equals the rate that the volume of the sphere increases]

(8.10.7) $f(x) = x^2 + 6x + 7$

(a) when is the rate of change of $f(x)$ zero?

(b) When is the derivative positive?

(c) When does the graph $y = f(x)$ have positive slope?

(8.10.8) Sketch the graph $y = f'(x)$ for each of the functions $f(x)$ shown below.

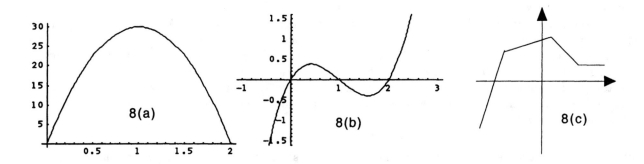

(8.10.9) (a) Sketch the graph of a function with derivative equal to 2 at all x.

(b) Sketch the graph of a function for $0 \le x \le 4$ which is negative and whose **derivative** is positive and increasing

(c) Sketch the graph of a function whose **derivative** is positive but decreasing.

(8.10.10) The price of IBM stock t months after you buy it is $\$p(t)$. Below is a graph for $p'(t)$ the **derivative** of $p(t)$. When should you sell for the most profit? Explain!!

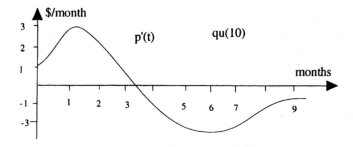

8.11   Viewing a graph under a Microscope.

A graph under high magnification looks straight.

This section describes a new way to think about the whole idea of derivative. The figure (I) shows a circle

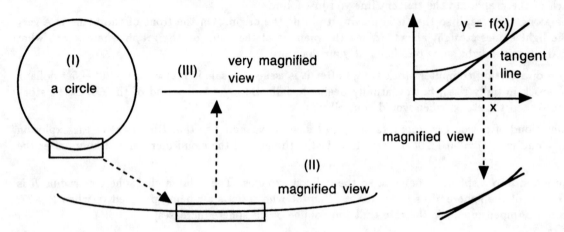

together with a magnified view, (II) and a highly magnified view (III), of portions of the circle. The magnified view does not look very curved and the highly magnified view looks completely straight. This is something from common experience: if we look at the Earth from space we see a sphere. But from the ground the Earth seems flat instead of round. Similarly, a small circle seems very curved, but a huge circle seems to curve very little. Looking at a highly magnified piece of a circle is like looking at a small part of an enormous circle. It seems to be almost flat. This is a fairly simple idea but it is really the **entire enchilada** when it comes to the derivative: a highly magnified view of a smooth curve looks almost straight.

Suppose you have a function $f(x)$ and you increase x by the small amount Δx. Think about the corresponding change $\Delta f = f(x + \Delta x) - f(x)$,   in $f(x)$. This is actually like looking at the graph of $y = f(x)$ under a microscope. We **focus in** on the portion of the graph near x. We draw the familiar diagram above showing Δx and Δf. Everything is so small that we magnify the diagram. The magnified graph looks very nearly straight and the tangent line almost coincides with this portion of the magnified graph. The slope of the tangent line is $f'(x)$. If we do not move far from x then the the tangent line provides a very good approximation to the function $f(x)$. We have also seen that

$$\Delta f = f(x + \Delta x) - f(x) \approx \Delta x \cdot f'(x).$$

We interpret this equation as saying "what the function $f(x)$ does near x is very nearly equal to what the tangent line does." It says "if you move a small distance Δx then the increase Δf is almost the same as if you move along the tangent line." It also says "if you want a good approximation to the complicated function $f(x)$ then use the tangent line." The equation of the tangent line is very simple:

$$y = m \cdot x + b.$$

Here the slope of the tangent line m is just the derivative of f.

The **slope** of the graph $y = f(x)$ can also be thought of this way. If we highly magnify the portion of the graph near x, it **looks** like a straight line, and the slope of this line is the derivative of the function at the point x.

A sophisticated way to think about derivatives is the following. An arbitrary function might be very complicated. But over a small range, we can **approximate** it by a straight line, which is very simple. This is an example of the motto: if you don't understand something (a complicated function), try to understand something simpler (the tangent line approximation.)

We have seen this idea in action before. The idea of linear interpolation and extrapolation is based on the idea of approximating a complicated function by a straight line (I should really say "linear function plus a constant").

In a way this is philosophy: it is a way to think about the meaning of the derivative. But, it is VERY important. The summary is

A smooth function is well approximated by a straight line over a small range.

Problems

(8.11.1) What straight line does the graph of $f(x) = x^3$ near $x = 1$ look like under a microscope?

(8.11.2) Draw a graph of $f(x) = 10^x$ for $0 \leq x \leq 0.1$. This graph is almost a straight line. What is the equation of this line?.

8.12 Acceleration and the Second derivative.

acceleration equals rate of change of velocity.

You are in a Ferrari on a race track. You put your foot on the gas pedal. Vroom! The car accelerates forward and you are pushed into your seat, blood draining to your feet. You are accelerating. This means you are speeding up. Acceleration tells you how quickly you are speeding up. If your speed increases from 50 *mph* to 60*mph* in one second then your acceleration is 10 **miles per hour per second.** In other words your speed increased by 10 *mph* in one second. An object that is falling speeds up. Every second that it falls its speed increases by 32 *ft/sec.* [30 *mph* equals 44 *ft/sec.*] This is described by saying that the acceleration due to gravity is 32 **feet per second per second.** It means that in one second of falling speed increases by 32 feet per second. In the metric system the acceleration due to gravity is 9.8 meters per second per second [ie ms^{-2}]. Often to make calculus questions easier this is changed slightly to 10 ms^{-2}.

Example: An object has an initial velocity of u ms^{-1} at time zero and has a constant acceleration of a ms^{-2}. Then the speed after t seconds is $u + at$. This is because every second the velocity increases by a ms^{-1}. So in t seconds the velocity increases by t times this amount, namely (at) ms^{-1}. Since the initial velocity was u, we add (at) to get the the final answer.

Example: An object has an initial velocity of u and constant acceleration of a. The distance, s, traveled by the object in t seconds is given by the famous formula

$$ s \;=\; ut \;+\; \frac{1}{2}at^2 . $$

We can check this formula is correct by differentiating it. The rate of change of distance traveled equals the velocity. So if we work out the derivative we should get the answer $u + at$ as in the example above.

$$ \frac{d}{dt}\left(ut \;+\; \frac{1}{2}at^2 \right) \;=\; u \;+\; \frac{1}{2}a(2t) \;=\; u \;+\; at. $$

This is indeed the correct answer.

Now acceleration equals rate of change of velocity and velocity equals rate of change of distance. Thus acceleration equals rate of change of (rate of change of distance). Two rates of change combined! Here is another example. Suppose that the distance an object has traveled after t seconds is $f(t) = t^3$ meters. Then the velocity, $v(t)$, after t seconds is the derivative of $f(t)$ so

$$ v(t) \;=\; \frac{d\,(t^3)}{dt} \;=\; 3t^2 . $$

Now we can work out the acceleration. Acceleration, $a(t)$, equals rate of change of velocity, so

$$ a(t) \;=\; \frac{d\,(3t^2)}{dt} \;=\; 6t. $$

We started with the distance, differentiated **twice** and ended up with the acceleration.

We often want to differentiate a function like $f(t)$ twice. This can be expressed mathematically by writing

$$ \frac{d}{dt}\left(\frac{df}{dt} \right) $$

which just means "take the derivative of df/dt." There is a shorthand notation for this:

$$\frac{d^2 f}{dt^2}.$$

Read this "dee two eff dee tee squared." Do not fall into the trap of thinking we are squaring something! The superscript ^{2} means "do it twice!". In other words, "differentiate twice." For example, if $f(t) = 6t^2$ then $df/dt = 12t$ and $d^2 f/dt^2 = 12$. There is another even simpler notation for differentiating $f(t)$ twice. This is written

$$f''(t)$$

and read aloud as 'eff double prime tee." So

$$f''(t) = \frac{d^2 f}{dt^2}.$$

The derivative $f'(t)$ tells you how quickly $f(t)$ is changing. It is the instantaneous rate of change of $f(t)$. So the derivative of $f'(t)$, which is just $f''(t)$, tells us how quickly the rate of change of $f(t)$ is changing. This is just like acceleration; acceleration tells us how quickly velocity (=rate of change of distance) is changing.

There are many situations in which the second derivative is something we really care about. For example, suppose that the amount of money a company is worth at time t is the function $p(t)$. It is good to know $p(3) = 10$ *million*. But, it's very important to know if the company is loosing money. This happens if the derivative $p'(t)$ is negative. We do not like this situation. But, there is a more subtle (and perhaps ominous from a financial standpoint) possibility. Maybe the derivative is positive: the company is making money. At the same time, maybe the sales are dropping off. The company is making less and less money. Said another way, "the company is going down hill!" This means that the rate at which it makes money is decreasing. In other words, the rate of change of the rate it makes money is negative. For company B shown below $p''(t)$ is negative, which means the rate at which the company is making money is declining. This is not good. For company A shown below $p''(t)$ is positive. So the rate it is making money is increasing.

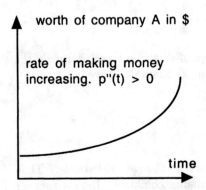

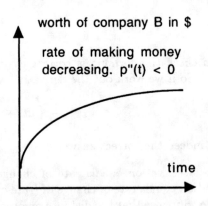

Look at the two graphs of company A and company B. Imagine instead that the graphs show the population of two countries. Both countries have increasing populations, but country A has a population which is growing more and more quickly, whereas country B has a population whose growth is slowing down. Now imagine that the graphs show the spread of two infectious diseases, which one are you more worried about? Imagine the two graphs show the demand for two different products. In which should you invest?

You see, the second derivative is often very important. You might remember the slogan "speed kills." Actually, this is wrong. Speed never killed anyone. Instead, you are in trouble when you hit a brick wall at 60 *mph*. All of the sudden your speed is 0 *mph*! Your speed changed very quickly from 60 to zero. So, it is acceleration that kills. Second derivatives are important.

On a graph of $y = f(x)$ we can see quite easily where the second derivative is positive. It is where the slope increases (becomes more positive) as you move along the graph to the right. If you look at a graph you can see where the graph curves upwards (this is called **concave up**) and where it curves downwards (this is called **concave down**). It is easy to remember this with "the smiley face and the frown." The + sign on the smiley face reminds you that the graph is concave up whenever $f''(x)$ is positive (+). The − sign on the frown tells you the graph is concave down whenever $f''(x)$ is negative (−). The reason this all works is that the **second derivative is positive means the derivative is increasing.**

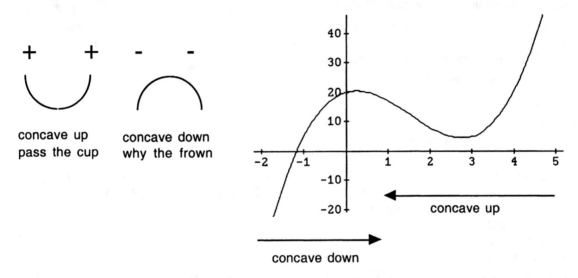

Problems

(8.12.1) Find the second derivatives of
(a) $3x + 7$ (b) x^2 (c) x^3 (d) x^n when $n \geq 2$.
(e) x^{-1} (f) $4x^5 - 3x^3 + x^2 + 2x + 7$ (g) $Ax + B$ where A and B are constants

(8.12.2) What is the derivative at $x = 2$ of the rate of change of x^3?

(8.12.3) Calculate $f''(2)$ and $f''(-1)$ when $f(x) = x^4 - 2x^2$.

(8.12.4) Copy the graph below and clearly indicate the part of the graph for which $f''(x) > 0$. Also indicate where the graph is concave down. Where is $f'(x)$ zero ?

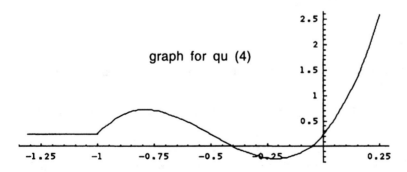

(8.12.5) The height of the water in a reservoir varies with time. After t days the height of the water is $h(t)$ meters.
(a) How would you explain to someone who does not know any calculus the meaning of $h'(30) = 1$ and $h''(30) = -0.5$?
(b) If h'' does not change, will the height of water be rising or falling on day 33? Explain. How quickly would you guess it is rising or falling?

(8.12.6) An object is dropped out of an airplane. The height of the object t seconds after being dropped is

$h(t) = 2000 - 5t^2$ meters.
(a) What was the height of the plane when the object was dropped ?
(b) What is the velocity of the object at $t = 3$
(c) What is the acceleration of the object ?
(d) When did the object hit the ground ?
(e) How fast was it going when it hit the ground?

(8.12.7) You are the mayor of a city. Last year the income taxes you collected increased by 2% from the previous year. You look at a graph showing the population of your city and see that it is concave down. Should you expect an increase of more or less than 2% this year? Explain.

(8.12.8) Find the derivative and second derivative of e^{3x}.

(8.12.9) Find the first 10 derivatives of e^{-x}.

(8.12.10) $f(x) = 2x^3 - 3x^2 - 36x + 2$.
(a) For which values of x is $f(x)$ increasing ?
(b) Where is $f(x)$ concave up ?

(8.12.11) I stood on top of Cheadle Hall and threw an egg upwards. It landed on the ground after 4 seconds. The height of the egg **above Cheadle Hall** t seconds after I threw it was $h(t) = 10t - 5t^2$ meters.
(a) What was the speed of the egg when it hit the ground ?
(b) When was the speed of the egg zero ?
(c) How high is Cheadle hall ?
(d) How high **above the ground** did the egg go?

(8.12.12) The position of an object on the x-axis after t seconds is $x(t) = t^3 - 4t^2 - 7t + 4$ cm to the right of the origin. At time $t = 1$:
(a) is the object moving left or right ?
(b) is the object speeding up or slowing down ?
(c) is the acceleration increasing or deceasing ?

(8.12.13) (a) At what point on the graph $y = x^4$ is the slope 32?
(b) What is the second derivative at this point ?

(8.12.14) The amount of profit a company makes is given by a function $f(t)$ where t is time measured in years. Write a short sentence explaining the meaning of each of the following (a) $f(3) < 0$ (b) $f'(3) > 0$ (c) $f''(3) < 0$

(8.12.15) Suppose that $f(t)$ is the price of an average bag of groceries at time t (measured in years).
(a) If the annual inflation rate is 4% after 5 years does this tell you anything about any of the derivatives of f ?
(b) Draw a sketch showing the average price of a bag of groceries if the inflation rate is increasing when $t = 4$.
(c) Does this tell you anything about any of the derivatives of f ?

8.13   Maxima and Minima.

The plural of maximum is maxima. Sometimes we want to find the largest or the smallest occurrence of something. Such a problem is usually called a max/min problem.

Example 8.13.1 *You have two numbers which add up to 20. What is the largest possible value for the product of the two numbers.*

We name the two numbers x and y. Since $x + y = 20$ we immediately discover that $y = x - 20$. Our job is to try to make $x \times y$ as large as possible. We now know that $y = 20 - x$. Therefore, we want to make $f(x) = x(20 - x) = 20x - x^2$ as large as possible. To solve this we could draw the graph of $f(x)$ which is an upside down parabola and we see that the maximum is at $x = 10$. This gives a maximum of $f(10) = 10(20 - 10) = 100$. But there is a way to do this without drawing the graph. The clue is to look at the graph and notice

$$\boxed{\text{The derivative is zero at a maximum}}$$

The tangent line is horizontal at a maximum, and the slope of the tangent line is the derivative at that point. So at the maximum the derivative is indeed zero. So now the trick is to differentiate $f(x)$ and set this derivative equal to zero. We then solve for x. Now,

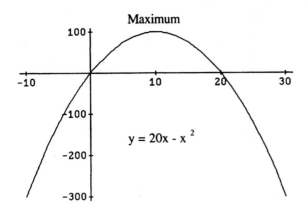

$$\frac{df}{dx} = \frac{d}{dx}\left(20x - x^2\right) = 20 - 2x.$$

Therefore, we set this equal to zero and solve for x. Thus,

$$20 - 2x = 0 \quad \Rightarrow \quad x = 10.$$

So we get the answer $x = 10$. We must remember that we are asked for the largest value of the product of the two numbers. Having found x we can find $y = 20 - x = 20 - 10 = 10$. Finally, the product $xy = 10 \times 10 = 100$ is the maximum. Magic!

To find the min/max the method is to "**kill the derivative**" ("kill" means set equal to zero):
(1) Solve $f'(x) = 0$.
(2) Substitute the value for x found in step (1) into $f(x)$. This is the minimum or maximum (depending on the problem).

Example 8.13.2 *A robot is navigating a hall in a maze. The hall is shaped like a long, thin rectangle. The robot must move down the hall in the right direction. To do this, it uses SONAR mounted on its swivelling head to find the wall farthest from it. The distance in feet from the nearest wall is reported by the SONAR to vary as $d(x) = 12 - 2x - x^2$, where $x \times 90°$ is the rotation angle of the robot's head. Find the angle which points to the farthest wall and the corresponding distance.*

Solution: we simply want to maximize the function $d(x)$. To do this, we compute $d'(x) = -2 - 2x$ and set this equal to 0. This gives us that $2x = -2$ or $x = -1$. This means the maximizing angle is $-1 \times 90° = -90°$ and the farthest wall is $d(1) = 12 - 2(-1) - (-1)^2 = 13$ *feet.*

To do a max/min problem here is the plan
(1) Find the function $f(x)$ that we want to maximize.
(2) Differentiate to get $f'(x)$.
(3) Solve $f'(x) = 0$. We will call the solution p.
(4) The maximum/minimum is $f(p.)$

The only hard bit is step (1). The other steps involve calculus but they are really easy. Step (1) is usually algebra (meaning a word problem) and can be a bit tricky.

Example 8.13.3 *A cardboard box is to be made from a rectangular sheet of cardboard by cutting out squares from the four corners of the cardboard as shown in the figure and then folding up what is left to make the box. The box has a bottom and sides but no top. The cardboard sheet is 80 cm by 30 cm. What dimensions produce the box with largest volume?*

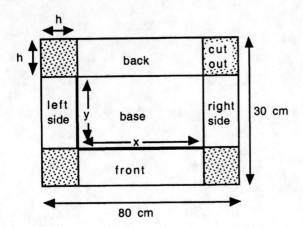

The first thing to do is draw a diagram. Next the goal is

$$\boxed{\text{Find the dimensions to maximize volume.}}$$

We must decide what this means. There are some quantities that can be varied. We must decide what these quantities are before we can hope to discover the solution. The figure has labeled three lengths x, y, and h. This is the hard part and it has been done for you. If the diagram did not have those sides labeled you would have to reason this way:

If I had this sheet of cardboard, how would I go about making a box? Any box: it does not matter for the moment if it has the biggest volume. Once I have figured out how to make a box, I will figure out its volume. There are many boxes I could make, and so the volume of the box I make will depend on some quantity (or quantities) that I choose during the construction. This means I will need to find a formula for the volume of the box I make. The formula will have some unknown (or several unknowns) in it. I will then try to find a maximum for the formula using calculus. So, how do I go about making a box? Well, I need to know how tall to make the box. I don't know how tall to make the box so I will call the height of the box h cm. Now I also need to know the lengths of the other sides of the box; that is, I need to know the length and width of the bottom of the box. Let

$$
\begin{aligned}
h &= \text{ height of box} \\
x &= \text{ width of bottom of box} \\
y &= \text{ length of bottom of box.} \\
V &= \text{ volume of the box in } cm^3
\end{aligned}
$$

All these are measured in centimeters except volume. So there are 3 unknowns in this problem. The game now is:
(a) express the volume of the box in terms of the unknowns
(b) get rid of 2 of the unknowns, so the volume only involves one unknown.

The volume of the box is area of bottom times height which is

$$(1) \qquad V = x \times y \times h.$$

Now we need to get rid of two variables. We will need some equations to do this. Looking at the diagram, we see that

$$(2) \qquad 2h + x = 80 \qquad 2h + y = 30$$

so we can use these to express everything in terms of just one of the variables. Which one? Well, it doesn't matter which one we choose; the final answer will be the same. By using h we avoid dividing by 2 so we express x and y in terms of h as follows:

$$(3) \qquad x = 80 - 2h \qquad y = 30 - 2h.$$

Now plug this into (1) to get

$$(4) \qquad V = x \times y \times h = (80 - 2h) \times (30 - 2h) \times h.$$

This is a formula for the volume with only one unknown, h, in it. Now we need to find the maximum volume. This means working out the derivative of V and setting this equal to 0. In order to work out the derivative of V we first need to multiply out the right hand side of (4):

$$(5) \qquad V = 2400h - 220h^2 + 4h^3.$$

Now find the derivative of this:

$$(6) \qquad \frac{dV}{dh} = 2400 - 440h + 12h^2.$$

This is the derivative. It must be zero for a maximum, so we set this equal to 0 and solve to find the height h. Once we know h we can find the other dimensions x and y of the box.

$$\begin{aligned} \text{Solve} \quad & 2400 - 440h + 12h^2 = 0 \\ \Rightarrow \quad & 600 - 110h + 3h^2 = 0 \qquad \text{divide both sides by 4} \\ \Rightarrow \quad & (3h - 20)(h - 30) = 0 \qquad \text{factoring} \\ \Rightarrow \quad & h = 20/3 \quad \text{or} \quad h = 30 \qquad \text{one of the factors is zero} \end{aligned}$$

Which one of these is correct? A height of 30 is impossible since the cardboard is not big enough. Therefore, the answer must be $h = 20/3$. Then $x = 80 - 2(20/3) = 200/3$ and $y = 30 - 2(20/3) = 50/3$.

Problems

(8.13.1) Find the maximum of $f(x) = 2 + 5x - x^2$.

(8.13.2) Find the minimum of $x + x^{-1}$ for $x > 0$.

(8.13.3) A rectangular area is to be fenced with 500 feet of chicken wire. Find the maximum area that can be enclosed. What are its dimensions?

(8.13.4) Use calculus to find the largest possible area for a rectangular field which can be enclosed with a fence 400 meters long.

(8.13.5) Consider $f(x) = 4ax - x^2$. Write down a formula in terms of a for the maximum of $f(x)$.

(8.13.6) A rectangular field is to have an area of $1000 \ m^2$ and is to be surrounded by a fence. The cost of the fence is \$20 per meter of length. What is the minimum cost this can be done for ?

(8.13.7) A box, with rectangular sides, base and top is to have a volume of 4 cubic feet. It has a square base. If the material for the base and top costs \$10 $/ft^2$ and that for the sides costs \$20 $/ft^2$ what is the least cost it can be made for.

(8.13.8) A farmer wants to make a rectangular field with a total area of $2000 \ m^2$. It is surrounded by a fence. It is divided into 3 equal areas by fences. What is the shortest total length of fence this can be done with ?.

(8.13.9) (*) A manufacturer wishes to make tee-shirts for the band **Indigo Girls.** They sell for \$20 each. She must deliver all the tee-shirts in 20 days time. The manufacturer will first set up some machines. Once all machines are set up she will turn them on. Each machine takes one day to set up. A machine produces 200 shirts in one day. All the machines must be set up before any of them are turned on. (a) Express the amount of money she will receive for the shirts in terms of the number of machines she decides to set up. (b) Use calculus to find how many machines should she set up to obtain most money.

(8.13.10) An airline sells **all the tickets for a certain route at the same price.** If it charges \$250 per ticket it sells $5,000$ tickets. For every \$10 the ticket price is reduced, an extra thousand tickets are sold. It costs the airline \$100 to fly a person. What price will generate the greatest profit for the airline?

(8.13.11) A commuter railway has 800 passengers per day and charges each one \$2. For each 2 cents that the fare is increased, 5 fewer people will go by train. What is the greatest profit that can be earned.

(8.13.12) A poster is to have a total area of 500 cm^2. There is a margin round the edges of 6 cm at the top and 4 cm at the sides and bottom where nothing is printed. What width should the poster be in order to have the largest printed area.

(8.13.13) A sports field is to have the shape of a rectangle with semi-circles put on the two ends. It must have a perimeter of 1000 m. What is the maximum area possible for the rectangular part.

(8.13.14) What point on the graph $y = \sqrt{x}$ is closest to $(1, 0)$. [hint: work out the **square** of the distance of a point on the curve from $(1, 0)$ and minimize the distance squared (this is a standard trick which makes the algebra easier.)]

(8.13.15) On a certain island there are two populations of deer. After t years the numbers of deer in the two populations are $p(t) = 100e^t$ and $q(t) = 10^4 e^{-t}$. When is the total population smallest?

(8.13.16) A cylindrical metal can is to have no lid. It is to have volume 8π in^3. What height minimizes the amount of metal used? (ie the surface area).

(8.13.17) A farmer can get \$2 per sack of potatoes on July 1'st and after that the price drops by 2 cents per sack per day. On July 1'st the farmer has 80 sacks of potatoes in the ground. She estimates that the crop is increasing by one sack per day. When should she harvest to get the most money?

(8.13.18) A wise old troll wants to make a small hut. Roofing material costs \$5 per square foot and wall material costs \$3 per square foot. According to ancient troll customs the floor must be square, but the height is not restricted.
(a) Express the cost of the hut in terms of its height and the length of the side of the square floor.
(b) If the troll has only \$1500 to spend, what is the biggest volume hut he can build?

(8.13.19) A manufacturer sells lamps at \$6 each and sells 3000 each month. For each \$1 that the price is increased, 1000 fewer lamps are sold each month. It costs \$4 to make one lamp. What price should lamps be sold at to maximize profit?.

(8.13.20) A rectangular field will have one side made of a brick wall and the other three sides made of wooden fence. Brick wall costs \$20 per meter. Wooden fence costs \$40 for 4 meters. The area of the field is to be 2400 m^2. What length should the brick wall be to give the lowest total cost of wall plus fence ?

8.14 Summary

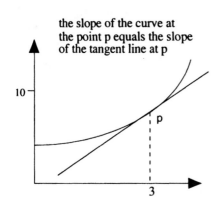

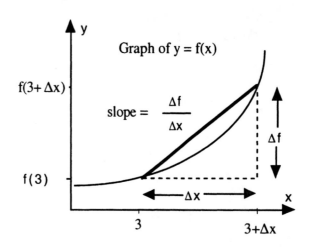

The derivative $f'(3)$ is
- the instantaneous rate of change of $f(x)$ at $x = 3$
- the slope of the tangent line to the graph of $y = f(x)$ at $x = 3$
- the limit as $\Delta x \to 0$ of $[f(3 + \Delta x) - f(3)]/\Delta x$
- how steep the graph of $y = f(x)$ is at $x = 3$
- how quickly $f(x)$ is increasing [or decreasing if $f'(3) < 0$] near $x = 3$
- has units (units of f)/(units of x)
- if $f(t) = $ (distance traveled at time t) then $f'(3) = $ (velocity at $t = 3$)
- if $f(t) = $ (velocity at time t) then $f'(3) = $ (acceleration at $t = 3$)
- $f'(3) = 0$ if $x = 3$ gives a max/min of $f(x)$
- is approximately $[f(3.0001) - f(3)]/.0001$
- is positive if f is increasing at $x = 3$ and negative if f is decreasing at $x = 3$
- is close to the average rate of change $[f(4) - f(3)]/[4 - 3]$

$$\frac{d}{dx}(x^n) = nx^{n-1} \qquad \frac{d}{dx}(constant) = 0 \qquad \frac{d}{dx}(e^{3x}) = 3e^{3x}$$

$$\frac{d}{dx}(\sqrt{x}) = \frac{1}{2}x^{-1/2} \qquad \frac{d}{dx}\left(\frac{1}{x}\right) = -x^{-2}.$$

review problems: You should attempt these problems BEFORE looking at the model solutions. (8.1.2) (8.2.6) (8.4.7) (8.4.9) (8.5.3) (8.6.1) (8.7.1) (8.7.8) (8.8.2) (8.10.1) (8.10.5) (8.12.1) (8.12.4) (8.12.11) (8.13.6) (8.13.17) (8.13.19)

Chapter 9

Integration.

It may seem surprising at first, but very often we want to calculate the area under part of a graph for reasons that have nothing to do with wanting to know about area. This is true because the area under the graph is a number which frequently **represents** something we want to know about. The process of finding this area is called **integration.** As explained in the introduction, integration is the reverse of differentiation. If we know how quickly a quantity changes with time, and we want to know what the quantity is at some particular moment, then integration gives the answer.

We will illustrate this idea by continuing the example we started in the introduction on page 1, which you should re-read now. If the rate at which the temperature changes is constant, then we can work out what the temperature will be 24 hours from now: multiply the rate of change by 24 and add this to the current temperature. However, if the rate at which the temperature changes is not constant, then this procedure no longer works. Thus, instead of computing the temperature change over 24 hours in one step, we break up the 24 hour time interval into one minute time intervals (for example). During each of these one minute time intervals, if we know how quickly the temperature is changing then we can work out the change in temperature during that minute. Then, by adding up all the changes for all the minutes in the 24 hour period, we get a very accurate picture of the total change in temperature. The big idea to understand is the following.

> The area underneath a graph represents adding up a lot of numbers.

This idea will become clearer as we proceed. We will discuss some examples in which the area represents: a distance traveled, the number of fish killed, an amount of water used, etc.

9.1   The Signed Area Under a Graph.

The area between the x-axis and the graph $y = f(x)$ between $x = a$ and $x = b$ shown shaded in the diagram is expressed mathematically by writing

$$\int_a^b f(x)\, dx.$$

This expression is called an **integral.** It means nothing more or less than this area. Read this as "the integral from a to b of $f(x)$ integrated with respect to x." The numbers a and b are called the **limits of integration.** This has nothing to do with the word limit in the sense of "taking a limit as x goes to zero." Here limit simply refers to the limits (meaning boundaries) of the region where we want to find the area. Thus,

$$\int_0^3 f(x)\, dx$$

is the area under the graph between $x = 0$ and $x = 3$. So, for example, if we take a **constant** function $f(x) = 2$ and find the area under the graph between $x = 0$ and $x = 3$ we get

$$\int_0^3 2\, dx = (\textit{area of a rectangle width 3 and height 2})$$

$$= 6.$$

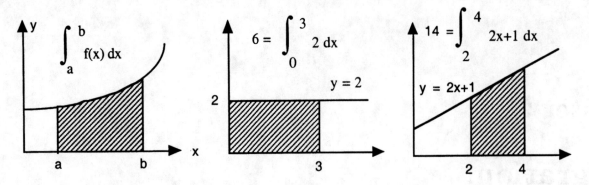

The Definite integral gives area under the graph

If the graph $y = f(x)$ is a **straight line** then it is easy to work out the area under the graph between any two points. For example,

$$\int_2^4 (2x + 1) \, dx \; = \; (area \; of \; trapezium \; shown)$$

$$= \; 2 \times \frac{5 + 9}{2} \; = \; 14.$$

Remember that the area of a trapezium is the **average height** times the width, and 5 and 9 are the heights of the two sides of the trapezium in this example.

We now come to an important idea: **signed area.** For reasons that will become clearer soon, we usually want to count area underneath the x-axis as **NEGATIVE.** For example,

$$\int_0^3 -2 \, dx \; = \; -6.$$

The graph in this case is **below** the x-axis. The shaded area shown is the the region between the x-axis and the graph $y = -2$. This shaded area is **beneath** the x-axis, so it counts negative.

  Thus, the rule is that

$$\int_a^b f(x) \, dx$$

is the **signed area** between the graph and the x-axis. Often people say "the integral is the area **under** the

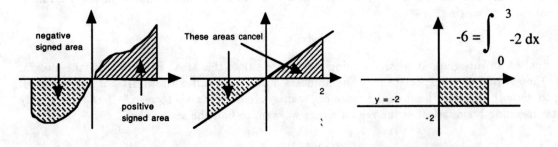

Signed Area: counts Negative below x-axis, Positive above

graph" but what they really mean is **signed area between** the graph and the x-axis. The only difference is that

---

area above the x-axis counts positive and area below the x-axis counts negative.

Now, frequently the graph of a function $y = f(x)$ is above the x-axis for part of the time and below the x-axis for part of the time. In this case, to work out

$$\int_a^b f(x)\ dx$$

we must find the area below the x-axis and subtract it from the area above the x-axis. Sometimes this is very easy to do. For example:

$$\int_{-2}^2 3x\ dx\ =\ 0.$$

If we look at the graph, then this integral gives the **signed area** of the shaded region. We notice that the region is made up of two equal sized triangles: one is below the x-axis while the other is above the x-axis. Therefore, their **signed** areas cancel out to give a total signed area of 0.

You should get clear in your mind the following: area is always positive, but **signed area** can be positive or negative. Signed area is not the same as area, just as Mel B. is not the same person as Mel C. It does not make sense to talk about a soccer field having negative area. Area can't be negative. But **signed area** is a bit trickier.

If the graph $y = f(x)$ is not a straight line, then one can find the area approximately by using graph paper and counting squares. Later we will see some other methods for working out integrals. But often the most important thing is not working out the integral but understanding what it represents.

Example 9.1.1 *Suppose that $f(t)$ is the rate at which people become infected with a disease. We will measure t in days, so that $f(5) = 220$ means that the rate at which people are being infected after 5 days is 220 new infections per day. What is the meaning of*

$$\int_0^{10} f(t)\ dt.$$

Solution: The integral is the number of people who become infected during the first 10 days. Here is how to think about this: The graph $y = f(t)$ is probably rather complicated, but to understand how to interpret the meaning of the integral use the following trick. **Pretend that $f(t)$ is constant C.** Now draw a diagram showing the graph $y = C$ which is a horizontal line. This way, it will look like the diagram below. Then

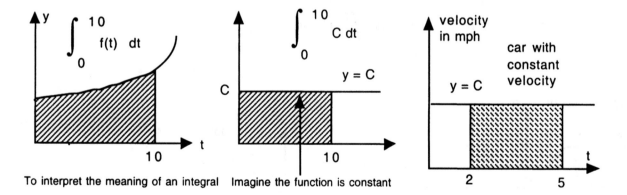

To interpret the meaning of an integral   Imagine the function is constant

$$\int_0^{10} C\ dt\ =\ 10 \times C.$$

Now think to yourself "this is the situation where the function $f(t)$ is constant." This means that **the rate at which people are getting infected is C people per day.** So what does $10 \times C$ mean?. It is how many people are infected in 10 days. Think about the **units** of the quantity that the area represents. The units on the y axis are **people per day**. The units on the horizontal axis are **days**. Thus, the units represented by the area is (*people per day*) × (*days*) = *people*. The area represents how many people get infected during 10 days. The important point to realize is that this is still what integral means when $f(t)$ is not constant: it is the number of people who become infected in 10 days.

Example 9.1.2 *Suppose $f(t)$ is the velocity in mph of a car driving along the freeway and t is measured in hours. What does*

$$\int_2^5 f(t)\, dt$$

represent?

Solution: the integral represents the distance the car travels between the times $t = 2$ and $t = 5$; that is, during the three hours starting at $t = 2$. To understand this, imagine the velocity of the car is constant at C mph. Now sketch the graph $y = C$ and draw the area that the integral represents as shown above. The area of this rectangle is $(5 - 2) \times C = 3 \times C$. The units this area represents are *hours* $\times$ (miles per hour) $= miles$. Thus, when the velocity of the car is the constant C then $3C$ is the number of **miles** the car travels in 3 hours. Which 3 hours are we talking about? Looking at the rectangle, the time starts at $t = 2$ and ends at $t = 5$. So it is the three hours between $t = 2$ and $t = 5$.

Problems

(9.1.1) For each of the following, sketch a diagram showing what signed area the integral represents and then calculate the integral from your diagram:

$$(a) \int_0^2 3\, dx \quad (b) \int_3^7 -5\, dx \quad (c) \int_{-2}^0 5\, dx \quad (d) \int_0^2 x\, dx \quad (e) \int_{-2}^2 x\, dx \quad (f) \int_{-2}^2 |x|\, dx.$$

(9.1.2) Find a such that

$$\int_1^a 3\, dx = 10.$$

[you will find it helpful to sketch the graph first]

(9.1.3) This question refers to the graph of the piece-wise linear function shown. Find

$$(a) \int_0^2 f(x)\, dx \quad (b) \int_{-2}^0 f(x)\, dx \quad (c) \int_{-2}^2 f(x)\, dx$$

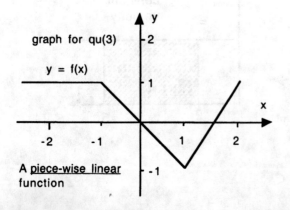

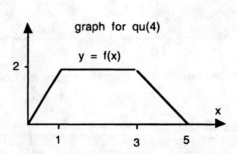

(9.1.4) Using the graph shown for what value of t does

$$\int_0^t f(x)\, dx = 6.$$

(9.1.5) Suppose that $f(t)$ is the rate that the world population is increasing in units of millions of people per year. What is the meaning of the statement

$$\int_{1970}^{1980} f(t) \, dt \; = \; 1200.$$

(9.1.6) Suppose that $q(t)$ is the rate at which the temperature of an oven increases in degrees celsius per minute. What is the meaning of

$$\int_{0}^{20} q(t) \, dt \; = \; 120.$$

(9.1.7) What does the statement

$$\int_{0}^{5} f(x) \, dx \; = \; 0$$

tell you about the function $f(x)$. Illustrate your answer with a diagram.

(9.1.8) The rate that a city uses water is $f(t)$ gallons per hour. Here t is the time measured in hours, starting at midnight at the start of July 4'th.
(a) How much water was used during July 4'th?.
(b) Write down an equation which expresses the fact that between midnight and 7am 200,000 gallons were used.
(c) Let T be the number of hours, starting at midnight, it takes for the city to consume 400,000 gallons. Write down a mathematical equation involving an integral with T in it that encodes this information.

(9.1.9) Draw a diagram showing the area the following integral represents. Using the formula for the area of a trapezium calculate this area:

$$\int_{0}^{a} (2x + 1) \, dx.$$

(9.1.10) (a) Sketch the graph $y = x^2$. Divide the area under this graph into two trapezia each of width 1/2. Show these trapezia on your sketch. Calculate the width of these trapezia and use this to find an approximate value for the area under the graph between $x = 0$ and $x = 1$. The exact answer is 1/3. What is the percentage error in your approximation?.
(b) Now use four trapezia of width 1/4 each. Again calculate their total area and the percentage error. Notice how small the percentage error is this time.

(9.1.11) Use the graph at the back to find

$$(a) \quad \int_{0}^{0.4} 10^x \, dx \qquad (b) \quad \int_{0.1}^{0.4} 10^x \, dx \qquad (c) \quad \int_{0.1}^{0.4} 2 \times 10^x \, dx.$$

(9.1.12) Explain why

$$\int_{0}^{1} 10 \times f(x) \, dx \; = \; 10 \times \left(\int_{0}^{1} f(x) \, dx \right).$$

Illustrate your answer with diagrams.

(9.1.13) Use the previous two questions and the fact that $10^{x+1} = 10 \times 10^x$ to find approximately

$$\int_{1}^{1.4} 10^x \, dx.$$

(9.1.14) Explain why

$$\int_{a}^{b} (3 + f(t)) \, dt \; = \; 3(b - a) + \int_{a}^{b} f(t) \, dt.$$

Illustrate your answer with diagrams.

9.2   The Definite Integral.

The number

$$\int_0^a f(x)\, dx$$

is the area under the graph between $x = 0$ and $x = a$. This number depends a. In other words, it is a **function** of a. Let's work out an example.

$$\int_0^a x\, dx.$$

This is shown on the diagram. The area it represents is a triangle with area $a^2/2$.

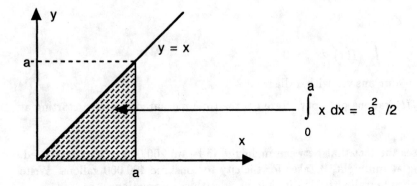

What we have here is a formula which tells us the area under the graph between 0 and a. We can actually use this formula to find the area under the graph between a and b. Looking at the diagram below, we see that

(area between 0 and a) + (area between a and b) = (area between 0 and b).

Translating this into integrals gives

$$\int_0^a f(x)dx + \int_a^b f(x)dx = \int_0^b f(x)dx$$

This formula just expresses the fact that the area of the two shaded regions added together is the area of the entire shaded region

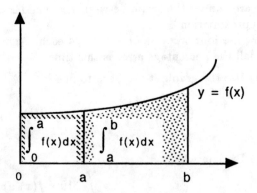

Moving things around gives

$$\int_a^b x\, dx = \left(\int_0^b x\, dx\right) - \left(\int_0^a x\, dx\right)$$

$$= \frac{b^2}{2} - \frac{a^2}{2}.$$

The key point to understand is this:

Once we have a formula that tells us the area under the graph between 0 and a, we can use it to get a formula for the area between a and b.

We found that the area under the graph $y = x$ between 0 and x is

$$\int_0^x x \, dx \; = \; \frac{x^2}{2}.$$

There is an interesting way to think about this formula. That is, we start with a function $f(x) = x$ and integrate to produce a new function $F(x) = x^2/2$. The new function $F(x)$ gives the area under the graph of the original function $f(x)$ between 0 and x. The function $F(x)$ is called a **definite integral** of $f(x)$. You might notice as an odd "coincidence" that the derivative of $x^2/2$ is x.

Here is another example of a definite integral:

$$\int_0^x 3 \, dx \; = \; 3x.$$

This just tells us that the area under the line $y = 3$ between 0 and x is $3x$. We know this is correct because the area in question is a rectangle of height 3 and width x. Again, we have the "coincidence" that the derivative of $3x$ is 3.

These "coincidences" are in fact examples of the most important idea in calculus:

> Differentiation is the reverse of integration.

This means that one "undoes" the other. As an example, suppose that our integration limits are 0 and x. Then integrating the function $f(x) = x$ gives $F(x) = x^2/2$. Differentiating $x^2/2$ gives x back again. The amazing fact is that this works for any function! If we integrate it and then differentiate the answer we end up back where we started. We can express this as an equation:

$$\frac{d}{dx} \left(\int_0^x f(x) \, dx \right) \; = \; f(x).$$

This is fantastically useful since it allows us to find integrals. For example, we know that the derivative of x^3 is $3x^2$. Therefore

$$\int_0^x 3x^2 \, dx \; = \; x^3.$$

Now if we divide by 3 we find the derivative of $(x^3/3)$ is x^2. So

$$\int_0^x x^2 \, dx \; = \; \frac{x^3}{3}.$$

We will see why integration is the reverse of differentiation soon. Before that let us examine the powerful tool

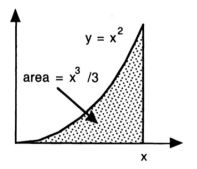
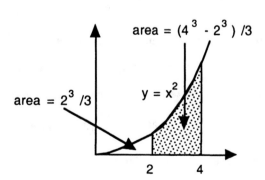

we have at our disposal. We start by considering this integral:

$$\int_0^x x^2 \, dx \; = \; \frac{x^3}{3}$$

If we set $x = 1$, we find that the area under the parabola $y = x^2$ between $x = 0$ and $x = 1$ is $1^3/3 = 1/3$. This is a **remarkable** achievement! There is no obvious way to find **exactly** the area under a parabola because it is not a simple shape like a rectangle. Yet we have done it! We can use also this to find the area between $x = 2$ and $x = 4$. For example, look at the figure (and refer back a couple of pages if you don't understand this formula):

$$\int_2^4 x^2 \, dx \;=\; \int_0^4 x^2 \, dx \;-\; \int_0^2 x^2 \, dx \;=\; \frac{4^3}{3} - \frac{2^3}{3}.$$

Since integration is the reverse of differentiation, we get a simple rule for finding the integral of any polynomial. We need only notice that we can use the rules for differentiating polynomials backwards. To begin, we know that

$$\frac{d}{dx}\left(\frac{x^{n+1}}{n+1}\right) \;=\; x^n.$$

The reverse of this rule is

$$\boxed{\int_0^x x^n \, dx \;=\; \frac{x^{n+1}}{n+1}}.$$

Next, if a is a constant then

$$\int_0^x (a \times x^n) \, dx \;=\; a\,\frac{x^{n+1}}{n+1}.$$

Finally to integrate any polynomial, we compute each term separately and then add up all the terms. For example:

$$\int_0^x (7x^3 - x^2 + 4x + 1) \, dx \;=\; \frac{7}{4}x^4 - \frac{1}{3}x^3 + \frac{4}{2}x^2 + 1x^1.$$

We can summarize this:

> To integrate a polynomial do each term separately.
> For each term add one to the exponent, and divide by the increased exponent.

It's actually fairly easy to memorize this rule. To differentiate, we **subtract** one from each exponent and **multiply** by the **old exponent**. The reverse of this process is to **add** one to each exponent and **divide** by the **new** exponent.

There is a special case which is very simple. By thinking about the area of a rectangle we know that

$$\int_0^x 1 \, dx \;=\; x.$$

Also $x^0 = 1$ and our general rule for integrating a power of x says

$$\int_0^x 1 \, dx \;=\; \int_0^x x^0 \, dx \;=\; \frac{x^1}{1}.$$

Thus, the general rule gives the right answer even in the case that $n = 0$.

Example 9.2.1 *Find the area under the graph $y = 3x^2 + 1$ between $x = 1$ and $x = 4$.*

The first thing to find is the definite integral using the rules above.

$$F(x) = \int_0^x (3x^2 + 1) \, dx \;=\; x^3 + x.$$

Then $F(x)$ is the area between 0 and x. Thus

$$\begin{aligned}
(area\ between\ 1\ and\ 4) \;&=\; (area\ between\ 0\ and\ 4) \;-\; (area\ between\ 0\ and\ 1)\\
&=\; F(4) - F(1)\\
&=\; (4^3 + 4) \,-\, (1^3 + 1)\\
&=\; 66.
\end{aligned}$$

The following notation is used for **definite integrals.** To find the area under the graph $y = 3x^2 + 1$ between $x = 1$ and $x = 3$.

$$\int_1^3 (3x^2 + 1) \, dx = \left[x^3 + x \right]_1^3.$$

The right hand side is basically what you get by writing down the integral of the function. The

$$[\cdots]_1^3$$

means plug $x = 3$ into what is inside the $[\cdots]$ and compute. Then plug in $x = 1$ and work that out. Now subtract what you got for $x = 1$ from what you got for $x = 3$. Thus

$$\left[x^3 + x \right]_1^3 = \left[3^3 + 3 \right] - \left[1^3 + 1 \right]$$
$$= 30 - 2$$
$$= 28.$$

The point here is that $\left[3^3 + 3 \right]$ is the area between 0 and 3 and $\left[1^3 + 1 \right]$ is the area between 0 and 1. Subtracting one from the other gives the area between 1 and 3.

Example 9.2.2 *A disease among fish is discovered on June 1'st. The rate at which fish are being killed is $200t + 1000$ fish per day t days after the outbreak was discovered.*
(a) How many fish are killed during June?
(b) How long until 200000 fish are dead?.

June has 30 days, we must find out how many fish are killed on each day and then add up the results. But there is an easier way. The answer is

$$\int_0^{30} (200t + 1000) \, dt.$$

Make sure you understand why this is the answer. The definite integral is

$$F(t) = \int_0^t (200t + 1000) \, dt = 100t^2 + 1000t.$$

We want to know $F(30)$ so this is

$$F(30) = 100(30)^2 + 1000(30) = 120000.$$

Thus 120000 fish die during June. To find out how long it will be until there are 200000 dead fish means to find out when $F(t) = 200000$. So we must solve

$$100t^2 + 1000t = 200000.$$

This is a quadratic and solving it gives $t = 40$. So by July 10'th (which is 40 days after June 1'st) 200000 fish will have died.

Problems

(9.2.1) Calculate the following

$$(a) \int_0^1 (3x^2 + 4x) \, dx \qquad (b) \int_{-1}^1 x^5 \, dx \qquad (c) \int_2^3 x^2 \, dx.$$

(9.2.2) What is the area under the graph of the function $f(t) = t^7 + t$ between $t = 0$ and $t = 1$.
(9.2.3) Find

$$\int_2^4 (y^3 + 1) \, dy.$$

(9.2.4) For what value of a is the area under the graph $y = x^2$ between $x = 0$ and $x = a$ equal to 72?.

(9.2.5) Find

$$(a) \int_0^1 (x^4 + x^3 + x^2 + x + 1)\, dx \qquad (b) \int_1^2 (x+1)^2\, dx \qquad \int_0^1 (a \cdot x^2 + b)\, dx.$$

(9.2.6) Find the following integrals

$$(a) \int_0^a x^2\, dx \quad (b) \int_a^1 x^3\, dx \quad (c) \int_a^b (x^2 + 1)\, dx.$$

(9.2.7) Find [hint: convert them all into the form e^{Kx}. Eg $10^x = e^{x ln(10)}$]

$$(a) \int_0^2 e^{7x}\, dx \quad (b) \int_{-1}^5 10^x\, dx \quad (c) \int_{-1}^0 (5 \cdot 2^x + x^2)\, dx.$$

(9.2.8) Consider the functions $f(x) = x^2$ and $g(x) = x^3$. Find the area of the region between $g(x)$ and $f(x)$ bounded on the left by the vertical line $x = 1$ and on the right by $x = 4$. [Hint: draw a diagram, subtract one area from the other.]

9.3   The Indefinite Integral and The Fundamental Theorem of Calculus.

In the last section we "discovered" that differentiation is the reverse of integration. This fact is used in the following way: There are simple rules which allow us to find the derivatives of many functions. We can use these rules in reverse to do integration. So, the question "what is the integral of $f(x)$" has the same answer as the question "which function has derivative $f(x)$." In symbols this says

$$\text{If} \quad \int_0^x f(x)\, dx \;=\; F(x) \qquad \text{then} \qquad f(x) \;=\; \frac{dF}{dx}.$$

Actually, there is a **complication.** The derivative of a constant is zero. So for example

$$\frac{d}{dx}(x^2 + 7) \;=\; \frac{d}{dx}\left(x^2\right).$$

This means that **there is more than one function with the same derivative!** It is not quite true that differentiation and integration are reverse processes. So we need to be a little bit careful. To help clarify this, we introduce a slight variation. The term **indefinite integral** of a function $f(x)$ means **the most general function whose derivative is** $f(x)$. This is expressed by writing

$$\int f(x)\, dx.$$

Notice that there are **no limits of integration.** When there are limits of integration the result is a number: the area under the curve between the limits of integration. But when there are no limits of integration the result is a **function** [something with an x in it]. As an example, the indefinite integral of x is

$$\int x\, dx \;=\; \frac{x^2}{2} + C \qquad\qquad \text{where } C \text{ is a constant.}$$

The first thing to realize is that the constant C, which is called a **constant of integration** disappears when we differentiate. Now, the fact is the most general function with the property that its derivative is x has the form

($x^2/2$ **plus some constant** C). The word **indefinite** refers to the **constant** C. We might say that "indefinite" here means "can't make up its mind" since a definite value for the constant C is unknown. That is, C might be 0, it might be 5, it might be -3.6178, or it might be any other value. Really, the indefinite integral can not make up its mind about C. In contrast, the definite integral $x^2/2$ has no unknown constant C in it to confuse the situation. It is definite because it **knows exactly** what function it is. The more accurate statement that integration and differentiation are (almost) reverse operations is summarized by a VERY FAMOUS THEOREM.

Theorem 9.3.1 *Fundamental Theorem of Calculus for Indefinite Integrals.*
Differentiation and Integration are Reverse Processes. This means

$$\boxed{\frac{d}{dx}\left(\int f(x)\, dx\right) = f(x) \qquad \int \frac{d}{dx}(f(x))\, dx = f(x) + C.}$$

This business about the constant C needs to be remembered when dealing with indefinite integrals. Actually, it does not complicate things very much. The usefulness of this theorem is that there are lots of rules (that we have not covered) for finding derivatives; differentiation is fairly "easy". Integration tends to be more difficult. If we want to find an indefinite integral

$$F(x) = \int f(x)\, dx.$$

the fundamental theorem tells us that

$$\frac{dF}{dx} = f(x).$$

So we can try to find the integral of $f(x)$ by thinking "what function, $F(x)$, can we differentiate to give $f(x)$?" Put differently, the theorem says "look, if you don't have a clue how to find integral of $f(x)$, just remember that it is something whose derivative is $f(x)$. So start trying to guess a function whose derivative is $f(x)$." Mathematicians call this **integration by trial and error.**

For example we know that even if n is negative that the derivative is

$$\frac{d}{dx}\left(\frac{x^{n+1}}{n+1}\right) = x^n.$$

Thus, the integration formula

$$\boxed{\int x^n\, dx = \frac{x^{n+1}}{n+1} + C}$$

works for negative exponents n also. However, there is **one important exception.** You **can not** integrate x^{-1} with this rule. Adding 1 to -1 gives 0 but **it is not true** that

$$\int x^{-1}\, dx = \frac{x^0}{0} + C.$$

This answer is clearly wrong because it involves dividing by zero! Although it's not at all obvious, it turns out that the integral of $1/x$ is the **natural logarithm:**

$$\boxed{\int \left(\frac{1}{x}\right) dx = ln|x| + C.}$$

We know the derivative of e^{Kx} is Ke^{Kx} so this means

$$\boxed{\int e^{Kx}\, dx = \frac{e^{Kx}}{K} + C.}$$

You can check this formula by differentiating the right hand side (which brings down a factor of K). At this point, you might ask "why do we bother with indefinite integrals? Why not just forget about C, make it zero every time?" Well, sometimes it is OK to forget about C; at other times it is not. It really depends on the context. When we are finding the area under a curve we can forget about C with impunity. It's easy to see that this is true by considering a typical example. Imagine working out

$$\int_4^7 x^2 \, dx.$$

We will use the indefinite integral $(x^3/3) + C$ and discover that the C's cancel out:

$$\int_4^7 x^3 \, dx = \left[\frac{x^3}{3} + C \right]_4^7$$
$$= \left[\frac{7^3}{3} + C \right] - \left[\frac{4^3}{3} + C \right]$$
$$= \frac{7^3}{3} - \frac{4^3}{3}.$$

However, there are occasions when we need to answer the question "a certain function has such and such a derivative. What are all the possibilities?" For example, this happens when solving a **differential equation**. In that case, failure to include C will inevitably lead to disaster. Just remember:

$$\boxed{\text{An indefinite integral has a } + C.}$$

Here is another explanation of this troublesome constant C. Suppose we have a function $f(x)$ and we integrate it from 0 to x to get that the integral is the function $F(x)$ where

$$F(x) = \int_0^x f(x) \, dx.$$

Instead we could integrate $f(x)$ from, for example, 3 to x and get a function $G(x)$ with

$$G(x) = \int_3^x f(x) \, dx.$$

Now there is a simple relationship between these two functions $F(x)$ and $G(x)$: **they differ by a constant.** Thus,

$$F(x) = G(x) + C$$

where C is the constant

$$C = \int_0^3 f(x) \, dx$$

which is **the area under the graph between 0 and 3**. The reason for this is simple:

$$\int_0^x f(x) \, dx = \int_0^3 f(x) \, dx + \int_3^x f(x) \, dx.$$

In other words, starting out with a function $f(x)$ we can make a new function $G(x)$ by computing the integral of $f(x)$ from some constant K (instead of using 3 as above) to x. This function $G(x)$ differs from $F(x)$ by a constant. However $F'(x) = G'(x)$. We will say more about this when we do the Fundamental Theorem of Calculus.

Here is an important example. The indefinite integral of a polynomial is just like the definite integral but with an arbitrary constant attached. In other words, we just write $+C$ at the end. So

$$\int (7x^3 - x^2 + 4x + 1) \, dx = \frac{7}{4}x^4 - \frac{1}{3}x^3 + \frac{4}{2}x^2 + 1x^1 + C.$$

Be advised: if you leave off the $+C$ and write instead

$$\int (7x^3 - x^2 + 4x + 1) \, dx = \frac{7}{4}x^4 - \frac{1}{3}x^3 + \frac{4}{2}x^2 + 1x^1,$$

then you will lose points!

Summary: The definite integral

$$\int_0^x f(x)\, dx \;=\; F(x)$$

equals the area under the graph between 0 and x. This is some definite number that depends on x. There is no constant C.

The indefinite integral of $f(x)$ is any function whose derivative is $f(x)$. Thus it is $F(x) + C$ where C is an arbitrary constant and $F'(x) = f(x)$. WARNING: you will lose points if you forget the dx, if you fail to put parentheses around the function being integrated, or if you leave off $+C$ in an indefinite integral.

Example 9.3.2 *Find the solution of the differential equation* $f'(x) = 3f(x)$ *for which* $f(0) = 7$.

We know that the derivative of an exponential function, is a **multiple** of itself:

$$\frac{d}{dx}\left(Ae^{Kx}\right) \;=\; K \times \left(Ae^{Kx}\right).$$

We seek $f(x)$ whose derivative is 3 times itself. So $K = 3$ works. In other words, no matter what the constant A is we have

$$\frac{d}{dx}\left(Ae^{3x}\right) \;=\; 3 \times \left(Ae^{3x}\right).$$

So $f(x) = Ae^{3x}$ works. But we are also told we need $f(0) = 7$ so that we must have

$$7 \;=\; f(0) \;=\; Ae^{3\times 0} \;=\; A.$$

So we discover that $A = 7$ and so the solution is $f(x) = 7e^{3x}$. In fact this is the **only** solution to this problem.

Problems

(9.3.1) Find the following indefinite integrals

$$(a)\ \int (2x^2 - 5)\, dx \qquad (b)\ \int x^5\, dx \qquad (c)\ \int x^2\, dx.$$

(9.3.2) Find

$$\int (y^3 + 1)\, dy.$$

(9.3.3) Find

$$(a)\ \int (4x^3 + 3x^2 + 2x + 1)\, dx \qquad (b)\ \int (x+3)^2\, dx \qquad \int (a\cdot x^2 + b\cdot x)\, dx.$$

(9.3.4) Find the indefinite integrals of [hint: convert them all into the form e^{Kx}.]

$$(a)\ e^{-3x} \quad (b)\ 2^x \quad (c)\ 10^{x+3}.$$

(9.3.5) What functions have derivative equal to x^4 ?

(9.3.6) Find a function $f(x)$ such that $f'(x) = 3x^2$ and such that $f(0) = 7$. [hint: think polynomials]

(9.3.7) If $F'(x) = x^3 - 1$ and $F(1) = 2$ what is $F(x)$?

(9.3.8) If $f'(x) = 0$ what are the possibilities for $f(x)$?

(9.3.9) What is the most general solution of the equation $g'(x) = e^{2x}$?

(9.3.10) Find a non-zero exponential function $h(t)$ so that $h'(t) = 7h(t)$ [hint: look back at the section on differentiating exponential functions]

(9.3.11) Find the solution of the differential equation $y'(x) = e^x + 4x^3$ for which $y(0) = 42$.

(9.3.12) The temperature of a cup of coffee is a function $h(t)$ where t is the time in minutes. The room temperature is $15°$ celsius. The **rate** at which the coffee cools down is proportional to the difference between the temperature of the coffee and the room temperature. Use this information to write an equation which sets the derivative of the coffee temperature equal to something.

(9.3.13) The number of bacteria in a pond after t hours is $p(t)$. The more bacteria there are, the more rapidly the number increases. This is expressed by the equation

$$p'(t) \; = \; 5p(t).$$

Find the most general exponential function of the form $A \cdot e^{Kt}$ which satisfies this equation. [A and K are constants. Hint: substitute this function into the equation and see what you get] If there were 2000 bacteria at $t = 0$ how many are there 5 hours later?

(9.3.14) It is known that the mass, $m(t)$, of a radio-active substance decreases as it decays, and that the equation governing this is

$$m'(t) \; = \; -0.005m(t).$$

What is the half life? [hint: find an exponential function which solves the equation. You will then have a formula for the mass, so you can find the half-life.]

9.4   The Integral is an Adding Machine: Riemann Sums.

Integration is very important because it is essentially the process of adding up many numbers. This section will explain this idea. To start though, here is an example. Integrating the function $f(x)$ from 0 to 10 is **approximately** the same as adding up the 10 numbers $f(1), f(2), \cdots, f(10)$. That is,

$$\sum_{n=1}^{10} f(n) \; = \; f(1) + f(2) + \cdots + f(10) \; \approx \; \int_0^{10} f(x)\,dx.$$

To help explain this idea, we will convert the summation into an area. Imagine 10 rectangles, each one of width one, so that the first one has height $f(1)$, the second one has height $f(2)$ and so on. Thus, the area of the first rectangle is $f(1) \times 1$, the area of the second is $f(2) \times 1$, etc. Hence the sum of the areas of the 10 rectangles is the sum above. Now compare this with the area under the graph $y = f(x)$ between $x = 0$ and $x = 10$. The two areas are approximately the same.

We have already seen that if $v(t)$ is the velocity of a car at time t (measured in hours) then

$$\int_0^1 v(t)\,dt$$

is the distance traveled by the car in the hour between time $t = 0$ and time $t = 1$. Here is another way to think about this. If the velocity of the car is constant, then the car travels $1 \times v(1)$. This is because $v(1)$ is the velocity of the car for the entire hour and the time of travel is 1 hour. So multiplying velocity by the time gives distance traveled. Imagine now that the car changes its velocity every minute. We could work out the total distance traveled by adding up how far it goes in the first minute, the second minute, and so on. We would find out how far it travels by adding up these 60 numbers. The velocity of the car for the first minute is $v(1/60)$ (one minute is $(1/60)$ of an hour). Notice that the time $t = 1/60$ is the end of the first minute. We could equally well have used the velocity at the start of the first minute by choosing $t = 0$. The distance the car goes in the first minute

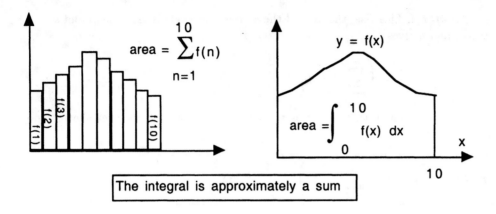

The integral is approximately a sum

is $(1/60) \times v(1/60)$: just the time times the velocity. The distance traveled by the car in the second minute is $(1/60) \times v(2/60)$ and so on. Adding all these up gives

$$\frac{1}{60} \times \left(\sum_{n=1}^{60} v(n/60) \right).$$

If the car changed its velocity every second, then we would figure out how far the car goes every second, and add up all these numbers. There are 3600 seconds in an hour. In the first second the car travels $(1/3600) \times v(1/3600)$. In the second second it travels $(1/3600) \times v(2/3600)$ and so on. Adding all these up gives

$$\frac{1}{3600} \times \left(\sum_{n=1}^{3600} v(n/3600) \right).$$

An expression like this is called a **Riemann sum** since Riemann invented this way of thinking about integration. We might imagine continuing this process of making the time interval shorter and shorter. In doing so, we are essentially taking a limit; that is, in the limit as the time interval gets shorter and shorter, we add up more and more numbers. What we are really saying can be expressed as follows:

$$\int_0^1 v(t)\,dt = \lim_{K \to \infty} \frac{1}{K} \left(\sum_{n=1}^{K} v(n/K) \right).$$

So we see that integration is basically like adding up a lot of numbers. This is the key point to understand.

Now, the same reasoning shows that when we integrate any function, the result can be thought of as obtained by adding up the values of the function at many points. That is, suppose we want to integrate $f(x)$ from a to b. Then we can imagine dividing up the range of integration from a to b into a large number K of small subintervals

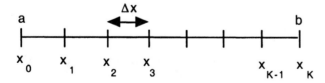

each of the same width

$$\Delta x = \frac{b-a}{K}.$$

The end points of these subintervals are

$$x_0 = a, x_1, x_2, \cdots, x_{K-1}, x_K = b.$$

The areas of the first rectangle is $\Delta x \cdot f(x_1)$. Likewise, the area of the second rectangle is $\Delta x \cdot f(x_2)$ and so on. Adding up the areas of all these rectangles gives

$$\Delta x \left(\sum_{n=1}^{K} f(x_n) \right).$$

This is a Riemann sum. Now we take the limit of these sums as the width of the subintervals get shorter and shorter. This gives the exact area:

$$\int_a^b f(x)\, dx = \lim_{K \to \infty} \Delta x \left(\sum_{n=1}^{K} f(x_n) \right). \qquad \Delta x = \frac{b-a}{K}$$

It is very important for you to understand this formula. It says that the the definite integral is a limit of Riemann sums.

The notation

$$\left(\sum_{n=1}^{K} f(x_n) \right) \Delta x$$

for a Riemann sum is what leads to the notation

$$\int_a^b f(x)\, dx$$

for an integral. The $\sum$ which is a Greek S becomes $\int$ which is an elongated S, and the Δ which is a Greek D becomes a Roman d.

Remember that the integral gives **signed** area, not just area. Area that is below the x-axis counts negative in an integral. We can now see why this should be so. An integral should be thought of as **adding things up.** The integral

$$\int_a^b f(x)\, dx$$

is thought of as adding up lots of values of $f(x)$ and then multiplying by Δx. The point is that when $f(x)$ is **negative** we are adding up negative numbers. Now $f(x)$ is negative when the graph $y = f(x)$ is **below** the x-axis. We **want** the integral to serve as an "adding machine." So we want area below the x-axis to count negative.

Problems

(9.4.1) Write out the following sum as 10 things added together but do not add them up

$$\frac{1}{10} \sum_{n=1}^{10} (n/10)^2.$$

(a) Make a sketch of the graph $y = x^2$ for $0 \leq x \leq 1$ and mark on the rectangles that this sum represents the area of. [hint: the first rectangle has height $(1/10)^2$] (b) How does this relate to this section ?

(9.4.2) (a) Write out the following sum as 10 things added together

$$\frac{1}{10} \sum_{n=1}^{10} f(n/10).$$

(b) Copy the graph shown and mark on it the 10 rectangles that this sum gives the area of. Write down an integral that this is a Riemann sum for. (c) What is this an approximate average of ?

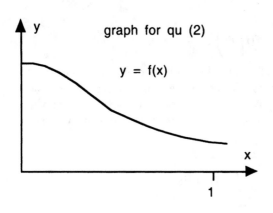

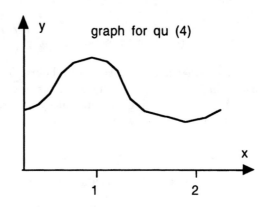

(9.4.3) Using a calculator, total up you answer to (1). The area under the graph of x^2 between 0 and 1 is 1/3. What is the percentage error in using the total you just calculated as an approximation to this area?

(9.4.4) Copy the graph, and mark on it the 5 rectangles you would use for a Riemann sum to approximate

$$\int_1^2 f(x)\ dx.$$

What is the width of each rectangle?. What are the x-coordinates of the left and right hand endpoints of the second rectangle on the x-axis?.

(9.4.5) Explain in your own words (about 100) what a Riemann sum is and what the relationship is between a Riemann sum and an integral. Include a diagram.

9.5 Integrals and Averages.

The **average** of the numbers $27, 14, 52, 10, 20$ is

$$(\text{average})\ =\ \frac{27 + 14 + 52 + 10 + 20}{5}.$$

If we have 100 numbers $y_1, y_2, \cdots, y_{100}$ we obtain their average adding them up and dividing by 100 :

$$(\text{average})\ =\ \frac{1}{100}\left(\sum_{n=1}^{100} y_n\right).$$

Now this looks like a Riemann sum! We can put this to practical use immediately. Suppose that $f(t)$ is the temperature at time t and we want to know the average temperature for the 5 hours between $t = 2$ and $t = 7$.

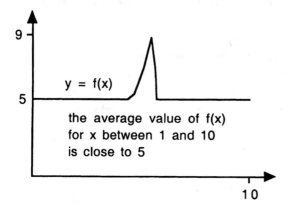

Then the answer is

$$(\text{average temperature}) = \frac{1}{5}\left(\int_2^7 f(t)\,dt\right).$$

We divide by 5 because the average is over a period of 5 hours. Think about this and remember the motto: "An integral is basically just adding things up." In this case, the integral is adding up the temperatures during the 5 hours between $t = 2$ and $t = 7$. So dividing the integral by 5 gives the average temperature. Now this is a fine way to remember what is happening. But we will look into why this works a bit more. First, here is the general formula you need to memorize:

$$\boxed{(\text{average of } f(x) \text{ between } a \text{ and } b) = \frac{1}{b-a}\left(\int_a^b f(x)\,dx\right).}$$

If we take the average of the numbers $5, 5, 5, 5, 5, 5, 5, 5, 5, 9$ then the answer is close to 5. Since there is only one 9 and all the rest are 5's the average is pulled down to be close to 5. Now look at the graph above. It is close to 5 most of the time between $x = 0$ and $x = 10$ but for a bit of time it rises up to 9. So we would expect that the average value of $f(x)$ between $x = 0$ and $x = 10$ is close to 5, but a little more. Now think about the area under the graph in the diagram. If the spike was not there, we would have a rectangle with height 5 and width 10. Thus, the area would be 50. But the spike adds a small bit of area. So perhaps the actual area is about 53. When we divide this area by the **width** of the region (which is 10) the result is $53/10 = 5.3$. This is the average height of the region.

Here is the key point:

$$\boxed{(\text{area under a graph}) = (\text{width of region}) \times (\textbf{average height}).}$$

The average height of the region is the same thing as the average value of $f(x)$.

Problems

(9.5.1) Use the integral formula to find the average value of the **constant** function $f(x) = 17$ over the interval $0 \le x \le 5$. [yes, I know there is an easier way to get the answer!]

(9.5.2) Use the integral formula to find the average of the function $f(x) = 2x$ over the interval $0 \le x \le 3$. Draw a diagram. Explain how you know this is the correct answer without doing any integrals.

(9.5.3) First use the integral formula to find the average value of $f(x) = x^3$ over the interval $-2 \le x \le 2$. Draw a diagram. Explain why the answer **had** to be zero without calculating the integral.

(9.5.4) What is the average value of $g(t) = t^2$ over the interval $1 \le t \le 2$. What is the largest and smallest value of $g(t)$ on this interval [just look at your diagram]. Is your average value between the min and max? Should it be?

(9.5.5) The number of megawatts supplied by a power station at time t is $p(t) = 100 + t^2$ during a 24 hour time interval $0 \le t \le 24$. What was the average number of megawatts supplied?

(9.5.6) (a) Find the average of e^t over the **time** interval $0 \le t \le 100$. (b) Find the value of t at which e^t takes this average value. [The answer is close to 100 because exponential functions get big so quickly that the average value is dominated by the values near the end. Thus the average human population (which has been growing exponentially) of the planet over the last 2000 years occurred since 1800.]

(9.5.7) Find the average value of $f(x) = 3 + (x^2/1000)$ during the interval $0 \le x \le 1$. Explain why the answer is close to 3.

9.6   The Fundamental theorem of Calculus for Definite Integrals

Theorem 9.6.1 *Fundamental Theorem of Calculus for Definite Integrals.*

$$\int_a^b f'(t)\ dt\ =\ f(b) - f(a).$$

$$\frac{d}{dx}\left(\int_0^x f(x)\ dx\right)\ =\ f(x).$$

By now we are quite convinced that the Fundamental Theorem of Calculus is extremely useful; the version we have seen in previous sections concerns **indefinite** integrals. We will now provide an analog for **definite** integrals. Actually, if we compare the two versions, we will see they are not really very different. By now, you may well wonder why the Fundamental Theorem in any form is true. Accordingly, we shall try to give some insight in this section.

The first part of the theorem tells us something **conceptually interesting,** a fact we already know and have used many times. Namely, if we know the **rate of change** of a function $f(x)$ then we can use the theorem to find how much the function $f(x)$ changes between two x values. We will explain why this works through an example. Suppose that an object moves along a line and that $f(t)$ is the distance of the object from the origin at time t. We will measure time in hours. Now $f'(t)$ is the rate of change of distance with time. Therefore $f'(t)$ is velocity. The quantity $f(1) - f(0)$ is the distance the object moves in 1 hour. If we know the velocity at each time t and we want to calculate how far the object moves in 1 hour, then according to the first part of the Fundamental Theorem

$$f(1) - f(0)\ =\ \int_0^1 f'(t)\ dt.$$

But how do we know this works? How do we know the first part of the Fundamental Theorem is correct?. The short answer is "Riemann sums." When we talked about velocity and Riemann sums before in section (9.4) we basically proved this works. "OK," you say, "I believe the first part of the Fundamental Theorem if we are dealing with velocity. But what about situations where $f(t)$ is not distance and $f'(t)$ is not velocity. Why does it work then?" Here is the answer: no matter what function $f(t)$ we have, we can always **pretend** that it is distance traveled in time t, and that $f'(t)$ IS velocity. After all, we could force an object to move with velocity given by the function. So, the discussion for velocity tells us the formula works. By pretending the function is distance traveled we can see that the Fundamental Theorem always works. In the next few pages there is another discussion of why the first part works. If you are happy with explanation just given, you need not bother with the next one. Really the first part of the Fundamental Theorem is common sense.

Incidentally, the variable t is used in the first part of the theorem and x in the second part. There is no significance to this. We actually could have used the same letter in both parts. But, for the explanation above, it was convenient to have the letter t in the first part. This section includes proofs of this theorem. These proofs are non-examinable. Read them only if you are interested.

Here is an explanation of:

$$\frac{d}{dx}\left(\int_0^x f(x)\ dx\right)\ =\ f(x).$$

The function we are differentiating is:

$$A\ =\ (\text{area under the graph between 0 and } x)\ =\ \int_0^x f(x)\ dx.$$

The derivative tells us the **rate of change** of this area as x increases. So the question is this: if we move x a small distance Δx to the right, how much does the area under the graph increase? In the diagram, the increase in area is shaded. This is a very thin shape, the sides and bottom are straight, but the top is not quite straight since it is part of the curve. However the area is almost exactly the area of a rectangle of width Δx and the height is $f(x)$. So

$$\Delta A\ =\ (\text{increase in area})\ \approx\ \Delta x \cdot f(x).$$

ΔA is the increase in area
due to increasing x by Δx.

So ΔA is the area of the
shaded region.

This is approximately a
rectangle of width Δx and
height f(x).

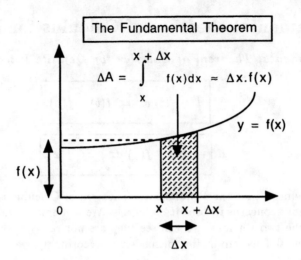

From this we find the average rate of increase of area:

$$\frac{\text{increase in area}}{\text{increase in x}} \;=\; \frac{\Delta A}{\Delta x} \;\approx\; \frac{\Delta x \cdot f(x)}{\Delta x} \;=\; f(x).$$

When Δx is very small this is almost exactly equal. In other words the derivative of area is the limit as $\Delta x \to 0$ of this

$$\frac{d}{dx}\left(\int_0^x f(x)\,dx \right) \;=\; \frac{dA}{dx} \;=\; \lim_{\Delta x \to 0} \frac{\Delta A}{\Delta x} \;=\; f(x). \qquad QED$$

The summary is that $f(x)$ determines the rate that the area under the graph between 0 and x increases as x increases. This is because $f(x)$ gives the height of the graph at the point where the region is expanding. If we move the vertical line in the diagram (the one at x) to the right, and ask "how quickly does the area under the graph up to this line increase?" then the answer is the height of that line.

This **proves** that differentiation is the reverse of integration. In other words, if we set

$$F(x) \;=\; \int_0^x f(x)\,dx$$

we have just proved that

$$\frac{dF}{dx} \;=\; f(x).$$

So we now **know** that the definite integral of $f(x)$ gives a function with derivative $f(x)$. We have already seen that one can find the derivative of x^n using algebra and limits. Putting these two facts together proves, for example, that

$$\int_0^x x^n\,dx \;=\; \frac{x^{n+1}}{n+1}.$$

This is an **astounding** accomplishment. We can find the **exact** area under a curve. Let us review the chain of reasoning. We can find the slope of a curve exactly using limits and algebra. We then established an unexpected relationship between the area problem and the problem of finding the slope of a curve: integration is the reverse of differentiation. Using this we found the exact area. This illustrates one of the most wonderful things in mathematics. Often the solution of one problem comes about in a surprising and unexpected way involving the solution to another entirely different problem. Finally, here is how one can deduce the first part of the fundamental theorem from the second part.

The second part of the Fundamental Theorem (which we have just proved) says

$$\frac{d}{dx}\left(\int_0^x f(x)\,dx\right) = f(x).$$

This works for **any** function $f(x)$. So, it certainly works for the function $f'(x)$ thus

$$(1) \qquad\qquad \frac{d}{dx}\left(\int_0^x f'(x)\,dx\right) = f'(x).$$

We know that

$$(\text{area under graph of } f'(x) \text{ between 0 and } x) = \int_0^x f'(x)\,dx.$$

Equation (1) then tells us that

(rate of change of [the area under the graph of $f'(x)$ between 0 and x]) = $f'(x)$

= (rate of change of f(x)).

Thus, the rate of change of one thing equals the rate of change of another thing. It follows (I lie a little bit) that the two things are equal:

(area under graph of $f'(x)$ between 0 and x) = $f(x)$.

This is not quite true because adding a constant to a function does not alter the derivative of the function. So, we should really write

(area under graph of $f'(x)$ between 0 and x) = $f(x) + C$

where C is some constant. We can work out this constant by plugging in $x = 0$. The area under the graph between $x = 0$ and $x = 0$ is 0. Thus $C = -f(0)$. Hence, we end up with

$$\int_0^x f'(x)\,dx = (\text{area under graph between 0 and } x) = f(x) - f(0).$$

Now that we have proved this fact, it follows from the discussion in (9.2) that

$$\int_a^b f'(x)\,dx = f(b) - f(a).$$

So the first version of the Fundamental Theorem is just a disguised version of the second version (and vice-versa). Really there is only one Fundamental Theorem that appears in several guises.

Problems

(9.6.1) The rate that water is flowing out of a reservoir is $R'(t) = 100 + t^2$ acre-feet per hour. How much water leaves the reservoir between $t = 0$ and $t = 3$.

(9.6.2) The velocity of a rat wearing a jet-pack is $v(t) = 3t^2$ m/sec. How far did the rat travel in the first 5 seconds?

(9.6.3) The rate of increase, as the radius increases, of the area enclosed in a circle of radius x cm is $2\pi x$ cm^2. How much does the area of the circle increase between a radius of 0 cm and a radius of 10 cm.

(9.6.4) The rate at which the temperature is dropping is $-t - 2t^2$ degrees celsius per hour t hours after sun-down. How much had the temperature decreased by 6 hours after sundown?

(9.6.5) A car accelerates from 0 mph to 60 mph. Its velocity at time t is $v(t) = 8t$ mph where t is in seconds.

Use an integral to find the total distance in miles it travels during its acceleration. [hint: work with seconds not hours]

(9.6.6) A valve in a full 6000 gallon water tank is slowly opening. Water flows out of the tank through the valve. The flow rate in gallons per hour is given by the function $f(t) = 300t^2$ where t is in minutes.
(a) How much water flows out of the tank during the first 2 minutes?
(b) How long does it take for the tank to empty?

(9.6.7) The rate of change of the price of a commodity after x years is 10^x dollars per year. January 1 is $x = 0$. Use the graph of 10^x to find the approximate change in price between the start of April and the start of July.

(9.6.8) The graph shows the rate that water leaves a tank, in gallons per hour
(a) How much water left the tank during the 6 hours shown.
(b) How long did it take for 9.5 gallons to leave the tank?

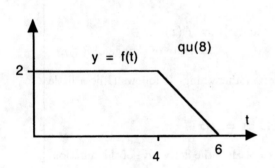

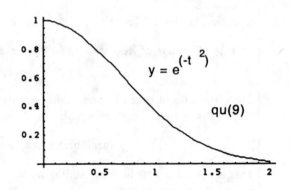

(9.6.9) The rate that heat leaves an object is e^{-t^2} Joules/minute. Use the graph shown to find approximately how much heat leaves the object during the time span from $t = 0.5$ to $t = 1.5$ minutes. [draw a trapezium]

(9.6.10) The acceleration of an object after t seconds is $a(t) = t^2 + 1\ ms^{-2}$. If the velocity at $t = 2$ was $7\ ms^{-1}$ what was the velocity at $t = 4$

(9.6.11) If an object has velocity $u\ ms^{-1}$ at time 0 and constant acceleration a use the fundamental theorem of calculus to show that the velocity after t seconds is $v(t) = u + a \cdot t$. Now use the fundamental theorem a second time to show that the distance the object travels between time 0 and time t is

$$ut + \frac{1}{2}a \cdot t^2.$$

(9.6.12) The table shows the rate that saline solution is entering a patient in cm^3/hour. Approximately how much saline entered between 1am and 3am. What does this have to do with the Fundamental theorem of calculus?

time	1:0	1:15	1:30	1:45	2:0	2:15	2:30	2:45	3:0
cm^3/hour	3	3.5	4.0	4.0	3.5	3.0	2.5	2.0	2.0

(9.6.13) The rate of change of temperature of a cup of tea is $-40e^{-t}$ degrees celsius per hour, t hours after it was made. If the initial temperature was $90^o\ C$
(a) what was the temperature 30 minutes later?
(b) what was the temperature t hours later?
(c) when was the coffee at $70^0\ C$?

(9.6.14) A water tank has a square base with each side of length $2m$. Water enters at a constant rate of 20 liters per minute. After t minutes, water leaves at a rate of t liters per minute. If the tank starts out filled to a depth of 3 meters, how long until the tank is empty?

(9.6.15) A particle is stationary at $t = 0$. The acceleration of a particle after t seconds is $1 + 2t$ ms^{-2}.
(a) Sketch a graph of the acceleration.
(b) Use your graph to find the velocity of the particle after 3 seconds
(c) What is the velocity after t seconds?
(d) Use your answer to (c) to write down an integral which gives the distance the particle went in the first 5 seconds. DO NOT CALCULATE THIS INTEGRAL.

(9.6.16) A water tank is a rectangular box with a base which is a square of side length $2m$. At $t = 0$ hours it has water to a depth of $3m$. The rate water flows into the tank after t hours is $1 + (t/2)$ m^3/hr. Sketch a graph of this. Use your graph to find the depth of water in the tank after 2 hours.

9.7 Summary

The integral

$$\int_a^b f(x)\ dx$$

represents the area under the graph of $y = f(x)$ between $x = a$ and $x = b$. It is the limit of adding up values of $f(x)$ between a and b in other words a Riemann sum. It is the average value of $f(x)$ between a and b times $(b - a)$. Integration is the opposite of differentiation.

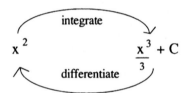

Integration is the reverse
of differentiation

$$\int f'(t)\ dt\ =\ f(t) + C.$$

Integrating the rate of change of $f(t)$ over the time period from $t = a$ to $t = b$ gives the total change in $f(t)$ during that time period.

$$\int_a^b f'(t)\ dt\ =\ [\ f(t)\ ]_a^b\ =\ f(b) - f(a).$$

The rules for integration are the opposite of those for differentiation. To integrate x^n add one to n and divide by $n + 1$:

$$\int x^n\ dx\ =\ \frac{x^{n+1}}{n + 1}\ +\ C.$$

The integral of velocity is distance traveled. The integral of acceleration is change in velocity. A definite integral has limits of integration, there is no $+C$. An indefinite integral has no limits of integration, there is a $+C$. You can quickly check an integral to see if you did it right by differentiating your answer. You should end up with the function you were asked to integrate in the first place.

Review problems: (9.1.1) (9.1.4) (9.1.6) (9.1.11) (9.2.1) (9.2.6)(c) (9.3.3) (9.4.1) (9.5.2) (9.6.6)

Chapter 10

Review of First 9 chapters.

10.1 Pre-Calculus Review

Use **parentheses** carefully. For example, to solve

$$\frac{2x+3}{x+1} = \frac{2x-4}{x+2}$$

we clear fractions by multiplying both sides by $x+1$ and $x+2$. Many students will write:

WRONG $2x + 3 \times x + 2 = 2x - 4 \times x + 1.$

If you can not see what is wrong with this, you are in serious trouble! Get help.

You should know:
- area of rectangle is length times width
- area of a triangle is half base times height
- area of a circle is πR^2
- circumference of a circle is $2\pi R$
- volume of a shape with vertical sides and horizontal base and top is (area of base) times height.

A line has equation $y = m \cdot x + b$. Here m is the **slope** which is

$$m = \frac{RISE}{RUN}.$$

The slope of the line going through the points (x_0, y_0) and (x_1, y_1) is thus

$$m = \frac{y_1 - y_0}{x_1 - x_0}.$$

To find the equation of a straight line, one method which always works is: first find the slope m. Then substitute for m into the equation for a straight line. Thus if you find $m = 3$ you write

$$y = 3x + b.$$

Finally, you find b by substituting into this equation the x and y values of any point which you know is on the line.

Pythagoras' Theorem says that for a right angled triangle with hypotenuse of length c and the other two sides of lengths a, b that

$$a^2 + b^2 = c^2$$

Rules for exponents	$a^{x+y} = a^x \times a^y$	$a^0 = 1$	$a^{-x} = \dfrac{1}{a^x}$	$(a^x)^y = a^{xy}.$

The logarithm function is the **inverse function** to the function 10^x.

$$\boxed{x = log(y) \qquad \text{MEANS} \qquad y = 10^x.}$$

What this means is the following:

$$\boxed{\text{10 to the power of } log(y) \text{ is } y. \qquad 10^{log(y)} = y}$$

This tells you what $log(y)$ IS by telling you what it DOES. The number $log(y)$ has a JOB to do: when you raise 10 to that power the result is y.

So $log(100) = 2$ because $10^2 = 100$, and $log(1000) = 3$ because $10^3 = 1000$. Also $log(0.1) = -1$ because $10^{-1} = 1/10 = 0.1$. The table shows some powers of 10 and the equivalent facts about logarithms.

x	1	2	3	6	0	-1	-2	-6
$y = 10^x$	10	100	1000	1000000	1	0.1	0.01	0.000001

y	10	100	1000	1000000	1	0.1	0.01	0.000001
$x = log(y)$	1	2	3	6	0	-1	-2	-6

Table 10.1: Table of values for $log(y)$

Logs convert multiplication into addition and **logs convert division into subtraction**.

$$\boxed{\begin{array}{c} log(x \times y) = log(x) + log(y) \qquad log(1) = 0 \qquad log(1/x) = -log(x) \qquad log(a^p) = p \times log(a). \\[2mm] log(10^x) = x \qquad 10^{log(x)} = x \qquad log(x/y) = log(x) - log(y) \quad log(10) = 1. \end{array}}$$

WARNING the following is a common mistake!

$$log(x + y) \neq log(x) + log(y).$$

There is NO WAY to simplify $log(x + y)$. Logs convert multiplication into addition, but they do not convert addition into anything at all!

In half life problems

$$m(t) = m_0 \left(\frac{1}{2}\right)^{t/K}$$

where K is the half life, t is the time, $m(t)$ is the mass at time t, and m_0 is the mass at time $t = 0$. For doubling-time problems

$$m(t) = m_0 2^{t/K}$$

here K is the time taken for the mass to double.

The number 2.718281.... is called e. The inverse function of e^x is $log_e(x)$. Instead of writing $log_e(x)$ the notation $ln(x)$ is used. This refers to **natural logarithms**. Thus

$$e^{ln(x)} = x$$

in other words $ln(x)$ is the power to which you raise the number e to obtain x. This is important in view of:

$$\frac{d}{dx}(e^x) = e^x \qquad \frac{d}{dx}(ln(x)) = \frac{1}{x} \qquad \int x^{-1} \, dx = ln|x| + C.$$

Example 10.1.1 *The half-life of a certain element is 40 years. If there was 17 grams in 1900 how much is there in 1950.*

Solution: The period from 1900 to 1950 is 50 years which is $50/40 = 1.25$ half lives. The initial amount is multiplied by $(0.5)^{1.25}$ in this time. So in 1950 there is

$$17 \times \left(\frac{1}{2}\right)^{1.25} \approx 7.15 \text{ grams.}$$

10.2 Calculus Review.

To differentiate a power of x the power rule says:

$$\frac{d}{dx}(x^n) = nx^{n-1}.$$

The derivative (rate of change!) of a constant is **zero**. The derivative of two things added together is just the sum of their derivatives:

$$\frac{d}{dx}(f(x) + g(x)) = f'(x) + g'(x).$$

The derivative of a constant times a function is the constant times the derivative of the function

$$\frac{d}{dx}(C \cdot f(x)) = C \cdot f'(x).$$

Putting these together we can differentiate a polynomial **term by term**

$$\frac{d}{dx}(3x^4 - 2x^3 + 3x + 5) = 12x^3 - 6x^2 + 3.$$

To differentiate an exponential function, first the derivative of e^x is itself

$$\frac{d}{dx}(e^x) = e^x.$$

And

$$\frac{d}{dx}(e^{K \cdot x}) = Ke^{K \cdot x}.$$

For example $10 = e^{ln 10}$ so $10^x = \left(e^{ln 10}\right)^x = e^{\ln(10) \cdot x}$ so

$$\frac{d}{dx}(10^x) = \frac{d}{dx}\left(e^{\ln(10) \cdot x}\right) = ln(10)e^{\ln(10) \cdot x} = ln(10) \times 10^x.$$

To differentiate natural logarithm $ln(x)$:

$$\frac{d}{dx}(ln(x)) = \frac{1}{x}.$$

If the derivative $f'(3) = 5$ then **rate of change** of $f(x)$ when $x = 3$ is 5 which means that a small change, Δx, in x results in $f(x)$ changing by approximately 5 times this amount, namely $5\Delta x$. Also the **slope of the graph** $y = f(x)$ at $x = 3$ is 5. This is the same as the **slope of the tangent line** to the graph at $x = 5$. This is the **limit** of the slopes of secant lines as they get closer to the tangent line. It is the limit of the **average rate of change** of $f(x)$ over a short interval near $x = 3$

$$\text{(average rate of change of } f(x) \text{ between 3 and } 3 + \Delta x) = \frac{f(3 + \Delta x) - f(3)}{\Delta x}.$$

$$f'(3) = \lim_{\Delta x \to 0} \frac{f(3 + \Delta x) - f(3)}{\Delta x}.$$

Indefinite integration, or anti-differentiation as some call it, is the process of finding a function $F(x)$ whose derivative is a given function $f(x)$. Thus

$$\int x^n \, dx = \frac{x^{n+1}}{n+1} + C \qquad \text{provided } n \neq -1.$$

Remember that C is an arbitrary constant (the constant of integration.) It appears because the indefinite integral is the most general function whose derivative is the function we started with. Under the process of differentiation, the constant C disappears. Then one integrates a polynomial one term at a time, for example

$$\int 3x^4 - 2x^2 + 7 \, dx = 3\frac{x^5}{5} - 2\frac{x^3}{3} + 7x + C$$

The other differentiation rules, when performed backwards give:

$$\int e^{Kx} \, dx = \frac{e^{Kx}}{K} + C.$$

$$\int \frac{1}{x} \, dx = \ln|x| + C.$$

Signed area is positive if it is above the x-axis and negative if it is below the x-axis. The signed area between the graph $y = f(x)$ and the x-axis between $x = a$ and $x = b$ is given by the definite integral

$$\int_a^b f(x) \, dx = [\, F(x) \,]_a^b = F(b) - F(a)$$

where $F'(x) = f(x)$ is an indefinite integral of $f(x)$. For example the area between the graph of $y = x^2$ and the x-axis between $x = 1$ and $x = 2$ is

$$\int_1^2 x^2 \, dx = \left[\frac{x^3}{3} \right]_1^2 = \frac{2^3}{3} - \frac{1^3}{3} = \frac{7}{3}.$$

The **average** value of the function $f(x)$ between $x = a$ and $x = b$ is

$$(\text{average value}) = \frac{1}{b-a} \int_a^b f(x) \, dx.$$

The **fundamental theorem of calculus** basically says that differentiation and integration are reverse processes:

Theorem 10.2.1 *The fundamental theorem for indefinite integrals says:*

$$\frac{d}{dx} \left(\int f(x) \, dx \right) = f(x) \qquad \int f'(x) \, dx = f(x) + C.$$

The fundamental theorem for definite integrals is what is used when finding a definite integral using symbolic integration.

Theorem 10.2.2 *The fundamental theorem for definite integrals says:*

$$\int_a^b f'(x) \, dx = f(b) - f(a).$$

It is also used for interpreting the meaning of certain integrals. For example

$$\int_a^b (\text{velocity}) \, dt = (\text{distance gone between times } a \text{ and } b)$$

because velocity is the rate of change of distance traveled.

To find the values of x where $f(x)$ has a **maximum or minimum** solve the equation $f'(x) = 0$. A solution is called a **critical point**. For example $f(x) = x^2 - 6x + 7$ has a minimum when $f'(x) = 2x - 6 = 0$. Thus the minimum is when $x = 3$ and the minimum value is $f(3) = (3)^2 - 6(3) + 7 = -2$.

The second derivative of $f(x)$ can be written as $f''(x)$ or

$$\frac{d^2 f}{dx^2}.$$

When $f''(x) > 0$ this means that the derivative $f'(x)$ is increasing. In other words the graph slopes upwards more as you move to the right. When $f''(x) < 0$ this means $f'(x)$ is decreasing. The graph $y = f(x)$ is concave up when $f''(x) > 0$ and concave down when $f''(x) < 0$.

Problems

(10.2.1) simplify the following

$$(a)\ \frac{x^5 \times x^7}{x^3.} \qquad (b)\ \frac{1}{y^{-7}} \qquad (c)\ \frac{y^3/x^4}{x^2 y^5} \qquad (d)\ (a^6 \times b^{-3})^4 \times (a^{-2}b^2)^3 \qquad (e)\ \frac{3}{4/(2x)^{-3}}$$

(10.2.2) multiply out and simplify

$(a)\ (x - y)(x + y)(x + 2y) \qquad (b)\ (1 - 2x)(1 + 2x + 4x^2) \qquad (c)\ (L + 2W)(L - 2W) - (L + 2W)(L + 2W)$

$$(d)\ (\frac{a}{b} - \frac{b}{a})\frac{ab}{a - b}$$

(10.2.3) Make the following substitutions into $x^2 - 2x + 1$, and simplify the answers

$$(a)\ x = 3 \qquad (b)\ x = a \qquad (c)\ x = y + 1 \qquad (d)\ x = 3y \qquad (e)\ x = y^{-1} + 1$$

(10.2.4) Factor then solve the equation $x^2 - 5x + 6 = 0$.

(10.2.5) What are the solutions of $3(x - 2)(x - 5)(x + 7) = 0$.

(10.2.6) Solve $\begin{aligned} x + 2y &= 12 \\ 2x + 2y &= 14 \end{aligned}$

(10.2.7) Write down two different solutions of $2x + 3y = 10$.

(10.2.8) Solve

$$\frac{3}{4x + 2} = 5$$

(10.2.9) Solve for x the equation $a \cdot x + b = 3x + 4$.

(10.2.10) Solve

$$\frac{2}{x - 4} = 5$$

(10.2.11) Solve

$$\frac{x + 2}{3x - 1} = \frac{x + 4}{3x - 5}$$

(10.2.12) Solve the following equation for w in terms of the other quantities

$$\frac{2}{w - 1} + \frac{a}{b + a} = -3$$

(10.2.13) Express a in terms of b given that

$$\frac{2a-1}{3a+2b} = -4$$

(10.2.14) Solve

$$\text{(a)} \ \ 2^x = 16 \quad \text{(b)} \ \ 10^x = .001 \quad \text{(c)} \ \ 3^{2-x} = 1.$$

(10.2.15) Find x satisfying the equation

$$\frac{3}{x+2} = \frac{7}{3x-1}$$

(10.2.16) For what value of x does

$$\frac{2}{x+b} - \frac{3}{c \cdot x + k} = 0.$$

(10.2.17) Find x and y so that

$$(x+y)^2 = 64 \quad \& \quad 2x + y = 13.$$

(10.2.18) Find x in terms of y if $(x - 3y)^2 - x^2 = 5$.

(10.2.19) Put over a common denominator and simplify

$$\frac{2x+3}{x-2} - \frac{x+1}{x+2}.$$

(10.2.20) Solve the equation $a^2 + b^2 = c^2$ for a.

(10.2.21) Students have written the following on exams. Find substitutions [like $x = 1$ and $a = 3$] which show these equations are wrong

$$\text{(a)} \ \ \frac{x}{a} + \frac{x}{b} = \frac{x}{a+b} \quad \text{(b)} \ \ (a+b)/x = a + (b/x) \quad \text{(c)} \ \ \frac{\sqrt{x}+1}{\sqrt{x}-1} = \frac{x+1}{x-1} \quad \text{(d)} \ \ \frac{1}{a+b} = \frac{1}{a} + \frac{1}{b}.$$

(10.2.22) If $f(x) = x^2$ what is

$$\text{(a)} \ \ f(3) \quad \text{(b)} \ \ f(a) \quad \text{(c)} \ \ f(1/x) \quad \text{(d)} \ \ f(x+2).$$

(10.2.23) The function $f(x) = x^3$. Find

$$\text{(a)} \ \ f^{-1}(8) \quad \text{(b)} \ \ f^{-1}(1) \quad \text{(c)} \ \ f^{-1}(y).$$

(10.2.24) What are the slopes of the following lines

$$\text{(a)} \ \ y = 3x - 5 \quad \text{(b)} \ \ 2x + y = 6 \quad \text{(c)} \ \ 3y = 4x - 2 \quad \text{(d)} \ \ 2x + 3y = 5.$$

(10.2.25) What is the slope of the line that goes through the two points $(2, 5)$ and $(17, 20)$.

(10.2.26) What is the equation of the line with slope 5 which contains the point with x coordinate 3 and y coordinate 2.

(10.2.27) What is the equation of the line going through the origin with slope 1/2.

(10.2.28) What is the equation of the line going through the two points $(2, 3)$ and $(5, 2)$.

(10.2.29) Where do the two lines $y = 3x + 2$ and $y = x - 1$ meet?

(10.2.30) Where does the line $y = x + 6$ intersect the parabola $y = x^2$?

(10.2.31) For each function $f(x)$ below, find the derivative $f'(x)$

(a) $3x + 5$ (b) $7x^2$ (c) $3x^5 - 2x^3 + 3x + 4$ (d) $(2x + 1)(x - 3)$ (e) $3x^{-5}$ (f) $7x^0$.

(10.2.32) Use calculus to find the maximum of $f(x) = 1 + 4x - x^2$.

(10.2.33) Where is the function $x^2 - 6x$ increasing? [ie where does the graph slope upwards?]

(10.2.34) What is the second derivative of $3x^3 + 5x$

(10.2.35) If $f(x) = x^4$ find $f'(x), f''(x), f'''(x)$.

(10.2.36) How many times must you differentiate x^7 before the answer is 0?.

(10.2.37) If $f''(x)$ is positive, what does this tell you about the graph of $y = f(x)$?

(10.2.38) Using log tables at the back, find:

(a) $log(400)$ (b) $log(50)$ (c) $log(0.002)$.

(10.2.39) Simplify

(a) $log(x^3) - 3log(x)$ (b) $log(x/y) - log(x \cdot y)$ (c) $log(\sqrt{x})$.

(10.2.40) Simplify

(a) $e^{ln(20)}$ (b) $e^{2ln(x)}$ (c) $\left(e^{ln(10)}\right)^x$ (d) $e^{x \cdot ln(a)}$.

(10.2.41) Express the following in the form e^w for some w.

(a) 37 (b) x (c) 10^x (d) 2^x.

(10.2.42) Find dy/dx for

(a) $y = 7e^x$ (b) $y = 3e^{2x} + 7$.   (c) $y = e^{x \cdot ln(10)}$ (d) $y = e^{x \cdot ln(2)}$.

(10.2.43) Use the results from the previous question to find

(a) $\dfrac{d}{dx}(10^x)$ (b) $\dfrac{d}{dx}(7 \cdot 2^x)$

(10.2.44) Find the indefinite integrals

(a) $\displaystyle\int x^2 \, dx$ (b) $\displaystyle\int 7 \, dx$ (c) $\displaystyle\int x^4 - 3x^2 + 5 \, dx$ (d) $\displaystyle\int 2x^{-3} \, dx$ (e) $\displaystyle\int x^{-1} \, dx$.

Now **check** your answers by differentiating them.

(10.2.45) What is the area between the graph of $y = 6x^2$ and the x-axis between $x = 1$ and $x = 2$.

(10.2.46) Find

$$\int_{-1}^{1} x^2 + x + 1 \, dx.$$

(10.2.47) Sketch on a single graph $y = x$ and $y = x^2$. Where do these graphs cross?. Shade the region lying between these two graphs in-between the crossing points. Find the area of this region using integrals. [hint: it is the difference between two areas that come from integrals.]

(10.2.48) Sketch the graphs $y = x^2$ and $y = 2 - x^2$. Find the area of the region enclosed between these two graphs.

(10.2.49) Evaluate

$$\int_{1}^{3} e^x \, dx.$$

(10.2.50) What is the average value of x^2 between $x = 0$ and $x = 2$? Make a sketch graph (plotting the points where $x = 0, 1, 2$), and use this to explain why the average value is bigger than 1.

(10.2.51) A circle must have the same area as a certain square. Express the radius of the circle in terms of the length of a side of the square.

(10.2.52) Two positive numbers have total 100. Use calculus to find the largest possible value for the product of these two numbers.

(10.2.53) A box has rectangular sides, top and bottom. The volume of the box is 5 m^3. The height of the box is half the width of the base. Express the total surface area of the sides of the box in terms of the height of the box.

(10.2.54) A tank of water has base a circle of radius 2 meters and vertical sides. If water leaves the tank at the rate of 5 liters per minute, how fast is the water level falling in centimeters per hour? [1 liter is 1000 cubic centimeters]

(10.2.55) A colony of bacteria doubles its mass every 5 hours. How long will its mass take to increase by (a) a factor of 8 (b) a factor of 20?

(10.2.56) Car A leaves the Grand Canyon at noon. It travels at 60 mph, but stops at 3 pm for an hour. Car B leaves the Grand Canyon at 2 pm and travels at 70 mph without stopping on the same route as car A. When does B catch up with A?.

(10.2.57) The temperature t hours after 6a.m. is $f(t)°F$.
(a) What is the meaning of the statement $f(5) = 17$.
(b) What are the units of $f'(t)$.
(c) What is the meaning of $f'(5) = 3$.

(10.2.58) Copy the graph $y = f(x)$ shown, then using a different color, sketch the graph of $y = f'(x)$ on the same diagram. What is $f'(1)$.

(10.2.59) A drug is injected into the blood of a patient. As the body absorbs the drug, the amount remaining in the blood declines. The table shows the amount of drug (in micrograms) in one cubic centimeter of blood at various times. What was the average rate of intake of the drug during the first ten minutes and during the last ten minutes. Assume the patient has 4 liters of blood. [1 liter is 1000 cm^3.]

time	1:00	1:05	1:10	1:15	1:20	1:25	1:30
micrograms/cm^3	37	30	24	19	14	10	7

Table 10.2: Qu (59)   Amount of drug in 1 cm^3 of blood.

(10.2.60) The graph shows the rate in grams/minute that a chemical is produced during a certain chemical reaction. How long did it take for 100 gms to be produced? Explain how you did it. What mathematical idea is involved?

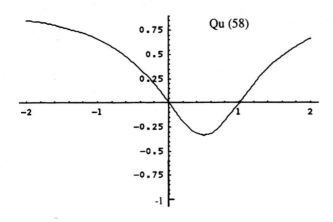

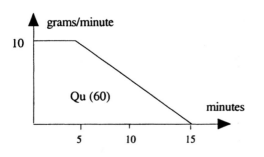

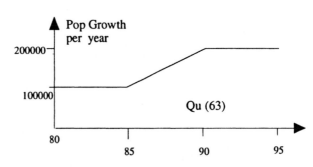

(10.2.61) What are the critical points of $f(x) = x + 2/x$.

(10.2.62) What is the minimum of $x^2 + 32x^{-4}$.

(10.2.63) The diagram shows the rate of population growth of a city during the period 1980 to 1995. The population of the city in 1980 was 2 million. What was the population in 1995?. When did the population first reach 2.5 million.

(10.2.64) The acceleration (rate of change of velocity) of an object is $2t + 1 \ ms^{-2}$ where t is the time is seconds. The velocity of the object at $t = 0$ is $3 \ ms^{-1}$.
(a) What is the velocity after t seconds?
(b) When is the velocity $23 \ ms^{-1}$.
(c) How far does the object move between $t = 0$ and $t = 4$.?

(10.2.65) A water tank has a square base of area $5 \ m^2$. Initially the tank contains $40 \ m^3$. Water leaves the tank, starting at $t = 0$, at the rate of $2 + 4t \ m^3/hr$. Here t is the time in hours. What is the depth of water remaining in the tank after 3 hours?

(10.2.66) In the previous problem, when will the depth of water remaining in the tank be $3 \ m$.

(10.2.67) Evaluate the integral

$$\int_0^{90} 2^{(t/10)} \ dt.$$

[hint $2^x = e^{x \cdot ln(2)}$.]

(10.2.68) Assume the amount of pollution entering the world's oceans grows exponentially. Starting in 1900 suppose that 10^7 tons enter per year, and that the amount doubles every 10 years. Express the **rate** that pollution enters, in units of tons per year, at a time t years after 1900. Write down an integral which expresses the total amount of pollution which has entered during the period 1900 to 1990.

(10.2.69) An hot object cools down according to Newton's law so that t hours after the start, the rate that heats leaves is $20e^{-0.3t}$ Joules per hour. How much heat leaves the object during the first 3 hours? [4.2 Joules is the heat required to raise 1 gram of water $1^\circ C$.]

(10.2.70) Find the coordinates of the points on the graph of $y = x^2$ at which the tangent line through that point also passes through the point $(0, -1)$. Start by drawing a diagram.

Chapter 11

Word Problems.

Problem Solving Strategies. Refer back to chapter 3, section (3.1) for problem solving strategies. Pay particular attention to **rule 3.** See also sections 16.4 and 16.11. If you have dont know where to start with word problems, do problems (3.1.1) to (3.1.8). Try to come up with a plan in the style of the plans given for those problems.

Problems

(11.0.1) The perimeter of a rectangle is 24 cm. If the area of the rectangle is 32 cm^2 find the length and width of the rectangle.

(11.0.2) (*) A tourist travels 1500 miles using two planes. The second plane averages 50 miles per hour faster than the first plane. The tourist uses the slower plane for the first 500 miles and the faster plane for the next 1000 miles. The total flying time is $6\frac{1}{2}$ hours. What is the speed of the first plane?

(11.0.3) It takes George 1 hour longer to mow the lawn than it takes Henry. Working together, using two mowers, they can mow the lawn in 1 hour and 12 minutes. How long would it take Henry to mow the lawn by himself? [hint: Let x be the time taken by George. How much of the lawn does George mow in 1 hour? How much does Henry do in 1 hour?. How much do they mow together in 1 hour?]

(11.0.4) The sum of the reciprocals of two consecutive odd integers is 8/15. Find the two integers.

(11.0.5) Let A,B,C and D be four points on the $x-$axis with coordinates $2, x, -3$ and 13 respectively. If the distance between A and C is the same as the distance between B and D, find the two possible values for the coordinate x of B.

(11.0.6) (*) A swimming pool can be filled by either or both of two pipes of different diameters. It takes the smaller pipe twice the time that the larger pipe takes to fill the pool. If water flows through both pipes it takes $2\frac{1}{3}$ hours to fill the pool. How long will it take the larger pipe to fill the pool? [hint: let x be the amount of the pool filled by the small pipe in one hour]

(11.0.7) Car A and car B both travel 300 miles. Car A drove at an average speed of 10 miles per hour faster than car B and made the trip in one hour less time. Find the average speeds of both cars.

(11.0.8) Four tasks, A,B,C and D, must be performed. They take 6 hours, 4 hours, 3 hours and 2 hours respectively. Tasks can be performed simultaneously but there are certain constraints. Task B can not be started until half of task A has been completed. Task C can not be started until 7 hours in total have been spent on tasks A and B combined. An hour before task C ends, task D is started. When does task D end? Draw a suitable diagram showing the time period each task takes.

(11.0.9) It is required to make 20 grams of certain chemical compound called Z. This is made from compounds W,X and Y in the ratio of 2 : 1 : 3. The compound Y is itself made from W and X. To make 6 grams of Y requires 4 grams of W and 2 grams of X. How much W and X is required to make the required amount of Z.

(11.0.10) [long] There are 3 beakers each of which contains saline solution.

Beaker A initially contains 3 liters of 5% salt solution,
beaker B initially contains 2 liters of 10% salt solution,
beaker C initially contains 4 liters of 0% salt solution.
Two liters are transferred from A to B and the result is thoroughly mixed. Then one liter is transferred from B to C and the result mixed. Finally two liters are transferred from C back to A. What is the percentage concentration of salt in A after all this?

(11.0.11) There are three people: A,B,C. They play a game in which they take turns. When it is a players turn, that player first gives half their money to each of the other two players. Having done this, that player then takes half of the money from each of the other two players. They start with a, b and c respectively. If the order of play is A,B,C, how much does each end up with? If you are the person who starts with the least money, do you want to go first, second, or last?

(11.0.12) The total number of people at a football game was 5600. Field-side tickets were $40 and end-zone tickets were $20. If the total amount of money received for the tickets was $154000 how many of each kind of ticket sold?

(11.0.13) 10 adults and 5 children can watch a movie for $75. But 8 adults and 10 children can watch the movie for $78. How much would it cost for 6 adults and 10 children to watch the movie?

(11.0.14) What amounts of 75% pure silver and 80% pure silver should be mixed to obtain 14 grams of 78% pure silver.

(11.0.15) A tank initially contains 1000 liters of pure water. Then water containing 4 mg of detergent per liter starts to enter the tank at the rate of 20 liters per hour.
(a) How long until the average concentration of detergent in the tank is 2 mg per liter?.
(b) How long until the average concentration of detergent in the tank is x mg per liter.
(c) Sketch a graph showing the function you obtained in (b). Put x on the horizontal axis and t on the vertical axis.
(d) What does you answer to part (b) give when $x = 5$. Do you notice anything strange?. Can you explain this?

(11.0.16) If 1000 liters of water with an unknown chlorine content are combined with 500 liters of water with 20 ppm (parts per million) of chlorine the result is 60 ppm chlorine. What is the concentration of chlorine in the unknown water?

(11.0.17) Car A starts in Sacramento at 11am. It travels along a 400 mile route to Los Angeles at 60mph. Car B starts from Los Angeles at noon and travels to Sacramento along the same route at 75 mph. The route goes past Fresno which is 150 miles along the route from Los Angeles. How far from Fresno are the cars when they meet?.

(11.0.18) A railroad and a highway intersect at right angles. A train is 10 miles from the intersection at the same moment that a car is 6 miles from the intersection. Both are traveling at 30 mph. How long until they are 4 miles apart.

(11.0.19) Express the length of the perimeter of a circle in terms of the area of the circle.

(11.0.20) A box with rectangular sides has width twice the length of the base. The volume is 24 in^3 and the total surface area of all six sides is 52 $inches^2$. Write down a (cubic) equation whose solution is the length of the box.

(11.0.21) A tin can has a circular base has a volume of 75π cm^3 and the surface area of the vertical (curved) side is 30π cm^2. What are the dimensions of the can?

(11.0.22) A tin can has a circular base and volume 100π cm^3. Express the diameter of the base of the can in terms of the height of the can.

(11.0.23) There are two positive numbers. Four times the small number plus 3 times the big number is 46. Two times the small number plus the big number is 19 what is 10 times the big number minus 6 times the small number?

(11.0.24) A car travels 365 miles in 6 hours. It travels at 50 mph for the first two hours and travels the last 165 miles at 60 miles per hour. It travels at constant speed for the remaining time.
(a) How long did it take to travel the first 150 miles?
(b) Where was the car after 3 hours?

(11.0.25) The graph shows the rate at which rain is falling in cm per hour. What is the total rainfall after t hours?

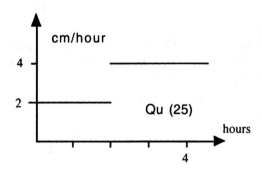

(11.0.26) A lake has a volume of 10^8 m^3. A river flows into the lake at the rate of 2×10^5 m^3/day. The same amount of pure water evaporates from the lake each day, so the volume of the lake does not change. The river contains 5% pollution. The lake is initially unpolluted.
(a) What is the percentage pollution in the lake after t days?
(b) When does the pollution in the lake reach 1%.

(11.0.27) (*) A plane flies at 200 mph for the first and last half hour of a flight. It flies at 400 mph the rest of the time. The route is 1000 miles long. The town of Erehwon is on the route 300 miles before the end. What is the distance of the plane from Erewhon after t hours of flying?.

(11.0.28) Disease A starts killing people in 1900 at the rate of 10000 people per year. Disease B starts killing people in 1910 at the rate of 20000 people per year.
(a) Sketch on one diagram for the period 1900 to 1950 how many people in total have been killed by each disease
(b) How many people have died from these two diseases t years after 1900.
(c) In what year does the total number of people killed by the two diseases become equal.
(d) In what year does the total number of deaths first exceed half a million?

(11.0.29) Colony A of Bacteria has a mass of 10 mg at noon on Jan 1, and doubles in mass every 12 hours. Colony B has a mass of 5 mg on noon of Jan 3 and doubles every 6 hours.
(a) Find the masses of the two colonies as functions of time (measured from noon Jan 3).
(b) Write down an equation whose solution is the time [measured in hours after noon Jan 3] at which the masses of the two colonies are equal.
(c) Solve the equation in (b) [you may wish to use logs]

(11.0.30) The graph below shows the total number of people who have or have had a certain flu virus on or before a given day. On what day of the epidemic did the **average** number of people who have had the flu
(a) equal 1000 [hint: draw the horizontal line on this graph at height 1000. You want as much area above this line as is missing below. why?] (b) equal 500 ?

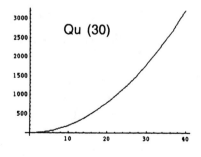

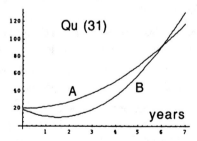

(11.0.31) The graphs above show the total worth of two companies.
(a) During what period of time is Company A worth more than Company B?
(b) During what period of time is the worth of A increasing faster than that of B ?

(11.0.32) The population of a country is growing exponentially. The population in millions was 100 in 1970 and

140 in 1980.

(a) What is the population t years after 1970 ?

(b) How long does it take the population to double?

(c) When will the population be 400 million?

(11.0.33) A crop-duster plane is spraying a circular field of radius 1 Km. It flies in a spiral inwards from the edge of the field. After half an hour the plane has sprayed all the field **outside** of a circle of radius 0.8 Km. Sketch a diagram showing the area sprayed. Approximately how long will it take to finish the field.

(11.0.34) Two cars are at the start line on a circular race track with a circumference of 7000 meters. Car A starts first and travels at 150 Km/hr. Car B sets out 20 minutes later and travels at 200 Km/hr.

(a) At what point on the circle does car B first pass car A.

(b) How long until car B has traveled the same distance as Car A ?

(c) At what point on the circle does the event in (b) happen?

(11.0.35) Plane A leaves New York at noon (EST) traveling to Los Angeles along a route which is 2500 miles. Plane A travels at 300 mph for the first and last half hour of flight, and at 600 mph for the rest of the flight. Plane B takes off from Los Angeles 1 hour after plane A takes off. Plane B flies along the same route. Plane B flies at 250 mph for the first half hour and last half hour and flies at 500 mph for the rest of the flight. The flight route goes over Salt Lake City, which is 700 miles along the route from Los Angeles. How far from Salt Lake city is the point where the two planes pass each other?

(11.0.36) A river flows at a speed of 5 mph towards the sea. A woman in a canoe can paddle through water at a speed of 8 mph. She wishes to paddle for a total time of 2 hours, first upstream (away from the sea) then turning round and paddling back to her starting point. How long should she paddle upstream?

(11.0.37) A plane has an average **air speed** (this is the speed the plane moves through air) of 550 mph. The plane flies a route of 5000 miles from Los Angeles to London. The average speed of the wind along this route is 50 mph blowing east from Los Angeles towards London. How much longer does it take the plane to fly from London to Los Angeles than the reverse direction?

(11.0.38) A reservoir is in the shape of a circle of radius 2000m. The water levels drops 2 cm over a period of a week. At what rate is the lake losing water water in m^3 per hour?.

(11.0.39) A building has a hemispherical roof. The roof is to be painted with a layer of paint one millimeter thick. The radius of the hemisphere is 50 meters. How many liters of paint are needed? [1 liter = 1000 cm^3. 1 meter is 100 cm is 1000 mm. The surface area of sphere is $4\pi R^2$.]

(11.0.40) The speed of car A after t minutes is $5t$ m/s. How long will it take the car to travel 100/6 meters?

(11.0.41) A water tank has a circular base of radius 3 meters.

(a) If water is entering the tank at the rate of 50 liters per minute, how quickly is the level of the water rising?

(b) If the tank is 5 meters high, how long until the tank is full?

(11.0.42) A spherical snowball is melting. The radius after t hours is $1 - t$ meters.

(a) What fraction of the initial mass of the snowball remains after half an hour.

(b) How quickly is the volume of the snowball changing after half an hour.

[volume of a sphere is $4\pi R^3/3$]

(11.0.43) How quickly a leaf grows is proportional how big [ie the surface area] the leaf is. If the area of the leaf grows from 2 cm^2 to 3 cm^2 in 5 days, how long will it take for the leaf's area to increase to 5 cm^2?

(11.0.44) The number of items sold at a price of $\$x$ per item is $2000 - 300x$. It costs \$5 to make the item. What price should be charged to make the most profit?

(11.0.45) In 1990 a fatal disease evolves to which 40% of a population of 5 million trees is susceptible. The proportion of susceptible trees which survive for a period of t years beyond 1990 is e^{-t}.

(a) Sketch a graph showing how many trees remain during 1990-1995.

(b) How quickly is the disease killing off trees at the start of 1992. Give units.

(c) When will the population be reduced to 80% of the level in 1990 ?

(11.0.46) The graph on the next page shows the number of a certain species of locust during the period 1950 to 1980. Write down a plausible function for the number of locusts as a function of time in years.

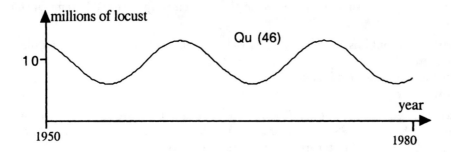

(11.0.47) The population of the planet Magrathea doubles every 40 years. In the year zero the population was 20 million. The total land area of Magrathea is 200 million square kilometers. One square kilometer of land can house 200 Magratheans. Land that is not used for housing is used for agriculture. Write down an equation whose solution is the year in which there are 30 m^2 of agricultural land available for each inhabitant. Solve this equation.

(11.0.48) Species A doubles every 2 hours and initially there are 5 grams. Species B doubles every 3 hours and initially there are 10 grams. How long until the two species have the same mass?

(11.0.49) Sketch the graph of $y = 2x$. Mark the point, P, on the x-axis where $x = 2$.
(a) What is the **square** of the distance of P from the point on the graph of $y = 2x$ at which $x = 4$.
(b) What is the **square** of the distance of P from the point on the graph of $y = 2x$ at which $x = a$.
(c) What point on the graph $y = 2x$ is closest to P ? [hint: it is the point which minimizes your answer to (b). Explain WHY]

(11.0.50) Using the method from the previous question, find the point on the graph $y = \sqrt{x}$ which is closest to the point $(3, 0)$.

(11.0.51) The height of water in a marina is shown on a vertical ruler fixed to a pier. The height t hours after noon is $h(t)$ inches. If $h(2) = 200$ and $h'(2) = -30$ what approximately is the height at 2 : 10 pm.

(11.0.52) A road starts at town A and goes into the mountains. When you are x kilometers from along the road from A, the height above town A is $h(x)$ meters.
(a) What does $h(12) = 400$ mean?
(b) What does $h'(12) = -60$ mean?
(c) If $h(12) = 400$ and $h'(12) = -60$ what, approximately, is the height of the road above A at a distance of 12.1 kilometers from A ?

(11.0.53) (a) In France you pay to enter a freeway and the ticket has a time on it. When you exit, they look at the ticket. If the distance you have traveled divided by the time taken is bigger than 160 Kilometers per hour, you are fined for breaking the speed limit of 160 Kilometers per hour. Question: how do they know you were speeding at any time in your journey?
(b) Use the reasoning from the previous question to answer this part. The temperature t hours after $6am$ is $f(t)$ degrees Fahrenheit. The temperature at 8 am was $50°F$ and the temperature at noon was $80°F$. How do you know that $f'(t)$ was bigger than 6 for some value of t between 2 and 6 ?

(11.0.54) How many liters of water containing 3 grams of salt per liter must be combined with x liters of water which contains y grams [y is less than 2] of salt per liter to yield a solution with 2 grams of salt per liter.

(11.0.55) Three people, A,B and C play a game. First A gives half her money to B then A takes half of C's money. Next B gives half his money to C and takes half of A's money. Finally, C gives half her money to A and takes half of B's money. A starts with $a and B starts with $b and C starts with $c. How much more money does C end up with than A .

(11.0.56) A plane flies at 200 mph for the first and last half hour of a flight. It flies at a higher constant speed for the rest of the flight. The route is 1950 miles. The plane is 850 miles from the destination 2 hours before the end of the flight. How many hours is the entire flight?

(11.0.57) A swimming pool contains x (less than 0.02) grams of chlorine per cubic meter. The pool measures 5 meters by 50 meters and is 2 meters deep. Some water will be drained and replaced by water containing 3 grams of chlorine per cubic meter. How much water should be drained so the pool ends up with 0.02 grams of chlorine per cubic meter.

(11.0.58) A full tank initially (at $t = 0$) contains 16 gallons. Then water is removed at a rate of $1 + t$ gallons per minute where t is the time in minutes.
(a) How much water remains in the tank after t minutes
(b) When is the tank half empty.

(11.0.59) A rectangular box has volume 20 cubic meters. The base of the box is a square. The top and the base of the box are made from material which costs \$30 per square meter. The material for the sides costs \$20 per square meter. Express the total cost of the box in terms of the length of edge of the base.

(11.0.60) Two planes A and B can each fly through still air at 500 mph. The distance from Paris to Los Angeles is 5000 miles. Wind blows from Los Angeles towards Paris at a speed of 100 mph. Plane A leaves Paris at noon flying towards Los Angeles. Plane B departs Los Angeles 1 hour later to fly to Paris. How far from Los Angeles are the two planes when they pass each other ?

(11.0.61) (*) Plane A flies at a constant speed from New York to Los Angeles along a route which is 2000 miles. Plane B flies in the opposite direction at a constant speed which is 100 mph faster than plane A. Plane B takes off one hour after plane A. They land at the same moment. How far are they from Los Angeles when they pass?

(11.0.62) Beaker A contains 1 liter which is 10% oil and the rest is vinegar, thoroughly mixed up. Beaker B contains 2 liters which is 50% oil and 50% vinegar, completely mixed up. Half of the contents of B are poured into A, then completely mixed up. How much oil should now be added to A to produce a mixture which is 60% oil?

11.1   Graphical Problems

Problems

(11.1.1) A car starts from rest and accelerates at a constant rate of acceleration for 2 minutes, then cruises at constant speed for 3 minutes, then decelerates at a constant rate of deceleration to a stop in one minute.
(a) Sketch a graph of speed against time
(b) sketch a graph of distance traveled against time

(11.1.2) A small colony of mold starts in a dish with plenty of food. Initially the colony grows exponentially until the dish starts to get crowded. The rate of growth slows down to almost nothing as the dish becomes full. After a while the nutrient becomes scarce and the colony starts to die off, slowly at first then more rapidly, Eventually all the mold is dead. Sketch a graph, marking significant features on the graph.

(11.1.3) The graph shows the populations of two species, one is a predator on the other. Which is predator and which is prey. Explain **clearly** your reasoning.

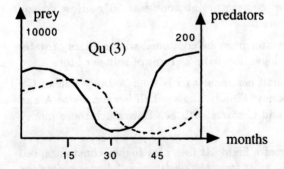
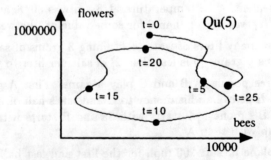

(11.1.4) Use the graphs for the previous question. Draw a curve in state space by plotting the number of prey on the horizontal axis and the number of predators on the vertical axis. Thus for each time t you will plot the point (a, b) where a is the number of prey at that time t and b is the number of predators at that time t. Put an arrow on the curve in the direction of increasing time. Mark the times of a few points on this curve.

(11.1.5) In this question you will do the reverse of the previous question. The graph above shows the number of bees and flowers over a period of 25 months. Use this to sketch graphs of the numbers of bees and flowers plotted against time t.

(11.1.6) Copy the graph from the next page for this question using different colors for $f(t)$ and $g(t)$,. Then, in two different colors, sketch the graphs $y = f(t) + g(t)$ and $y = f(t) - g(t)$ on the same diagram.

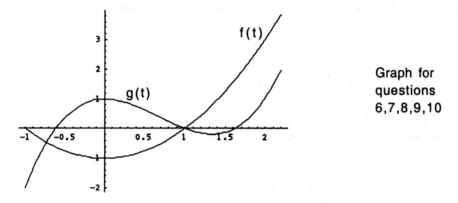

Graph for
questions
6,7,8,9,10

(11.1.7) Copy the graph for the previous question using different colors for $f(t)$ and $g(t)$, then in a different color, sketch the graph of $y = f(t) \times g(t)$ on the same diagram.

(11.1.8) Copy the graph of $f(t)$ from the previous question. Then sketch, using different colors, the graphs $y = 2f(t)$ and $y = -f(t)$ and $y = 1 + f(t)$ on the same diagram.

(11.1.9) Copy the graph of $f(t)$ from the previous question. Then sketch, using a different colors, the graph $y = 1/f(t)$ on the same diagram.

(11.1.10) Copy the graph for the previous question, then sketch on the same diagram the graph of $y = [f(t)]^2$ using a different color.

(11.1.11) The table of data shows the population of a city from 1900 to 1945. Plot a graph showing the **logarithm** of the population against time. Draw the best straight line you can through these data points. What do you think the population was in 1948?. The straight line provides an exponential function $p(t) = A2^{t/K}$ which approximates the original data. Find A and K from the straight line you found. What is the doubling time of this population?

0	5	10	15	20	25	30	35	40	45
275	342	378	488	559	795	824	1155	1281	1734

Qu (11)   Population in thousands from 1900 to 1945

(11.1.12) On graph paper, draw the following graphs for $-5 \le x \le 5$

$$y = 2^x, \quad y = 2^{-x}, \quad y = 3 - 2^{-x}, \quad y = 3 + 2^{-x}.$$

(11.1.13) On graph paper draw the graphs of $y = sin(t)$ and $y = 2sin(t - \pi/2)$ for $0 \le t \le \pi$.

(11.1.14) Draw the graph $y = log(e^x)$ on graph paper. What do you notice? Explain why this happened.

(11.1.15) Sketch a graph showing the population of the state of California over the past 100 years. (Make an educated guess about what it should look like) What is the significance of the slope of this graph for the year 1955?

(11.1.16) Sketch a graph showing the average daytime temperature in Santa Barbara throughout the year. Where do you think the slope is greatest?. Why?

(11.1.17) Carefully draw a reasonably accurate graph of the function 2^x for the range of values $-100 \leq x \leq 100$. Using the same diagram, sketch the graph of $y = x^{10}$. What do you conclude about these two functions from your graph? [hint in this question, you may either use a calculator to work out a table of values for 2^x or else use logarithms.]

(11.1.18) Use the graph for qu 6. Using different colors, sketch on a single diagram the graphs of $y = f(t) + 1$ and $y = f(t) - 2$.

(11.1.19) Use the graph for qu 6. Using different colors, sketch on a single diagram the graphs of $y = 2f(t)$ and $y = 3f(t)$.

(11.1.20) Use the graph for qu 6. Using different colors, sketch on a single diagram the graphs of $y = f(t)$ and $y = f(2t)$.

(11.1.21) Use the graph for qu 6. Using different colors, sketch on a single diagram the graphs of $y = f(t)$ and $y = f(t + 0.5)$ and $y = f(t - 0.5)$.

(11.1.22) The graph shows the rate of increase of money in circulation during 1995. What was the average increase per month in the amount of money in circulation during the year? **Explain clearly.**

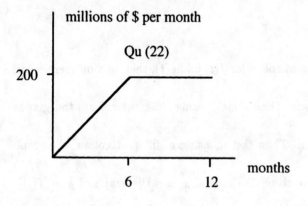
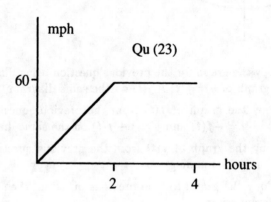

(11.1.23) The graph shows the velocity of car during 4 hours.
(a) how far did car go in the 4 hours ?
(b) what was the average velocity during the 4 hours ?
(c) what was the acceleration after one hour ? include units.

Chapter 12

Further Calculus.

12.1 Sine Waves.

The graph shows a familiar curve: the **sine wave.** This curve appears in many situations, for example sound waves, ocean waves, fluctuating populations, earthquakes and many more.

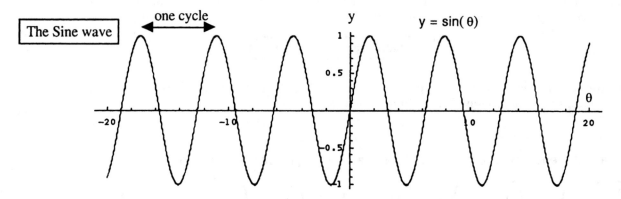

You first met the sine function in trigonometry. For calculus it is most convenient to measure angles using **radians** instead of degrees. There are 360 degrees round a circle but only 2π radians. In other words

$$\boxed{\text{one radian equals } 360/2\pi \approx 57.3 \text{ degrees.}}$$

The reason for this apparently strange unit of measurement is that it is in fact quite natural. If you want to know what the angle measured in radians between two straight lines is, draw a circle of radius 1 centered at the point where the lines cross. Then measure the length of the part of the circumference between the two lines. This length equals the number of radians in the angle.

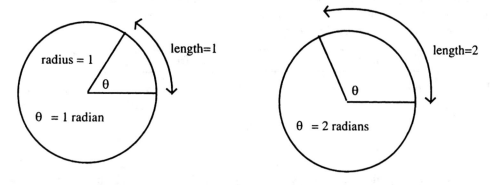

To find the sine of an angle θ, draw a right-angled triangle with an angle of θ. Then the length of the side **opposite** the angle divided by the length of the **hypotenuse** gives $sin(\theta)$.

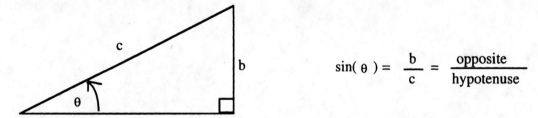

$$sin(\,\theta\,) = \frac{b}{c} = \frac{\text{opposite}}{\text{hypotenuse}}$$

This is usually where the sine function first makes its appearance. However for many of the further applications, this connection with triangles is not so important. The **key feature** of the sine function is that it **repeats** over and over again. The graph shows a wave which repeats. The **period** of repetition is $2\pi \approx 6.3$. The period is the time taken for one complete wave, or **cycle** as it is often called. It is the amount of time after which the function begins to repeat. It is the repetition that makes the sine wave so useful. Many phenomena have a repetitive nature, and can be described in terms of sine waves.

We can produce variations on the basic sine wave by changing three properties: the **amplitude, frequency** and **phase** of the sine wave. These terms will now be explained. The **amplitude** of a sine wave is "how high" it is. For example the amplitude of $sin(t)$ is 1 but the amplitude of $3sin(t)$ is 3. If you graph $y = 3sin(t)$ the result looks just like the graph of $y = sin(t)$ except that it is 3 times as tall. The amplitude of $0.2sin(t)$ is 0.2.

$$\boxed{\text{The amplitude of } A \cdot sin(t) \text{ is } A.}$$

The **frequency** of a sine wave is how rapidly it repeats. A high frequency means it repeats very quickly, a low frequency means it repeats slowly. The diagram shows examples of this.

a high frequency sine wave (solid) and a   low fequency sine wave (dashed)

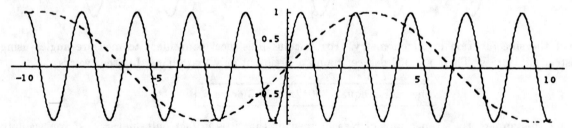

The function $sin(3t)$ has 3 times the frequency of $sin(t)$. To understand this, $3t$ increases 3 times as rapidly as t. The $sin(t)$ function goes through a complete cycle as t is increased from 0 up to 2π. Now $3t$ increases from 0 to 2π three times as quickly: as t increases from 0 to $2\pi/3$. Since it is $3t$ that we take $sin(.)$ of, this means that $sin(3t)$ goes through one complete cycle as t increases from 0 to $2\pi/3$. In other words it goes through one complete cycle in $1/3$ of the time it takes for $sin(t)$.

$$\boxed{\text{The frequency of } sin(Kt) \text{ is } K \text{ times the frequency of } sin(t).}$$

Frequency is how many complete cycles occur in one unit of time. For example if 3 cycles take one hour then the frequency is 3 cycles per hour. A frequency of 60 cycles per second is called 60 *hertz*. One hertz means one cycle per second. Thus

$$\boxed{\text{frequency} \; = \; \frac{1}{\text{period}}}$$

Now the function $sin(t)$ has a period of 2π so its frequency is $1/(2\pi) \approx 0.15$. Thus the frequency of $sin(Kt)$ is $K/(2\pi)$.

Different sine waves might be identical in amplitude and frequency, but still differ because they "start" at different times. In other words two sine waves might be the same except that they are "out of step." The technical term is that they have different **phases**. The two sine waves shown on the next page are $sin(t)$ and $sin(t + 2)$. The effect of taking sine of $(t + 2)$ instead of (t) is that $(t + 2)$ does everything that t does but 2 units of time **earlier**.

two sign waves that are out of **phase**

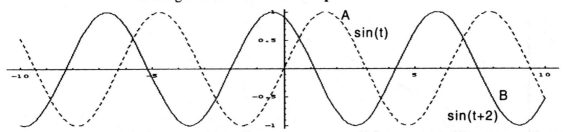

In other words, $t + 2$ is always 2 units of time bigger than t. So $sin(t + 2)$ is always 2 units of time ahead of $sin(t)$. Whatever $sin(t + 2)$ does, $sin(t)$ does 2 units of time later. The phase of $sin(t + \phi)$ is ϕ ahead of $sin(t)$.

> If you move sine wave A to the right by the **fraction** k of one complete cycle, the result is a sine wave B which has **phase** $2\pi k$ compared to A.

In the diagram, the dotted sign wave, A, must be moved $2\pi - 2 \approx 4.3$ units right along the t-axis to get the solid sine wave A. One complete cycle takes $2\pi \approx 6.3$ units. So the **fraction** of one complete cycle that A must be moved is $4.3/6.3$. This gives a phase difference of $2\pi(4.3/6.3) \approx 4.3$.

> $sin(Kt + \phi)$ has a phase lead of ϕ compared to $sin(Kt)$.

These variations can be combined. For example $3sin(4\pi(t + .1))$ has amplitude 3 and frequency $4\pi/(2\pi) = 2$. The phase shift is a bit trickier to work out. The idea is to compare $sin(4\pi(t + .1))$ to $sin(4\pi t.)$ If we write $4\pi(t + .1)$ as $4\pi t + .4\pi$ then we can see that this corresponds to increasing the thing you take **sine** of by 0.4π. Thus the phase shift between these two sine waves is 0.4π.

Beware that a phase shift of 2π is the same as a phase shift of 0. (why? : sketch a diagram!)

The cosine function, $cos(t)$, is a close relative of the sine function. In terms of right angled triangles, it it the length of the side **adjacent** to the angle divided by the length of the hypotenuse. However the feature that is most important here is that $cos(t)$ is a sine wave with a phase shift of $\pi/2$.

$$cos(t) = sin(t + \pi/2).$$

This can be seen easily from the diagram using right angles triangles. The graph of $y = cos(t)$ is therefore just the same as the graph $y = sin(t)$ but shifted sideways by $\pi/2$ units. The derivatives of $sin(t)$ and $cos(t)$ are given by a simple formula

$$\frac{d}{dt}sin(t) = cos(t) \qquad \frac{d}{dt}cos(t) = -sin(t).$$

For the most general sine wave the derivative is given by:

$$\frac{d}{dt}(A\,sin(Kt + \phi)) = AK\,cos(Kt + \phi)$$

Since integration is the reverse of differentiation we see that

$$\int sin(t)\,dt = -cos(t) + C \qquad \int cos(t)\,dt = sin(t) + C.$$

And

$$\int A \sin(Kt + \phi)\, dt = -\frac{A}{K}\cos(Kt + \phi) + C.$$

Problems

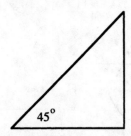 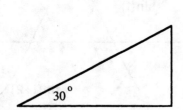 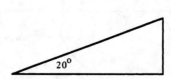

(12.1.1) Measure the lengths of sides of the triangles shown to work out the sine of the angles
(a) 30° (b) 60° (c) 45^0 (d) 20° (e) 70^0.

(12.1.2) What are the amplitude, frequency and period of the following
(a) $7\sin(2t)$ (b) $\sqrt{2}\sin(5\pi t)$ (c) $-3\sin(4\pi t)$ (d) $\sin(t/3)$.

(12.1.3) What is the phase difference between
(a) $\sin(t)$ and $\sin(t + \pi.)$ (b) $\sin(2t)$ and $\sin(2t + \pi)$ (c) $\sin(t + 1)$ and $\sin(t + 3)$.

(12.1.4) Find the derivatives of
(a) $\sin(5t)$ (b) $3\sin(2t - \pi)$ (c) $7\cos(3(t - 4))$ (d) $2\sin(4 - t)$.

(12.1.5) Sketch the following graphs between $t = 0$ and $t = 6$
(a) $y = 2\sin(2\pi t/2)$
(b) $y = -\sin(2\pi t/3)$
(c) $y = 3 + \sin(2\pi t/6)$
(d) $y = \sin(\pi(t - 1))$

(12.1.6) Sketch on the same diagram graphs of $y = x$ and $y = x \cdot \sin(\pi x)$.

(12.1.7) Sketch on the same diagram graphs of $y = x^2$ and $y = x^2 \cdot \sin(\pi x)$.

(12.1.8) Find

$$(a)\ \int \sin(x)\, dx \quad (b)\ \int \sin(2x)\, dx \quad (c)\ \int \cos(3x)\, dx.$$

In each case check your answer is correct by differentiating it.

(12.1.9) Calculate

$$\int_0^{\pi/2} \sin(t)\, dt.$$

Sketch a diagram showing the region that this integral represents the area of. How can you tell from your sketch if your answer is approximately correct?

(12.1.10) What is the area under one arch of the graph $y = \sin(2x)$. Begin by sketching the region your are finding the area of.

(12.1.11) The diagram shows the functions labeled (a), (b) and (c). These are the functions $\sin(2\pi x) + 2, 3\sin(2\pi x + (\pi/2)), 1 + \sin(2 + 2\pi x)$ Which is which? Say how you know.

(12.1.12) The diagram shows three functions labeled (d),(e) and (f). Each one is a constant plus a sine wave. Determine for each one the frequency, amplitude and phase of the sine wave. Then write down the function that gives each graph. Check your answers by plotting a few carefully chosen points.

(12.1.13) An artery has a circular cross-section of radius 2 millimeters. The speed which blood flows along the artery fluctuates as the heart beats. The speed after t seconds is $20 + 8\sin(2\pi t)$ meters per second.

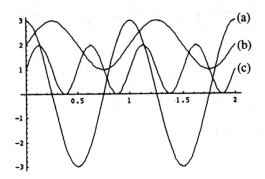
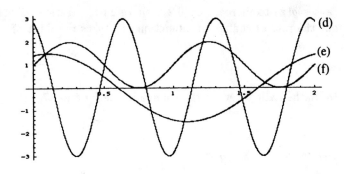

(a) Sketch a graph of the speed over a 3 second time span.

(b) What volume of blood passes along the artery in one second?

(12.1.14) The number of rabbits in a forest varies with time. During the period of 1980 to 1990 the number is given approximately by the formula $8000 + 2000sin(2\pi t/5)$ where t is the number of years after 1980.

(a) Sketch the population during 1980-1990

(b) What years was the population greatest?

(c) When was the population lowest?

(d) How quickly was the population changing at the start of 1985?, increasing or decreasing?. What are the units of your answer?

(12.1.15) The rate at which a power station provides electrical power during a 24 hour period is given by the function $f(t) = 200 + 100sin(2\pi t/24)$. The units are megawatts. Time is measured in hours after midnight. Sketch a graph of this. When is energy production greatest and when lowest?. How much energy is provided by the power station during a twenty four hour period period [answer will be in megawatt-hours]

12.2   The Product Rule.

This section is about how to find the derivative of two functions multiplied together. The **product rule** says:

$$\frac{d}{dx}\left(f(x)\cdot g(x)\right) = f'(x)\cdot g(x) + f(x)\cdot g'(x).$$

For example

$$\frac{d}{dx}\left(x^3 e^{2x}\right) = 3x^2 e^{2x} + x^3(2e^{2x}).$$

The rest of this section is an explanation of this rule. First it is important to look at a very common mistake. Many students "forget" the product rule and instead use the

WRONG FORMULA $\dfrac{d}{dx}\left(f(x)\cdot g(x)\right) = f'(x)\cdot g'(x).$

This formula is simpler and "seems" like it ought to be correct. If you try it out, you will find it gives the wrong answer. The rule for differentiating two functions **added** together is

$$\frac{d}{dx}\left(f(x) + g(x)\right) = f'(x) + g'(x).$$

In other words when functions are added, you just differentiate each one and add the results. So students tend to think a similar thing should work with multiply also. It does not!

To understand the product rule, we will first look at the simple case that $f(x)$ is the **constant** function 7. What is the derivative of $7g(x)$? The derivative is the **rate of change** of the function. How quickly does $7g(x)$ change ? The answer is 7 times as quickly as $g(x)$ changes. This is common sense. For example, if increasing x by Δx

causes $g(x)$ to increase by 0.2 then $7g(x)$ increases by 7×0.2. Any change in $g(x)$ is multiplied by 7. To work out the rate of change of a function one takes the limit of

$$\frac{\text{change in function}}{\Delta x}.$$

In mathematical terms, the change in the function $7g(x)$ is just

$$\Delta(7g) \;=\; 7g(x + \Delta x) - 7g(x) \;=\; 7\left(g(x + \Delta x) - g(x)\right) \;=\; 7\Delta g.$$

Divide this by Δx to get

$$\frac{\Delta(7g)}{\Delta x} \;=\; 7\frac{\Delta g}{\Delta x} \qquad \text{thus} \qquad \frac{d}{dx}\left(7g(x)\right) \;=\; 7g'(x).$$

Now think about the rate of change of $f(x) \cdot g(x)$. If x is increased then **both** $f(x)$ and $g(x)$ change. Thus there are two contributions to the change of $f(x) \cdot g(x)$. One from each of $f(x)$ and $g(x)$. Let us think first about the contribution from $g(x)$. Suppose that changing x by Δx causes $g(x)$ to increase by 0.2 The effect on $f(x) \cdot g(x)$ is to create a change of $f(x) \times 0.2$. It is similar to the change in $7g(x)$ caused by a change of Δx, any change in $g(x)$ is multiplied by 7. The only difference here is that the constant 7 is replaced by the function $f(x)$. A similar argument applies to the contribution to a change in $f(x) \cdot g(x)$ resulting from a change in $f(x)$. Whatever the change, Δf in $f(x)$ is, it is multiplied by $g(x)$ to give a contribution of $\Delta f \times g(x)$. These two contributions are

$$f(x) \cdot \Delta g \;+\; (\Delta f) \cdot g(x)$$

We thus see that the two terms in the product rule, $f'(x) \cdot g(x)$ and $f(x) \cdot g'(x)$ are the contributions to the rate of change of the product $f(x) \cdot g(x)$ resulting from the rates of change of $f(x)$ and $g(x)$ respectively.

(rate of change of $f(x) \cdot g(x)$) =
(contribution from rate of change of $f(x)$) + (contribution from rate of change of $g(x)$.)

The product rule involves the derivatives of three quantities: f, g and $f \times g$. If you know any two of the derivatives, you can use the product rule to find the third. For example if you know f' and $(f \times g)'$ then

$$(f \times g)' \;=\; f'g + fg' \qquad \Rightarrow \qquad g' = \frac{(f \times g)' - f'g}{f} \qquad \text{solving for } g'$$

Problems

(12.2.1) (a) Use the product rule to find the derivative of $(3x + 2)(5x - 1)$.
(b) Now multiply out and work out the derivative again and check the answers agree.
(c) Use the wrong formula so often used by students. What does this give?

(12.2.2) Use the product rule to find the derivatives of

$$(a) \;\; (x^2 + 7)(x^3 - 3x) \qquad (b) \;\; (3x^4 - x^{-2})(x^2 - 3x) \qquad (c) \;\; x^{-4}(2x^4 + 1) \qquad (d) \;\; (x^5 + 3x)^2$$

(12.2.3) Use the product rule to find the derivatives of

$$(a) \;\; (x^2 + 7)\sin(3x) \qquad (b) \;\; (x^2 + 1)e^x \qquad (c) \;\; x\ln(x) \qquad (d) \;\; \sin(3x)\ln(x)$$

(12.2.4) Find the derivatives, with respect to t, of

$$(a) \;\; e^t\ln(t) \qquad (b) \;\; (t^2 + 1)\sin(t) \qquad (c) \;\; \sin(t)\cos(t) \qquad (d) \;\; (3t^2 - 2t)\sin(7t) \qquad (e) \;\; e^{2t}\sin(3t).$$

(12.2.5) When a commodity is sold for $\$x$ per item, the number of items sold is $f(x)$. The item costs $\$5$ to make.
(a) Express the total profit in terms of x
(b) What is the rate of change of total profit as x changes.
(c) If 1000 items sell when the price is $\$15$ and if the number of items that are sold decreases by 200 for every $\$1$ increase in price, find the rate of change of profit when the item price is $\$15$.
(d) What is the practical significance of your answer to (c)?

(12.2.6) If x million dollars are spent in research to make a new computer, then the number of new computers sold is a function $n(x)$. The profit made from selling each computer is also a function $p(x)$.
(a) What is the total profit? (ignore the money spent on research for this)
(b) What is the rate of change of total profit as x is increased?

(12.2.7) In this problem you will work out a formula for the derivative of the product of three functions $f(x), g(x)$ and $h(x)$ multiplied together. To do this, think of the function $f(x) \cdot g(x) \cdot h(x)$ as the result of multiplying two functions together; namely $f(x)$ and $K(x) = g(x) \cdot h(x)$.
(a) What is the derivative of $K(x)$?.
(b) What is the derivative of $f(x) \cdot K(x)$?
(c) What is the derivative of $f(x) \cdot g(x) \cdot h(x)$. ?
(d) Use your answer to (c) to work out the derivative of $x \cdot x \cdot x$.
(e) Use your answer to (c) to find the derivative of $(f(x))^3$.

(12.2.8) Use the result (e) of the previous question to find the derivative of $(x^2 + 1)^3$ and of $[sin(x)]^3$.

(12.2.9) What would you guess the derivative of $[f(x)]^4$ is ? Use the product rule with $f(x)$ and $[f(x)]^3$ multiplied together to see if your guess is correct.

(12.2.10) In a certain country, the population is growing by 5% per year. The average income per person is growing by 3% per year. Show how the product rule can be used to show that the total income of the entire population is growing by (approximately) 8% per year. [hint: 5% pop. growth means $p' = .05p$]

(12.2.11) A city is in the shape of a rectangle. In 1995 the width of the city was 2 miles and the length of the city was 3 miles. The width of the city is growing at a rate of 1 mile in 3 years. The length of the city is growing at the rate of 1 mile in 8 years. Use the product rule to find how quickly the area of the city is growing in 1995.

(12.2.12) Use the product rule to show the derivative of $f(t) = sin(t)sin(t) + cos(t)cos(t)$ is zero. This means that $f(t)$ is constant. What is this constant?.

(12.2.13) We know that the derivative of $f(x) = x$ is 1. The function $g(x) = 1/x$. Pretend that you do not know what the derivative of $1/x$ is. Differentiate $f(x) \cdot g(x)$ using the product rule, but leave $g'(x)$ in the answer (because you are pretending you do not know what it is). Since $f(x) \cdot g(x) = 1$ the derivative of this product is 0 (why?) Use this fact to "discover" what $g'(x)$ is.

(12.2.14) The number of trees in a forest is 50 million in 1997 and increasing at a rate of .5 million trees per year. The total mass of wood in the entire forest at the start of 1997 is 2×10^{10} kilograms. In 1997 the total mass was increasing at a rate of 2% per year. (a) What is the average mass of a tree in 1997? (b) Is this average increasing or decreasing in 1997? (c) How quickly?

(12.2.15) The population of the country Dnalgne is 50 million in 1997 and increasing at a rate of .5 millions per year. The average annual income of a person in Dnalgne during 1997 was $\$15000$ per year, and increasing at a rate of $\$500$ per year. How quickly is the total income of the entire population rising in 1997?

(12.2.16) The annual profit a company makes by selling computers is 10 million dollars and increasing at a rate of 20% per year. The number of computers it sells is 2000 and increasing by 500 per year. (a) What is the average profit per computer? (b) Is it rising or falling? (c) How much ?

(12.2.17) A rectangle initially has width 5 meters and length 3 meters and is expanding so that the area increases at a rate of 2 m^2 per hour. If the width increases by 20 cm per hour how quickly does the length increase initially? [hint: Using the product rule, if you know **any** two of the derivatives you can work out the third one. Here you know $A = L \times W$ and A' and W'.]

(12.2.18) A tree trunk is approximated by a circular cylinder of height 100 m and diameter 1 m. The tree is growing taller at a rate of 2 *meters* per year and the diameter is increasing at a rate of 2 *cm* per year. The density of the wood is $2000Kg$ per cubic meter. How quickly is the mass of the tree increasing?

(12.2.19) In this problem you will discover the derivative of $1/g(x)$ is $-g'(x)/[g(x)]^2$. The function $f(x) = g(x) \times (1/g(x))$ is of course just the constant 1. Use the product rule to find the derivative

$$0 \; = \; f'(x) \; = \; g'(x)\frac{1}{g(x)} \; + \; g(x)\frac{d}{dx}\left(\frac{1}{g(x)}\right).$$

Now solve this equation for $\frac{d}{dx}\left(\frac{1}{g(x)}\right)$ to get

$$\frac{d}{dx}\left(\frac{1}{g(x)}\right) \; = \; \frac{-g'(x)}{[g(x)]^2}.$$

Try this out using $g(x) = x^2$. Does it give the correct answer?

(12.2.20) Use the product rule, and the result of the previous question to show that

$$\frac{d}{dx}\left(\frac{f(x)}{g(x)}\right) \; = \; \frac{f'(x)g(x) \; - \; f(x)g'(x)}{[g(x)]^2}.$$

This is called the **quotient rule** because it is what is used to calculate the derivative of the quotient of two functions (ie one function divided by another one.)

(12.2.21) Use the quotient rule from the previous question to find the derivatives of the following functions

$$(a) \; \frac{x}{2x+1} \quad (b) \; \frac{x}{1+e^x} \quad (c) \; \frac{1}{1+x^2}.$$

12.3   More Maxima and Minima: The Second Derivative Test.

If you stand on the top of a small mountain which is next to a larger mountain, you are at a **local maximum** of height. All the places near you are lower than you are. But you are not at the (global) maximum height because the other mountain is higher. We have seen that if a function $f(x)$ has a local maximum or minimum at $x = -1$ then $f'(-1) = 0$. This is shown in the diagram. At the maximum, the slope of the graph is seen to be horizontal, so the slope is zero at that point. We also can see something else. The graph is **concave down** at the maximum. This means that $f''(-1) < 0$. The function shown has a minimum at $x = 2$ there the graph is **concave up** so $f''(2) > 0$. This gives a useful method to distinguish maxima from minima. If you solve $f'(x) = 0$ the solutions are called **critical points**. Each critical point might be a maximum or a minimum, or in fact neither (see later).

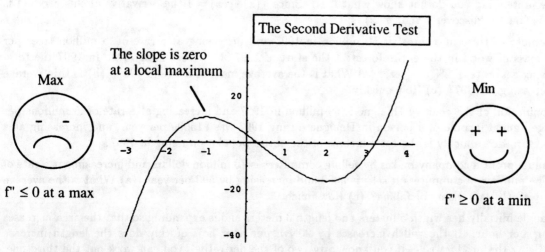

The second derivative can be used to decide whether a particular critical point is a local maximum or minimum.

SECOND DERIVATIVE TEST   A critical point x is a **local** max if $f''(x) < 0$ and a min if $f''(x) > 0$.

This can be easily remembered with the smiley face and frown shown.

Here is an example. The function $f(x) = 2x^3 - 3x^2 - 12x + 4$ has derivative $f'(x) = 6x^2 - 6x + 12$. So the critical points of $f(x)$ are the solutions of

$$6x^2 - 6x - 12 \;=\; 6(x+1)(x-2) = 0.$$

Thus there are two critical points $x = -1$ and $x = 2$. To find out which is a maximum and which a minimum we calculate the second derivative $f''(x) = 12x - 6$. The critical point $x = -1$ gives $f''(-1) = 12(-1) - 6 = -18$. So $x = -1$ is a maximum. The critical point $x = 2$ gives $f''(2) = 12(2) - 6 = 18$. So $x = 2$ is a minimum.

There are sometimes critical points which are not a maximum or a minimum. Such a point is called a **inflection point.** The diagram shows one. When the second derivative test is done, it might happen that $f''(a) = 0$. In this case a might be a maximum or a minimum or a inflection. The second derivative test fails to tells us what

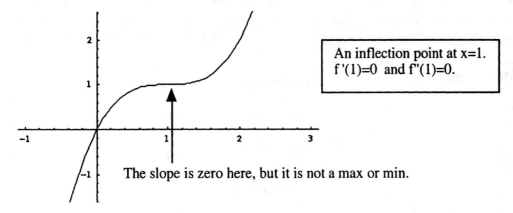

An inflection point at x=1.
f '(1)=0 and f''(1)=0.

The slope is zero here, but it is not a max or min.

is going on in this situation. There are more complicated tests to decide between the three possibilities, but we will not discuss them.

Problems

(12.3.1) For each of the following functions, find the critical point(s) and determine their type (max, min or inflection). If the second derivative test is inconclusive, state this.

(a) $x^2 + 5x + 3$　　(b) $x^3 + 6x^2 + 9x + 3$　(c) $x^3 + 3x^2 + 3x + 1$　　(d)$x + 2x^{-1}$　(e)$e^x + e^{-x}$.

(12.3.2) Find any local maxima or minima of the following

(a) $x^3 - 12x + 2$　　(b) $(x+1)^2$　　(c) $(x+1)^4$　　(d) $2x - 27x^{-2}$　　(e) $ln(x) - 2x^2$.

(12.3.3) Sketch the graph of a function which has local maxima at $x = 2$ and $x = 4$ and a local minimum at $x = 3$. Try to sketch the graph of a function which has local maximum at $x = 2$ and $x = 3$ and a local minimum at $x = 4$ **but no other local max/min.** Do you think this is possible? Give reasons.

(12.3.4) Where are the local max/min for the function $2sin(\pi x)$. Do this problem by sketching the graph and thinking instead of using calculus.

(12.3.5) Same as previous question but for $e^x sin(x)$. You do not need to find exactly where the max/min are, but you should say approximately where they are.

(12.3.6) Sketch the graph of a function which has no max or min but has inflections at $x = 1, 2, 3, \cdots$.

(12.3.7) Find any local maxima or minima of the following

(a) $4(x-1)(x+3)$　　(b) $1/x$　　(c) $2x^3 + 3x^2 - 36x + 5$　　(d) $2x^3 + 15x^2 + 24x - 7$.

12.4   Orders of Smallness.

In this section we will be concerned with making approximations in the situation where certain quantities are much smaller than others. For example $0.01 + 0.000001$ is the sum of two small numbers, but one of them is much smaller than the other. If this number is the solution to some problem, then a good approximation is given by taking 0.01 and neglecting the other term. We say that the second term has a **smaller order of magnitude** than the first term. Now suppose that Δx is a small number, for example $0 \leq \Delta x \leq 0.1$ Then $(\Delta x)^2$ is much smaller than Δx. For example if $\Delta x = .01$ then $(\Delta x)^2 = 0.0001$. When making a mathematical model, in order to keep things simple, one often neglects terms of smaller order. Thus for example one might calculate something as

$$3\Delta x \ + \ 2(\Delta x)^3$$

and then, if we know Δx will be very small, approximate this by taking $3\Delta x$. This has the great advantage of replacing something complicated, involving cubing Δx by something much simpler. The price is a small error.

Example. A square had side length 5 cm. If the length of the side is increased by 0.1 cm how much does the area increase by? The area of the larger square is $(5 + 0.1)^2 \ = \ 5^2 + 2(5)(0.1) + (0.1)^2$. The area of the original

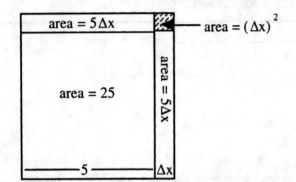

Increasing the length of the side of the square by Δx gives an increase in area of approximately $10\Delta x$

$(\Delta x)^2$ is small compared to $5 \Delta x$ so it is neglected

square is 5^2 so the increase in area is $2(5)(0.1) + (0.1)^2$. The first term is 1 and the second term is $(0.1)^2$ which is much smaller than the first term. So a good approximation to the answer is 1. (The exact answer is 1.01)

We can extend this problem a bit. Suppose the new square has the length of a side increased by Δx. Then the area of the new square is $(5 + \Delta x)^2 = 25 + 10\Delta x + (\Delta x)^2$. The area of the original square is 25 so the increase in area is $10\Delta x + (\Delta x)^2$. If Δx is much smaller than 1 then the increase in area is $10\Delta x$ and the error (which is $(\Delta x)^2$) is small compared to this answer. This is shown in the diagram.

A similar example is to take a circle of radius R, increase the radius by ΔR and calculate the increase in area as

$$(\text{area of circle of radius } R + \Delta R) - (\text{area of circle of radius } R) \ = \ \pi(R + \Delta R)^2 - \pi R^2.$$

This simplifies to $2\pi R\Delta R + \pi(\Delta R)^2$. If ΔR is much smaller than 1 then we neglect $(\Delta R)^2$ which is small in comparison to ΔR and get the increase in area is approximately $2\pi R\Delta R$.

Example. A box with rectangular sides has a square base. It is 3 meters high and the length of a side of the base is 2 meters. If the height is increased by a small amount Δh and the length of the side of the base is increased by a small amount Δx how much does the volume increase by?. Let x be the length of the side of the base and h the height, both measured in meters. Then the volume, V, of the box is the area of the base, x^2, times the height so:

$$V(x,h) \ = \ x^2 h.$$

The increase in volume will now be calculated. We will neglect any term which involves multiplying two small terms together.

$$
\begin{aligned}
V(x + \Delta x, h + \Delta h) \ - \ V(x,h) \ &= \ (x + \Delta x)^2(h + \Delta h) \ - \ x^2 h \\
&= \ (x^2 + 2x\Delta x + (\Delta x)^2)(h + \Delta h) - x^2 h \\
&\approx \ (x^2 + 2x\Delta x)(h + \Delta h) - x^2 h \qquad \text{neglecting } (\Delta x)^2 \\
&= \ (x^2 h + 2hx\Delta x + x^2\Delta h + 2x\Delta x\Delta h) - x^2 h \\
&= \ 2hx\Delta x + x^2\Delta h + 2x\Delta x\Delta h \\
&\approx \ 2hx\Delta x + x^2\Delta h \qquad \text{neglecting } \Delta x\Delta h.
\end{aligned}
$$

We are interested in the increase in volume when $x = 2$ and $h = 3$ then

$$\Delta V = 2(3)(2)\Delta x + (2)^2\Delta h = 12\Delta x + 4\Delta h.$$

We see that an increase in x has a larger affect on volume than the same increase in h would, since Δx is multiplied by 12 but Δh is only multiplied by 4.

> ### Problems

(12.4.1) (a) Find the increase to first order in Δx of $f(x) = 4x^2 + 3x + 2$ when x is increased by Δx.
(b) In particular, if x is increased from 3 to 3.02 what does your formula give for the approximate increase in $f(x)$.
(c) What is the exact increase?
(d) What is the percentage error in your approximation?

(12.4.2) The volume of a sphere of radius R is $V(R) = 4\pi R^3/3$. If the radius is increased by ΔR, what is the increase in volume to first order?

(12.4.3) A mass of gas at pressure P and temperature T (degrees Kelvin) and volume V satisfies the equation $T = KPV$ where K is a constant. The initial temperature is 300^oK, the initial volume is $2\ cm^3$ and the initial pressure is 5 [The units are N/cm^2. which is Newtons per square cm. This is force per unit area, which is what pressure is.] If the pressure increases by 2% and the volume increases by .03 cm^3 approximately how much does the temperature change by? [hint: use the initial information to find K]

12.5 Exponentially growing and damped oscillations.

In many situations a periodic phenomena is combined with an element of growth or decay. For example when a tuning fork is struck, one hears a pure musical note which starts loud and gradually dies away. The basic note is a single frequency sound wave. For example the A note above middle C is 440 cycles per second (**hertz**), and so is given by the function $A \cdot sin(440 \times 2\pi t)$. However the amplitude gradually decreases. Thus the amplitude A is a function of time. It starts out quite large at $t = 0$ and decreases to 0 as time passes. An example of such a function is the exponential function $A(t) = 17e^{-.2t}$. The initial amplitude is $A(0) = 17$. As t increases, the factor $e^{-.2t}$ decreases towards zero, but never quite get there. This is shown in the diagram. In fact such an exponential function does provide a good approximation to the way a musical note fades away. So that the musical note is described by

$$17e^{-.2t}sin(440 \times (2\pi t)).$$

The graph looks something like the one shown, except the frequency of the note shown is much lower than 440 hertz, in order to be able to see the sine waves.

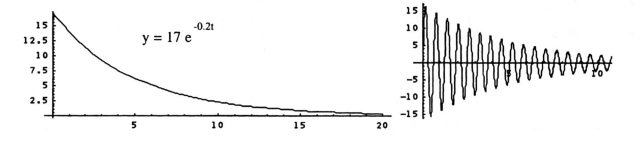

a musical note fading away

Problems

(12.5.1) The graph [on the next page] (1) shows an exponential function. It can be written as Ae^{-kt}. What are A and k ?

(12.5.2) The graph (2) shows an exponentially decaying sine wave. What is the equation. [hint: find the frequency, and now compare the amplitude at successive peaks. Think how you did the previous question.]

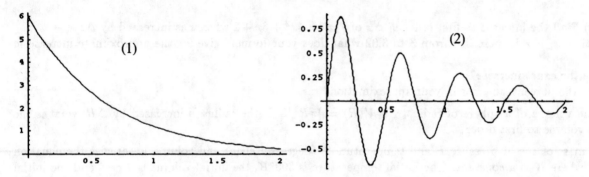

12.6   Power Series, Approximations and Linearization.

Often we can not get exact answers to a problem, and we must be satisfied with approximate answers. If the exact answer to a certain problem is 1.5 then approximate answers of 1.498 or 1.502 are fairly accurate. What does it mean to have an approximate answer when we are looking for is a **function** rather than a number?. It means that the approximating function should give almost the same answers as the real function. For example, an economist might try to find predict the unemployment rate during the next year. This means predicting how many people will be unemployed on any given day. An example of such a prediction is shown together with the real unemployment rate.

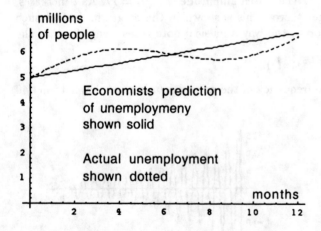

The function e^x is given by the power series

$$e^x = \sum_{n=0}^{\infty} \frac{x^n}{n!} = {}' 1 + \frac{x}{1!} + \frac{x^2}{2!} + \frac{x^3}{3!} + \frac{x^4}{4!} + \frac{x^5}{5!} + \cdots$$

This series goes on for ever. However if you wish to work out $e^{0.3}$ then an answer accurate to within 1% is obtained by using just the first 4 terms

$$e^{0.3} \approx 1 + \frac{0.3}{1!} + \frac{0.3^2}{2!} + \frac{0.3^3}{3!} + \frac{0.3^4}{4!} \approx 1.35.$$

The function

$$f(x) \;=\; 1 + \frac{x}{1!} + \frac{x^2}{2!} + \frac{x^3}{3!} + \frac{x^4}{4!}$$

is a pretty good approximation to e^x for small values of x. For values $-1 < x < 2$ the relative error is at most about 5%. This is shown on the diagram. The two graphs are very close together when x is between -1 and 2. Outside that range, the close agreement rapidly disappears.

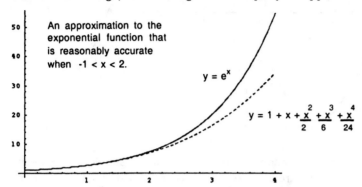

Often one is concerned with the values of a function, $f(x)$, over some limited range of x-values. In this situation it is often possible to give some simple formula for a function which is a good approximation to $f(x)$ over this range. The simplest kind of approximation is called a **linear approximation**. This means an approximation to $f(x)$ using a function like $g(x) = 3x + 7$. Such a function should really be called a **linear plus constant** approximation, or even **affine approximation**. But we, along with common usage, will call it a linear approximation.

As an example, the function $g(x) = 1 + x$ is a linear approximation to e^x which produces a relative error of less than 5% over the range $-0.3 < x < 0.3$. A linear approximation to the function $f(x) = x^2$ is given by $g(x) = 1 + 3(x - 1)$. This is accurate to within 12% over the range $1 \le x \le 2$. The graphs of these two functions are shown on the diagram.

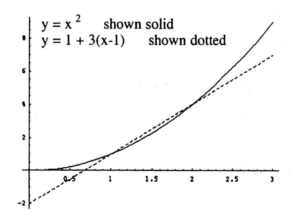

Another example is given by the function $sin(x)$

$$sin(x) \;=\; \sum_{n=1}^{\infty} (-1)^{n-1} \frac{x^n}{n!} \;=\; x - \frac{x^3}{3!} + \frac{x^5}{5!} - \frac{x^7}{7!} + \cdots .$$

This is approximated by the function

$$f(x) \;=\; x - \frac{x^3}{3!} + \frac{x^5}{5!} - \frac{x^7}{7!}$$

to an accuracy of about 0.001 over the range $-2 \le x \le 2$. This is shown on the diagram. The two graphs are so close together over this range that they appear as a single graph.

$y = \sin(x)$ shown solid

$$y = x - \frac{x^3}{3!} + \frac{x^5}{5!} - \frac{x^7}{7!}$$

shown dotted

The functions agree to
within 0.001 over the
range $-2 \le x \le 2$

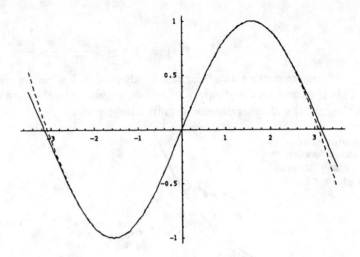

Sometimes when a difficult problem involves a complicated function, we make the problem simpler by replacing the complicated function by a simpler approximation. If we are able to solve this simplified problem, we hope that it's solution will be close to the solution of the original problem.

Here is an example. Solve the equation

$$e^x = 2 - x.$$

We will approximate e^x by using the first three terms of the power series:

$$e^x \approx 1 + x + \frac{x^2}{2}.$$

Then using this approximation we get

$$1 + x + \frac{x^2}{2} = 2 - x.$$

This gives a quadratic equation

$$\frac{x^2}{2} + 2x - 1 = 0.$$

Solving this gives

$$x = \frac{-2 \pm \sqrt{4+2}}{1} = -2 \pm \sqrt{6} \approx -2 \pm 2.45 \approx 0.45, -4.45.$$

If we try $x = 0.45$ we find $e^{0.45} \approx 1.57$ and $2 - 0.45 = 1.55$. So the equation is nearly satisfied. But if we try -4.45 this does not nearly solve the equation.

The most common kind of approximation is a linear approximation, because it is also the simplest and so the easiest to work with. A common linear approximation is

$$sin(x) \approx x.$$

This works quite well provided x is small. The relative error is less than 1% over the range $-0.2 \le x \le 0.2$. The relative error is less than 4% over the range $-0.5 \le x \le 0.5$. In some problems x is very small, and this linear approximation works quite well.

In economics the **rule of** 72 says that if you invest money and earn interest at a rate of $r\%$ per year (which is compounded annually) then the number of years it takes your initial investment to double is approximately $72/r$. Thus at 7% is takes approximately $72/7 \approx 10$ years for the money to double. If the interest rate is 10% then it takes approximately $72/10 \approx 7$ years for the money to double. This rule can also be used for working out interest accumulating on an unpaid debt (like a credit card) or the effects of inflation. For example with inflation running at 3% it takes approximately $72/3 = 24$ years for money to devalue to half of what it is currently worth.

Problems

(12.6.1) Use a calculator to make a table showing the difference between the function $ln(1 + x)$ and $f(x) = x - x^2/2 + x^3/3$ for the x values from 0 to 1 spaced 0.2 apart. When is the error greatest. What is the largest percentage error?.

(12.6.2) Find a linear approximation to the function $f(x) = \sqrt{x}$ that agrees with this function at $x = 4$ and at $x = 9$. [hint: you are looking for the equation of a straight line] Use this approximating function to find approximate value for $\sqrt{4.5}$ and $\sqrt{5}$. What are the errors with these approximate values?.

(12.6.3) Find the equation of the tangent line to the graph $y = \sqrt{x}$ at $x = 4$. Sketch the graph and the tangent line using the x-values $3.5, 4, 4.5$. The tangent line provides a linear approximation to $\sqrt{x}$ near $x = 4$. Use this approximation to find approximate values for $\sqrt{4.5}$ and $\sqrt{5}$. Is this approximation better or worse than the one from the previous question. Explain your answer.

(12.6.4) The area of a square whose sides have length $3 + x$ is given by the function $A(x) = (3+x)^2$. Thus $A(0)$ is the area of a square with sides of length 3. The function $f(x) = 9 + 6x$ is a linear approximation to $A(x)$. What is the error in this approximation when $x = 0.1$. Explain what this has to do with the diagram shown.

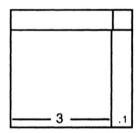

(12.6.5) What is the percentage error when the linear approximation for $sin(x)$ used above is applied at $x = 0.04, 0.1, 0.2$?

(12.6.6) What is a reasonable linear approximation to the function $f(x) = (3 + 2x)(x - 4)$ for the range of x values $-0.1 \le x \le 0.1$ [hint: how big is x^2 in this range, compared to x ?]

(12.6.7) Find a linear approximation to the function

$$f(x) = e^{x/500}$$

for the range of values $0 < x < 100$. Do this by making the linear approximation equal the function at the end points $x = 0$ and $x = 100$. Find the percentage error in the approximation when (a) $x = 25$ and (b) $x = 50$.

(12.6.8) Make a table showing for annual interest rates of $4\%, 8\%, 12\%, 16\%$ (a) how long it really takes an investment to double (b) how long the rule of 72 predicts. (c) The percentage error the rule of 72 makes for each of these interest rates.

Chapter 13

Differential Equations.

A **differential equation** is an equation that involves derivatives, for example

$$\frac{dy}{dt} = 3y.$$

(This particular equation is discussed in the section on Growth/decay equations) One solution of this differential equation is

$$y(t) = e^{3t}.$$

It is easy to check this is indeed a solution by substituting e^{3t} for y. The point is that

$$\frac{d}{dt}\left(e^{3t}\right) = 3e^{3t}.$$

In other words the derivative of the function e^{3t} is just three times the function e^{3t}.

To **solve** a differential equation means to find a **function** which satisfies the given equation. When you solve an ordinary equation involving x such as $3x + 5 = 2$ the solution is a **number** $x = 1$. Differential equations are much more complicated because the solution is a function, not a number. The quadratic equation $x^2 - 4 = 0$ has two solutions $x = 2$ and $x = -2$. A differential equation usually has infinitely many solutions. For example $y(t) = 7e^{3t}$ is another solution of the differential equation above, as you can check by plugging it in. In fact if C is any constant at all then $y(t) = C \cdot e^{3t}$ is a solution. Furthermore these are the **only** solutions of this particular differential equation.

Some differential equations can be solved easily, like the one above. Some can be solved by experts with a lot of work. But most can not be solved using basic calculus techniques. It is usually possible to use a computer to get good numerical approximations to a solution, and lots of computer software exists for this purpose. An **ordinary differential equation**, abbreviated to **ODE**, is an equation that involves (ordinary as opposed to partial) derivatives. The term **partial differential equation or PDE** means an equation involving partial derivatives. This book will not discuss PDE only ODE.

It is also often possible to understand the important features of a solution without actually having to solve the equation. For example if one thinks about the equation

$$x^3 + 3x = 10^6$$

then even without solving it in detail one realizes that x is probably going to have to be quite big. In fact plugging in $x = 100$ nearly gives the right answer: the left hand side becomes $10^6 + 300$, which is a bit too big. We notice that the x^3 term gives $100^3 = 10^6$ and $3x$ term $3x$ becomes 300 which is quite small compared to the x^3 term. By reducing x to 99 and plugging in we get $99^3 + 3(99) = 970596$ which is quite a lot too small. So we conclude that between 99 and 100 there is a solution to the equation, and in fact the solution is much closer to 100 than to 99. In a similar way one can often say important things about the solution of a differential equation without solving it precisely.

The **derivative** of a function $f(t)$ tells us **how quickly the function changes.** So a differential equation is an equation that expresses mathematically

$$\text{(how quickly a quantity changes)} = \text{(some other quantity)}.$$

The point is that very often this is the kind of thing we know about a situation, and so very often the mathematical model of the situation ends up being a differential equation.

We are going to explore various situations that lead to differential equations. We will develop some methods for predicting important features about the solution of some simple differential equations. We will look at simple numerical techniques for solving these equations, and finally we will see how sometimes seemingly minor changes in a differential equation can cause amazing changes in behavior of the solution.

13.1   Differential Equations, a brief discussion.

Imagine you are flying an airplane. You are going to fly along a route that follows a straight freeway along the ground, but you will go up and down, varying your height above the ground. You might be given some instructions on how to do this. Let us consider two imaginary flight plans. The first flight plan tells you what height the plane should be at each moment along the route. You start on the ground at height 0 and the instructions tell you that you will be at 300 feet after 1 minute, then 600 feet after 2 minutes, and so on. This information might be displayed on a graph showing your height above the ground against time.

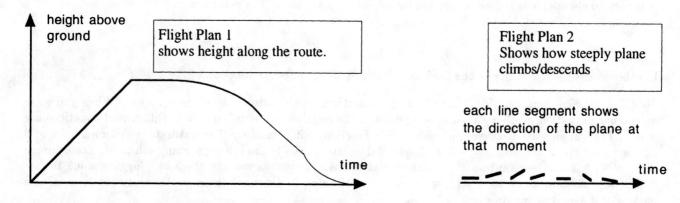

The second flight plan has instructions which say things like: after take-off you will climb at a rate of 300 feet per minute for the first 5 minutes of flight. Then you will level-off and fly at constant height for 3 minutes, then you will begin a gentle descent, gradually increasing the rate you are losing height until you are losing height at the rate of 300 feet per minute. When you are 2000 feet above ground you will begin to descend more slowly, so that when you land you are losing height at the rate of 100 feet per minute.

The second flight plan is easier to follow. It tells you what to **do** at each stage. The first flight plan tells you where you have to be, and when you have to be there, without telling you how to get there. The second flight plan contains information about **how quickly your height should change.** The first flight plan simply tells you what your height must be along the route. The second flight plan is like a differential equation because the instructions are about the **rate of change** of the planes height along the route.

Clyde Wild (of Griffith University, Queensland) wrote the following in answer to a question in *New Scientist* Nov. 1 1997.
The Argentinian moth *Cactoblastis cactorum* is quite specific to plants in the genus *Opuntia* (the "prickly pear" cacti), although it prefers some of these species to others and will hardly attack some at all. Several North American species of these cacti were an enormous problem in Queensland and New South Wales. When the moth was released, much of the cactus was destroyed and hoards of hungry caterpillars starved without suitable food. Later the cactus recovered, but the few remaining moths built up in numbers again, and once more

devastated the cactus, starting a long series of oscillations in which the cactus grows, the moths catch up, the cactus dies and the moth numbers fall. This cycle goes on to this day. Clearly, the cactus never dies out completely, nor do the moths. If they did the biological control would fail. What does *Cactoblasis* eat, then (when most of the the cacti have died) ? It eats prickly pear cactus which remains present, although in low density, throughout its former range.

> Problems

(13.1.1) What quantities are changing in the moth example above?. Make a sketch graph (using different colors) showing how those quantities might change over time. What derivatives are relevant?

(13.1.2) In the discussion of flight plan 2, what function and what derivative is important. What form does the ODE take?

13.2   Differential Equations and Constants of Integration.

The differential equation

$$y'(t) = 2t$$

can be solved by integration. The equation tells you that $y(t)$ is a function whose **derivative** is the function $2t$. This means that $y(t)$ is the **indefinite integral** of $2t$. Thus

$$y = \int 2t \, dt.$$

This integral is $t^2 + C$. The constant C might be anything. It disappears when you take the derivative. This is the **general solution**. Examples of **particular solutions** are $t^2 + 3$ and $t^2 - 7$ and t^2. It is a general feature of differential equations that there are **arbitrary constants** in them. Sometimes C is called a **constant of integration**. Often we have both a differential equation and an extra piece of information, an **initial condition** such as $y(1) = 3$. This can be used to determine the constant C. In this particular case we find C by using $y(1) = (1)^2 + C = 3$ from which we discover that $C = 2$.

The differential equation

$$y' = f(t)$$

can be solved in this way by integrating to get

$$y = \int f(t) \, dt.$$

Thus one can solve $y' = t^3 + 2$ this way, but not $y' = y + t$. The latter example **involves y on the right hand side**, and this complicates the problem. You can't just integrate $y + t$ because you don't know what y is, so you can't integrate it.

The above differential equation is called **first order** because it only involves taking **one** derivative. An example of a **second order** equation is

$$y'' = 4t + 2.$$

We can solve this by integrating **twice**. The Fundamental theorem of Calculus will now be used. Integrating once we get

$$y' = \int \frac{dy'}{dt} \, dt = \int (4t + 2) \, dt = 2t^2 + 2t + C.$$

Now integrating a second time gives

$$y = \int \frac{dy}{dt} \, dt = \int (2t^2 + 2t + C) \, dt = \frac{2}{3}t^3 + t^2 + Ct + K.$$

Here K is the constant that appears with this indefinite integral. The final answer has **two** arbitrary constants, C and K. One appeared each time we integrated. To check this really is a solution of the original equation, we differentiate **twice** to find y''. The first derivative gets rid of K. The second derivative gets rid of C leaving us with $4t + 2$.

A differential equation is **third order** if it involves a third derivative, but no higher derivative. It is n'th order if the most number of times a function is differentiated is n times. For example

$$\frac{d^4 y}{dt^4} + 3y\frac{dy}{dt} + y = sin(t)$$

is a fourth order equation. When one solves a differential equation of order n, there will almost always be n arbitrary constants. Even though we are not always able to solve the equation by integrating n times, we might think of the constants as arising this way. It is a fact that most of the important differential equations are first order or second order. Equations of higher order are comparatively rare.

> **Problems**

(13.2.1) What are the orders of the following differential equations.

(a) $y'' + (y')^2 = 3$ (b) $y''' - y' = 7$ (c) $\left(\frac{d^5 x}{dt^5} + 2\right)^2 = sin(t).$ (d) $7(y')^2 - 3y^3 = t^2$

(13.2.2) Find the general solution of the following equations by integrating

(a) $y' = 4t^3 + t$ (b) $y'' = 4t^3 + t$ (c) $\frac{d^3 y}{dt^3} = 4t^3 + t.$ (d) $2y' = 3t + 2$

(13.2.3) Find the solution of the differential equation $y' = (2t + 1)^2$ satisfying the initial condition $y(0) = 3$.

(13.2.4) Find the solution of the equation $y'' = 6t$ satisfying the initial conditions $y(0) = 2$ and $y'(0) = 3$.

(13.2.5) Find the general solution of the equation $y' = 2t$ and sketch on a single diagram, in different colors, the three particular solutions satisfying $y(0) = 0$ and $y(0) = 1$ and $y(0) = 2$.

(13.2.6) Find the general solution of the equation $y'' = e^{2t}$.

(13.2.7) The rate that a chemical reaction produces the molecule W after t seconds is $3t + 2$ gms/sec. Write down a differential equation involving the mass of W. Find the general solution of this equation. If there was 5 gms of W initially, how much is there after 10 seconds.

(13.2.8) The acceleration of a rocket fired vertically upwards t seconds after launch is $20 + 4t$ ms^{-2}. [as a rocket burns fuel, it becomes lighter, so accelerates more quickly.] Write down a second order differential equation for the height of the rocket. Find the general solution. Use the fact that at $t = 0$ the rocket was on the ground and not moving to find the particular solution that gives the height of the rocket. How high was the rocket after 10 seconds. How fast was it moving then? [hint: acceleration is rate of change of velocity. The velocity of the rocket is rate of change of what ?]

(13.2.9) The number of bees in a forest is growing at a rate of $200 + 10t$ bees per day, t days after being introduced into the forest. Express this as a differential equation. If initially 20000 bees we introduced, how many bees are there after 100 days?

(13.2.10) What is the general solution of $y''(t) = 12t - 2$. What is the particular solution for which $y(0) = 3$ and $y(1) = 5$

13.3   The Growth/Decay Equation.

The differential equation

$$y' = ky$$

is called **a growth equation** if $k > 0$ and a **decay equation** if $k < 0$. As an example the equation

$$y' = 3y$$

is a growth equation. The function $y = e^{3t}$ is a solution. To check this we just substitute into the equation. The left hand side becomes

$$y' = \frac{d}{dt}\left(e^{3t}\right) = 3e^{3t}.$$

The right hand side becomes $3y = 3\left(e^{3t}\right)$. These are equal so the equation is satisfied. Looking at the equation we see that it says that y is a function whose derivative is $3y$. We know that differentiating e^{3t} brings the 3 factor from the exponent out in front, resulting in 3 times the original function. This is not the only solution: $y = 7e^{3t}$ also works. In fact any constant y_0 instead of 7 gives a solution. Some of these solutions are shown in the diagram.

$$\boxed{y = y_0 e^{3t}}$$

is the **general solution** of this growth equation.   The constant y_0 is just the value of y when $t = 0$ since

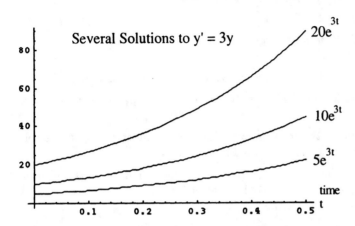

$y(0) = y_0 e^0 = y_0$. We say that $y(0)$ is the **initial value** of $y(t)$. A differential equation is often given with an additional piece of information: **initial conditions.** This means information about what is happening initially at time = 0. The solution of the differential equation has some constant in it (maybe more than one) and the initial conditions are used to determine what this constant is. The following example should help clarify this.

Example 13.3.1 *The mass of a colony of bacteria satisfies the growth equation*

$$m' = 0.3m.$$

Here time is measured in hours and mass in grams. If there was initially 2 grams of bacteria, how much is there after 4 hours?

Solution: Let $m(t)$ be the mass in grams of the colony after t hours. The mass satisfies the growth equation $m' = 0.3m$ and so $m(t) = m_0 e^{0.3t}$ where m_0 is a constant (the mass at time 0). To determine m_0 we use the fact that at time $t = 0$ the mass is 2 grams and plug in to get $m(0) = m_0 e^{0.3(0)} = 2$. This simplifies to $m_0 = 2$. So the mass after t hours is $2e^{0.3t}$ grams. After 4 hours the mass is $m(4) = 2e^{0.3(4)} = 2e^{1.2} \approx 6.64$ grams.

The growth equation in this example tells us how quickly the the colony is growing. But unless we know how big the colony is initially, we can not hope to know how big it is at some later time. What we can do is is find the solution of the differential equation which starts with an initial mass m_0. We can do this without knowing what m_0 is. If we are given additional information (initial conditions) we can then determine what m_0 is.

The diagram only shows three possible solutions of the infinitely many possible ones.   Each solution is represented by a curve starting on the y-axis. The initial value $y(0) = m_0$ is just the point on the y-axis from which the curve starts. The curves never cross. There is a single curve passing through each point in the plane (although if we attempt to draw all of them, we end up with a solid black picture!) There is a simple reason why there is a single curve passing through each point. A point (t, y) represents a colony of bacteria of mass y grams at time t. This colony will grow in a definite way (governed by the differential equation.) So there will be only one curve passing through that point representing the population of **that particular** colony as time progresses.

We can reason out the general features of the graph of a solution of the equation $y' = 3y$ without doing any calculus. The equation tells us that the slope of a solution curve is 3 times the y coordinate. So the further such a curve is above the x-axis, the larger y is, and so the bigger the slope of the curve. So not only does the curve slope upwards, but the slope gets bigger and bigger as you move to the right along the curve. In other words the curve is concave up.

If we do similar reasoning for the decay equation $y' = -2y$ then when the y coordinate is positive, the slope of the solution curve is negative. So this time, a solution curve which starts out above the x-axis, will decrease as you move to the right. But as y gets closer to zero, the slope, which is $-2y$ also gets closer to zero, so the curve slopes less and less. It must be asymptotic to the x-axis.

> [**Problems**]

(13.3.1) On a single diagram, sketch and label the curves $y = 0.1e^x, y = e^x, y = 2e^x, y = 4e^x$ and also the curves $y = -0.1e^x, y = -e^x, y = -2e^x, y = -4e^x$ for $-1 \leq x \leq 1$.

(13.3.2) On a single diagram, sketch the curves $y = 3e^{-x}, y = 2e^{-x}, y = e^{-x}, y = 0.5e^{-x}$ and also $y = -3e^{-x}, y = -2e^{-x}, y = -e^{-x}, y = -0.5e^{-x}$ for $-1 \leq x \leq 1$.

(13.3.3) Find the solution of $y' = 4y$ which has initial condition $y(0) = 3$. What happens to y as $t \to \infty$?

(13.3.4) Find the solution of $y' = -y/2$ which satisfies initial condition $y'(0) = 4$. What happens to y as $t \to \infty$?

(13.3.5) The function y satisfies a differential equation of the form $y' = ky$ for some number k. If you are told that when $t = 3$ that y is 2 and the rate of change of y is 3 then what is k? What is $y(6)$?

(13.3.6) A colony of bacteria is growing at a rate of 0.2 times its mass. Here time is measured in hours and mass in grams. The mass of the bacteria is growing at a rate of 4 grams per hour after 3 hours.
(a) Write down the differential equation the mass of bacteria satisfies.
(b) Find the general solution of this equation
(c) Which particular solution matches the additional information?
(d) What was the mass of bacteria at time $t = 0$.

(13.3.7) A certain isotope decays radioactively so that the mass satisfies the equation $m' = -0.04m$. Here the time is measured in days. Why is there a negative sign?. What is the half life? [hint: solve the equation. then find at what time your solution gives half the initial mass]

(13.3.8) The graph shows the mass of a certain isotope remaining after a time t. From the graph, determine the differential equation satisfied by this mass. [hint: you are looking for an equation $m' = -km$. To find k use that derivative equals slope]

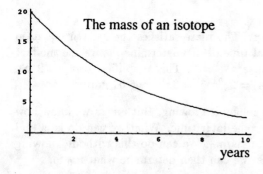

(13.3.9) The half-life of isotope W is 100 years. Use this information to determine the differential equation that describes the mass as a function of time. In other words $m' = km$ where k is a constant and $m(t)$ is the mass after t years. Use the information given to find k. Now solve this equation. Use the solution to determine how long it takes for an initial amount to decay to 15% of the original. This can also be determined [as in 34A] by solving the equation

$$\left(\frac{1}{2}\right)^{t/100} = 0.15.$$

Check that this gives the same answer.

(13.3.10) The graph shows an exponential function $f(t) = Ae^{kt}$. What are A and k ?

(13.3.11) The mass of a colony of bacteria is growing at a rate proportional to its mass.

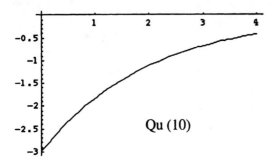

Qu (10)

(a) What is the differential equation which describes this?

(b) Initially the mass is 10 grams. Initially the mass is increasing at a rate of 5 grams per day. What is the mass after t days ?

(c) When will the mass be 60 grams? [you may leave ln(...) in your answer]

13.4 Long term behavior.

Often we want to know what will happen in a given situation over a long period. In studying an ecosystem one may wish to discover if a certain species will become extinct. If wind blows across a bridge in a certain way, the bridge might vibrate. If these vibrations become larger and larger, eventually the bridge will collapse (as happened on one famous occasion). These questions involve a function $f(t)$ of time t. We are asking what happens as t becomes very large, ie as $t \rightarrow \infty$. Here are some possible behaviors. They are illustrated in the diagram on the next page.

(1) $f(t)$ grows larger without bound

(2) $f(t)$ increases and approaches a limiting value

(3) $f(t)$ oscillates a bounded amount

(4) $f(t)$ oscillates more and more without bound.

(5) $f(t)$ oscillates less and less approaching a limiting value.

There are of course many other possibilities. We shall see that it is often possible to say what the long term behavior of a solution of a differential equation is, even without finding a formula for a solution.

Problems

(13.4.1) For each of the figures (a) to (j) shown, describe the long term behavior suggested by the graph. Describe any differences and similarities in the various behaviors.

(13.4.2) Produce a graph with a long term behavior different to any of those shown. Describe the difference.

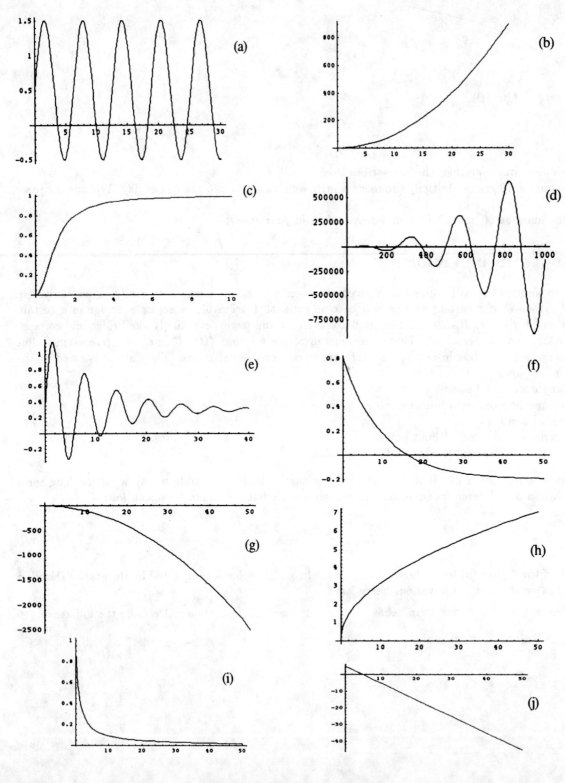

Some Long Term Behaviours

13.5   Graphing Solutions.

Imagine a map of the ocean showing which way the wind is blowing. At different places on the ocean, the wind blows in different directions and with different speeds. This is shown on the map by lots of arrows showing the direction of the wind at various places. Suppose you get into a sailing boat and do not steer, but just go whichever way the wind blows you. Your path across the ocean will probably not be a straight line, but weave this way and that way. Your boat is solving a differential equation.

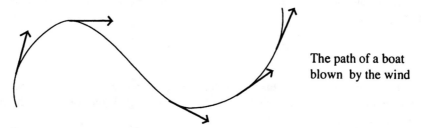

The path of a boat
blown by the wind

The wind changes the position of the boat. The rate of change of position depends on which way the wind blows, and how quickly. The path of the boat across the ocean is a curve which is always moving in the direction of the wind at each point along it. In other words, at each point along this curve the direction of the wind is tangent to the curve. This idea is the basis for sketching solutions of differential equations.

Often we do not need to know a precise formula for a solution of a differential equation, but would like to know some general properties of the solution. For example, does the solution increase or decrease? Does it become very large or limit on some value? Does it oscillate? How does the long term outcome change as the initial condition changes? These questions can be answered using a technique for sketching a graph of a solution of a differential equation of the form

$$y' = f(y, t).$$

There are two steps:
- Sketch the **slope field**
- Sketch a curve tangent to the slope field

The slope field for the equation

$$y' = y + t$$

is shown in the diagram, Fig 1. It consists of many short lines. Each line shows the **slope** of a solution curve to the equation at that point.

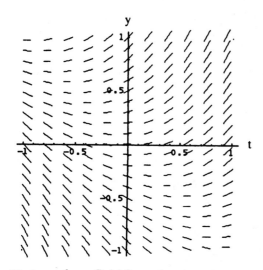

Fig 1   slope field for   y' = y+t

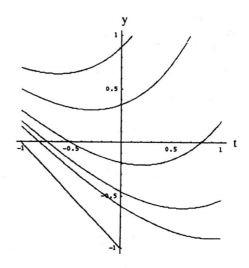

Fig 2   some solutions of y' = y+t

For example at the point $t = 0.5$, $y = 0.25$ then $y' = t + y = 0.75$ is the slope of a solution curve at that point. So the line segment at $(t, y) = (0.5, 0.25)$ has slope 0.75. The diagram is produced by taking a grid of points in the plane, and working out, for each point in the grid, the value of y' there, which is $y + t$ in this example. Once this is done, a short line of that slope is drawn through the grid point.

Once the slope field has been drawn, solution curves can be sketched. Some are shown in Fig 2. First one chooses a start point and then one draws a curve starting at that point so that the slope of the curve always agrees with the slope field at each point on the curve. This process is not very accurate. However quite a lot of information can be obtained this way.

Problems

(13.5.1) In this question use the two slope fields shown for this question. For each one, place your paper over the slope field. Trace the axes and draw seven solution curves, one passing through each of the points on the y axis where $y = -1, -.5, -.2, 0, .2, .5, 1$.

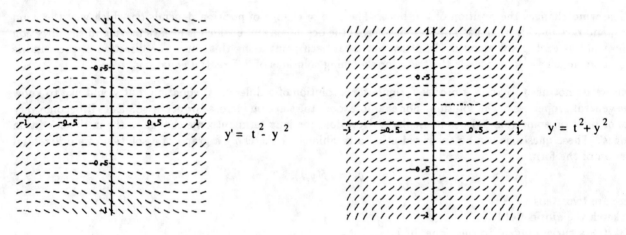

$y' = t^2 - y^2$ $y' = t^2 + y^2$

(13.5.2) In this question, you will determine the long term behavior of some solutions to the equation $y' = y + t$. Use the diagrams at the start of this section that describe the slope field for this equation. Use them to sketch the solutions with initial value $y(-0.5)$ each of $0.8, 0, -0.6$. Describe the long term behavior of these solutions.

(13.5.3) Sketch the slope field for the equation $y' = 2$ in the region $-1 \le t \le 1$ and $-1 \le y \le 1$. Use a grid spacing of 0.5 in each direction. Describe in words what the solutions to this equation are.

(13.5.4) Sketch the slope field for the equation $y' = y$ in the region $-1 \le t \le 1$ and $-1 \le y \le 1$. Use a grid spacing of 0.2 in each direction. Sketch the solution curve for which $y(-1) = 0.2$ What do you think is the long term behavior of this solution ?

(13.5.5) The diagram shows several solutions to an equation of the form $y' = f(t, y)$. Sketch the slope field for this equation.

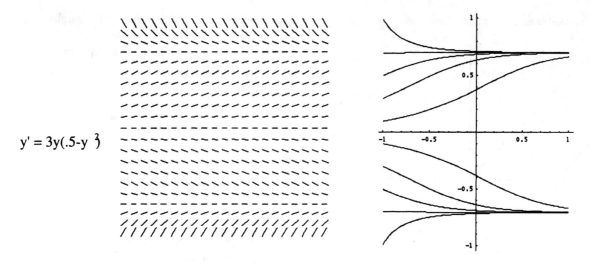

$$y' = 3y(.5-y^3)$$

(13.5.6) A slope field together with some solution curves are shown for the equation $y' = 3y(.5 - y^2)$.

(a) Say where the slope field has positive, zero, negative slope.

(b) Describe the solution with $y(0) = 1$.

(c) What is the long term behavior of the solution with $y(0) = 0.5$?

(d) Describe the long term behavior of the solution with $y(0) = y_0$ depending on what the value of y_0 is.

13.6   Isoclines

The process of sketching the slope field is straightforward, but it takes a long time. There is a way to sketch the slope field more quickly. The idea is to first figure out all the places where the slope field has slope 0.5 (for example). Here is an example $y' = -(t + y)/2$. The slope field has slope 0.5 exactly when $-(t + y)/2 = 0.5$. This is the equation of the line $y = -t - 1$. So at points on this line the slope field has slope 0.5. We sketch (very lightly) this line. Then draw line segments of slope 0.5 spaced along this line.

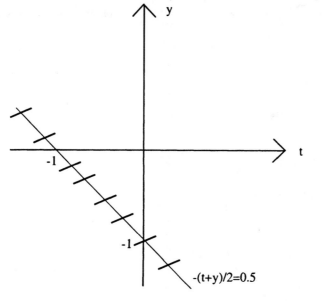

Having found the slope field has slope 0.5 along the line -(t+y)/2=0.5, sketch the slope field along this line.

An **isocline** is a curve on which all the points of the slope field have the same constant value (*iso* means "same" as in the word "*isotope*" and *cline* means "slope" as in the word in*clined*). For example $y = -t - 1$ is an isocline for the equation $y' = -(t + y)/2$. On this isocline the slope field consists of line segments all of slope 0.5. For each constant C there is an isocline along which the slope field has the constant value

C. One first sketches several of these isoclines. In this example we get a family of parallel straight lines.

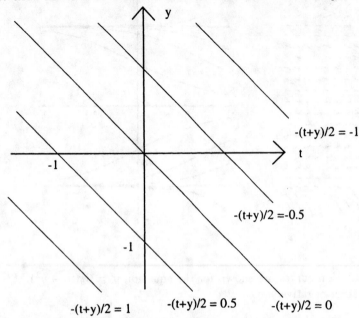

Some isoclines for the equation
y' = -(t+y)/2.

The slope field has slope C
when -(t+y)/2 = C.

This sketch shows the isoclines
for C = 1,.5,0,-.5,-1

In this example each isocline
is a straight line.

Then for each isocline, draw the slopes.

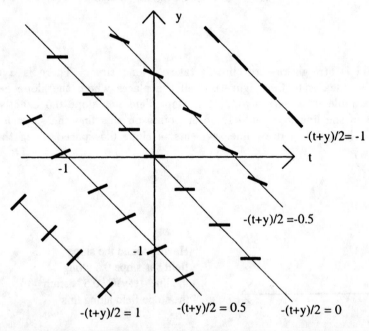

For each isocline
sketch some slopes
along it.

On the isocline -(t+y)/2=C
the slope field has
slope C.

Here is a summary of the method applied to the equation $y' = -(t + y)/2$.

(1) Sketch the isoclines $-(t + y)/2 = C$ for each of the values $C = -1, -.5, 0, .5, 1$

(2) For each isocline $-(t + y)/2 = C$ found in step (1), sketch short lines of slope C spaced along this isocline.

This method works very well when the differential equation has only y's on the right hand side, but not t's. For example the equation $y' = 2 - y$. For such an equation the isoclines are **horizontal lines**. For example the isocline along which the slope field has slope -1 is $2 - y = -1$ in other words the horizontal line $y = 3$. A differential equation is called **autonomous** if it is of the form

$$y' = f(y).$$

This means that the slope field does not depend on t but only on y. For example $y' = y - y^2$ is autonomous but $y' = y + t$ is not autonomous. (The word *autonomous* means "self governing." The idea is that since the rate of change of y depends only on y [and not on the whims of the **foreign** variable t] it's future is determined entirely by itself.)

> For an autonomous equation, once you have drawn one slope, every other point with the same y coordinate has the same slope.

This means you can draw parallel slopes at all points with the same y coordinate. You should look at the examples and decide which ones are autonomous and which are not. Look at the slope fields for the autonomous ones and see that as you move left or right, parallel to the t-axis, the slope does not change.

Problems

(13.6.1) (a) Write down the equations of the isoclines for the differential equation $y' = (y + t)/2$ for each of the slopes $C = -1, -.5, 0, .5, 1$.
(b) Sketch each of these isoclines in the region $-1 \le t \le 1$ and $-1 \le y \le 1$.
(c) Sketch the slope field in this region using the isoclines.
(d) Sketch the solution with $y(-1) = 0.5$.
(e) What do you think the long term behavior of this solution is ?

(13.6.2) (a) Write down the equations of the isoclines of the equation $y' = y - (t/2)$ for each of the slopes $C = -1, -.5, 0, .5, 1$.
(b) Sketch each of these isoclines in the region $-1 \le t \le 1$ and $-1 \le y \le 1$.
(c) Sketch the slope field in this region using the isoclines.
(d) Sketch the solution with $y(-1) = 0$.
(e) What do you think the long term behavior of this solution is ?

(13.6.3) (a) Write down the equations of the isoclines of the equation $y' = 1 - y^2$ for each of the slopes $C = -3, 0, 1$
(b) Sketch each of these isoclines in the region $-2 \le t \le 2$ and $-2 \le y \le 2$.
(c) Indicate on your diagram where the slope field has positive slope and where it has negative slope.
(d) Sketch the slope field in this region using the isoclines.
(e) Sketch three solutions with $y(-1) = 0, \quad y(0) = -1.1 \quad y(0) = 1.1$
(f) What do you think the long term behaviors of these solution are ?

(13.6.4) (* VERY long *) Plot graphs of the solutions to the following differential equations in the region $-2 \le t \le 2$ and $-2 \le y \le 2$. Use six isoclines and graph three different solutions.

$(a) \quad y' = 3y + 2t$	$(b) \quad y' = 1 - y$
$(c) \quad y' = 2y - t$	$(d) \quad y' = 1 + 2t$
$(e) \quad y' = y/(2 + t)$	$(f) \quad y' = (.5 + y)(.5 - y)$
$(g) \quad y' = t(1 - t)$	$(h) \quad y' = ty$

(13.6.5) Which of the equations in the previous question are autonomous?.

(13.6.6) In this question, you will determine the long term behavior of some solutions to the equation $y' = y(1 - y^2)$. On a diagram, indicate where the slope field has positive slope, zero slope and negative slope. Use this to sketch the solutions with initial value $y(0)$ each of $-2, -1, 0, 1, 2$. Use this to describe the long term behavior of the solution with initial value $y(0) = y_0$ in terms of the value y_0.

13.7 Exponential Decay towards a limiting value.

The equation

$$y' = k(M - y)$$

with $k > 0$ occurs frequently. The right hand side involves $M - y$. If $y < M$ then this is positive, but if $y > M$ then it is negative. This means that if y starts out less than M then y increases. However as y gets closer to M the number $M - y$ becomes smaller, and so y increases more slowly. In fact y increases more and more slowly and never quite reaches M. The diagram shows the slope field for this equation $y' = 0.5 - y$ together with some solutions. Notice that the solutions approach 0.5 at t increases.

This equation expresses the idea that

> (the rate of increase of y) is proportional to (how much smaller y is than some limiting value M.)

The constant k is the constant of proportionality.   There is a simple algebraic trick for solving this equation.

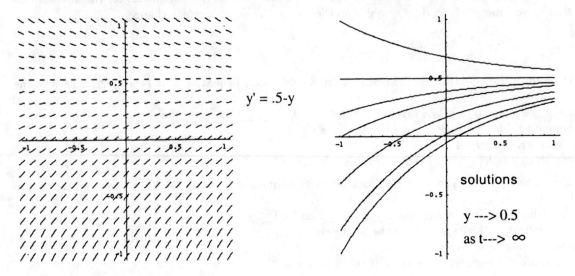

Introduce a new variable $u = y - M$. Then $u' = y'$ because M is constant and so has zero derivative. So the equation can be written in terms of u as

$$u' = -ku.$$

This is a decay equation for the quantity u. We know that the general solution is

$$u(t) = Ae^{-kt}.$$

We now wish to get a formula for $y(t)$ from this. Since $y = M + u$ this means that the general solution is therefore

> $y(t) = M + Ae^{-kt}$ A is an arbitrary constant

As t gets larger, $e^{-kt} \to 0$ and so y gets closer and closer to M. In other words,

> the limiting value of $y(t)$ as $t \to \infty$ is M.

When an object falls, the force of gravity causes it to speed up. However as it falls faster, it has to push more air out of the way more quickly. This is **air resistance.** The faster the object moves through the air, the greater is the air resistance. In reality the relationship between how fast the object moves and how strong the air resistance is can be rather complicated. Sometimes one assumes as an approximation that the air resistance is proportional to the speed. Using **down as the positive direction**, the acceleration due to gravity is a positive constant which we will call g. The force (weight) of a mass m is then mg. Air resistance is $-B \cdot v$ where v is the velocity downwards and B is a positive constant that depends on (among other things) how large the object is. The negative sign is because this force is **upwards**. Thus the total **downwards** force on the object is $mg - B \cdot v$. Since force equals mass times acceleration (Newton's second law) we get

$$mv' = mg - B \cdot v$$

Divide by m to get

$$v' = g - C \cdot v$$

where $C = B/m$. This equation can be re-arranged to be in the form we are looking at:

$$v' = C\left(\frac{g}{C} - v\right).$$

From this we see that the limiting velocity is g/C. This is to be expected for the following reason. The limiting velocity is when the force of gravity is balanced out by the air resistance. This happens when $mg - B \cdot v = 0$. Solving this gives $v = mg/B$ which equals g/C (since $C = B/m$.)

Newton's law of cooling states that the rate that an object gains or loses heat to the surroundings is proportional to the difference in temperature between the object and the surroundings. Let $f(t)$ be the temperature of the object at time t. Let M be the temperature of the surroundings, which we will assume is constant. The difference between the temperature of the object and that of the surroundings is then $M - f(t)$. Then Newton's law of cooling says $f'(t)$ is proportional to $M - f(t)$. Thus

$$f'(t) = k(M - f(t)).$$

The constant k is **positive.** This is because if the object is **colder** than the surroundings then $f(t) < M$ and so $M - f(t) > 0$. We know the object will warm up as it absorbs heat from the surroundings. So $f'(t) > 0$. This means that the constant k **must** be positive.

Problems

(13.7.1) What is the general solution of the equation $y' = 5(1 - y)$. What is the long term behavior of solutions to this equation? Do this graphically.

(13.7.2) Put the following equations into the form used in this section. (a) $y' = 7 - 4y$ (b) $y' = 3(1 - y/2)$.

(13.7.3) Find the solution of the equation $h' = 0.3(2 - h)$ with initial condition $h(0) = 1$. Sketch the slope field. Use this to sketch the solution you just found algebraically.

(13.7.4) If $y' = 3(6 - y)$ and $y(1) = 10$ what is $y(0)$?

(13.7.5) Find the general solution of the equation $f'(t) = 7 - 3f(t)$. [hint: first do some algebra to put the equation into the form used in this section]

(13.7.6) Describe the long term behavior of the temperature of a glass of water in terms of the initial temperature of the water and of the surroundings. Illustrate your answer with a sketch showing some solutions where the temperature of the water is (a) hotter (b) equal to and (c) lower than the surroundings.

(13.7.7) Solve the equation $y' = 15 - 2y$ with initial condition $y'(0) = 0$.

(13.7.8) Solve the equation $x' = 3(100 - x)$ with the initial condition $x(0) = 90$. Sketch this solution. For what value of t is $x(t) = 50$.

(13.7.9) The terminal velocity of a person falling through air is about $200\ Km/hr$. The acceleration due to gravity is $10\ ms^{-2}$. Use this information to find what C is in the equation above.
(a) How long does it take for a falling person to reach 90% of this terminal velocity ?
(b) How long to reach 95% ?
(c) How long to reach 99% ?

(13.7.10) A lake of water is at a temperature of $65°F$. The air temperature drops to $30°F$. Assume that Newton's law of cooling applies to the lake. If the water temperature drops $10°F$ during the first day, how long will it take till the temperature of the lake is $40°F$.

(13.7.11) A cup of coffee was made at a temperature of $90°C$ and cools according to Newton's law of cooling. The room temperature is $20°C$. If the temperature of the coffee 20 minutes after being made was $30°C$
(a) What was the temperature of the coffee 10 minutes after being made.
(b) When was the temperature of the coffee $80°C$.

(13.7.12) In the discussion of a falling object, what factor(s) other than how large the object is affects the constant B ? In the discussion of Newton's law of cooling, what factor(s) affect the constant k ?

(13.7.13) (hard algebra) The general solution of the equation in this section involves three constants M, A and k. Thus given three pieces of information, it should be possible to determine these constants. Suppose that an object is cooling in accordance with Newton's law. The temperatures at times $t = 0, 1, 2$ were $100, 80, 70$ respectively. Find the temperature as a function of time. What is the limiting temperature? What is the temperature at $t = 3$?

(13.7.14) The solution to this problem is used in the chapter **Pollution of the Marine environment.** Show that the equation

$$\frac{dL}{dt} = A - BL$$

has a solution

$$L(t) = L(0)e^{-Bt} + \frac{A}{B}[1 - e^{-Bt}].$$

Do this by substituting into the equation and checking that it works. Here $L(0)$ is an arbitrary constant. How does this fit in with the equation we studied in this section?

13.8 Logistic Equation.

The differential equation

$$y' = k \cdot y(1 - y/M)$$

is called the logistic equation. Here k, M are positive constants. This equation describes growth with limited resources. The idea is that there is a **maximum population** M which can live in the given environment. The population will grow to towards this limiting value, but never quite get there. The diagram show a solution of this equation.   For example if a colony of mold is growing on a piece of cheese in your refrigerator, then while

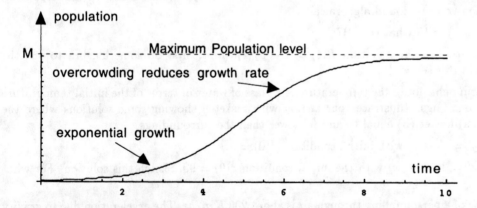

the colony is small, it can multiply as fast as it is able to reproduce. But when the cheese is nearly covered by mold, there are far fewer opportunities for young mold to go off and make the big time. There is crowding and a limited food supply. This slows down the growth. Initially, while the population is small compared to the amount of cheese, the population grows (almost) exponentially as would be described by the equation

$$y' = ky.$$

This can be seen mathematically since when y is small then $1 - y/M \approx 1$ and so the equation is nearly the same as $y' = ky$. But as y increases towards the limiting value of M (the maximum population the cheese can hold) the term $1 - y/M$ goes to zero, so that

$$y' = k \cdot y \times (\text{a very small number}).$$

Thus $y' \approx 0$ which simply means the rate of change of y is almost zero, or that y no longer increases very much.

The summary is

> Initial exponential growth decreases as y approaches a limiting value of M.

There is a formula giving the general solution of this equation

$$y = \frac{M \cdot e^{kt}}{A + e^{kt}} = \frac{M}{Ae^{-kt} + 1} \qquad A \text{ is an arbitrary constant.}$$

When t is very large $y(t) \approx M$. (This is because when t is large $e^{-kt} \approx 0$.) In other words as $t \to \infty$ we see that $y(t)$ approaches a limiting value of M.

There are distinct phases in this story. The first was unlimited growth, at a speed determined only by how much mold there was. As the amount increased, so also how quickly the amount grows increased. In fact

(rate of increase of mold) is proportional to(amount of mold.).

To put his mathematically, if $y(t)$ is the of mold at time t measured in minutes then

$$y'(t) = Ky(t).$$

Here K is a constant of proportionality. It depends on the type of mold. If the mold reproduces quickly then K is bigger than if it reproduces more slowly.

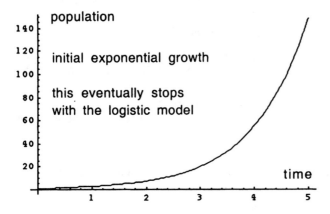

But as crowding sets in the rate of increase is not as big as before. There is competition for scarce food. There is overcrowding. The actual situation is very complicated. However this is sometimes modeled by a simple mathematical formula which seems to predict the right kind of answer. There is some maximum population of bacteria that can live on the dish. Suppose that this maximum number is M. Then how quickly the number of bacteria grows slows down when that number gets near M. Now a very simple function that becomes smaller the closer the population gets to M is $1 - (p(t))/M$. When the population is actually equal to M [that is when $p(t) = M$] then $1 - (p(t)/M) = 0$. If $p(t)$ is only a bit smaller than M then $1 - (p(t)/M)$ is nearly zero. If, on the other hand, $p(t)$ is small compared to M then $1 - (p(t)/M) \approx 1$. So we modify the growth equation $p' = Kp$ by writing

$$p'(t) = K\left(1 - \frac{p(t)}{M}\right)p(t).$$

When the population, $p(t)$, is much smaller than the maximum population, M, then this gives

$$p'(t) \approx Kp(t).$$

This is what we had before. So we would expect exponential growth of population as before. But as the population size grows towards the maximum size, M, the multiplying factor $1 - (p(t)/M)$ gets closer and closer to 0. This means that

$$p'(t) \approx (\text{very small number})p(t).$$

This in turn tells us that the rate of increase, $p'(t)$, of the number of bacteria is much smaller than before. So the closer the population gets to the maximum, the more slowly the population grows. This seems to fit with experiment. One might have imagined a different scenario. One where the population grew faster and faster until suddenly the population hits the maximum, and immediately stops growing. This sounds unlikely, and does not seem to happen in practice.

The logistic equation can be thought of as a hybrid between the growth equation $y' = ky$ and the exponential decay to a limiting value equation $y' = k(M - y)$. The solution of the logistic equation is also a hybrid, of the solution $y = e^{kt}$ of the first equation and the solution $y = M + Ae^{-kt}$ of the second equation.

Problems

(13.8.1) Use the formula for the general solution given in the text to sketch the solutions of the logistic equation $y' = y(1 - y/10)$ with initial condition $y(0) = 1$ and another with $y(2) = 1$.

(13.8.2) Sketch a slope field for the equation $y' = y(1 - y)$ and use this to sketch several solution curves. [hint: it is autonomous]

(13.8.3) Use the formula from the text to sketch, on the same diagram, the solutions to $y' = y(1 - y)$ and $y' = 0.3y(1 - y)$ both satisfying the initial condition $y(0) = 0.5$. Explain how the difference in the equations explains the difference in these two curves. [hint: which one rises more quickly? why?]

(13.8.4) On a single diagram, in different colors, sketch slope fields for $y' = 0.3y(1 - y)$ and $y' = y(1 - y)$. Does this help answer the second part of the previous problem?

(13.8.5) If $y' = ky(1 - (y/10))$ and $y'(0) = 2$ and $y(0) = 5$ what is k ? What is $y(1)$?

(13.8.6) The population of a certain planet is believed to be growing according to the logistic equation. The maximum population the planet can hold is 10^{10}. In year zero the population is 50% of this maximum, and the rate of increase of the population is 10^8 per year.
(a) Write down the logistic equation satisfied by the population [hint: previous question]
(b) How many years until the population reaches 90% of the maximum? [hint: use the general solution given in the text]
(c) How many years since the population was 10^9

(13.8.7) A colony of bacteria is growing in a petri dish with a maximum capacity of 100 mg. The mass of bacteria is increasing at a rate given by the logistic equation. Initially there was 2 mg of bacteria and the rate of increase was 1 mg per day.
(a) Write down the logistic equation satisfied by the mass
(b) When will the mass of bacteria be 50 mg ?
(c) What is the mass of bacteria 10 days after the mass was 2 mg ?

13.9   Numerical Solutions: Euler's method.

Consider the differential equation

$$\frac{dy}{dt} = \frac{1 - y}{10}.$$

Here t stands for time measured in seconds. What this tells us is that how quickly y increases depends on y itself. For example if at time $t = 5$ the value of y is $y(5) = 3$ then

$$y'(5) = (1 - 3)/10 = -0.2.$$

This means that at $t = 5$ that y is decreasing at the rate of 0.2 units per second. So if you want to know what the value of y at $t = 6$ then it is **approximately**

$$y(6) \approx 3 - (0.2).$$

This is simply because during the second between $t = 5$ and $t = 6$ the value of y decreases by 0.2 from it's value of 3 at $t = 5$. The reason this is only approximate and not quite exact is that as y changes so does

$$\frac{1 - y}{10}$$

which means that how quickly y changes **also changes**. Nevertheless, we can be certain that y does in fact decrease, and the decrease is about 0.2.

This is the basis of **Euler's method.** This is a simple numerical technique which will find numerical solutions to almost any differential equation. Euler's method was used to create the diagrams in this text. The idea is to do what we did above many times. There, we knew the value $y(5)$ of y when $t = 5$ and used the differential equation to calculate $y(6)$ (approximately). Now we know what $y(6)$ is we can repeat the method and find $y(7)$. Then $y(8)$ and so on. Since the method is only approximate, there are errors, and these errors become progressively worse the more steps we do. (When doing many steps the errors from each step can add up and become rather large). To get more accurate answers one uses smaller **time steps.**

Euler's Method. We have a differential equation $y' = f(y, t)$. We wish to find numerically the solution with initial value $y(0) = y_0$. We decide on a **time step** Δt. A small Δt will give more accurate answers, but involve more calculation. We first calculate $y(\Delta t)$. This is done using

$$y(\Delta t) \approx y(0) + \Delta t f(y_0, 0).$$

The reason this works is simply that we know how quickly y is changing when $t = 0$. This is just $y'(0)$. It is given by the differential equation and equals $f(y_0, 0)$. So during the short time interval Δt the change in y is approximately $\Delta t f(y_0, 0)$. However the change in y is just $y(\Delta t) - y(0)$. Thus

$$y(\Delta t) - y(0) \approx \Delta t f(y_0, 0).$$

That is it! We now know (approximately) the value $y(\Delta t)$ of y at time Δt. Next we calculate $y(2\Delta t)$. Then $y(3\Delta t)$. And so on. It is convenient to give the times at which we calculate y [namely $0, \Delta t, 2\Delta t, 3\Delta t \cdots$] names

$$t_0 = 0 \quad t_1 = \Delta t \quad t_2 = 2\Delta t \quad \cdots t_{17} = 17\Delta t \cdots$$

and similarly give the values of y which we calculate at these times names:

$$y_0 = y(t_0) \quad y_1 = y(t_1) \quad y_2 = y(t_2) \quad \cdots y_{17} = y(t_{17}) \cdots$$

Then Euler's method is given recursively by the formula

$$\boxed{\textbf{Euler's Method} \quad y_{n+1} = y_n + \Delta t f(y_n, t_n) \quad t_{n+1} = t_n + \Delta t.}$$

We start out knowing $t_0 = 0$ and $y_0 = y(0)$. Using these formulae, we can work out y_1 and t_1. Then y_2 and t_2. And so on. The sequence $y_0, y_1, y_2 \cdots$ are the approximate values of y at the times $t_0, t_1, t_2 \cdots$.

Example 13.9.1 *Use Euler's method with a time step of 0.2 to find the solution for the time period from 0 to 1 of $y' = y + t$ with initial condition $y(0) = 0.5$*

Solution: the time step $\Delta t = 0.2$. We are told $y_0 = 0.5$ It will take 5 steps to get $y(1)$. We now make a table as shown above. The rows are filled in one by one, starting with the first row. We find $y_5 \approx 1.73$.

step n	time t_n	$f(y_n, t_n) = y_n + t_n$	$y_{n+1} = y_n + \Delta t f(y_n, t_n)$.
0	$t_0 = 0$	$0.5 + 0 = 0.50$	$y_1 = 0.5 + (0.2)(0.5) = 0.6$
1	$t_1 = 0.2$	$0.6 + 0.2 = 0.80$	$y_2 = 0.6 + (0.2)(0.8) = 0.76$
2	$t_2 = 0.4$	$0.76 + 0.4 = 1.16$	$y_3 = 0.76 + (0.2)(1.16) = 0.99$
3	$t_3 = 0.6$	$0.99 + 0.6 = 1.59$	$y_4 = 0.99 + (0.2)(1.59) = 1.31$
4	$t_4 = 0.8$	$1.31 + 0.8 = 2.11$	$y_5 = 1.31 + (0.2)(2.11) = 1.73$

Table 13.1: Euler's Method for $y' = y + t$ with time step $\Delta t = 0.2$ and $y_0 = 0.5$

Problems

(13.9.1) Use Euler's method with $\Delta t = 0.1$ for the equation $y' = y$ with initial condition $y(0) = 1$. Complete the table shown.

t	0	0.1	0.2	0.3	0.4	0.5	0.6	0.7	0.8	0.9	1.0
Euler's Method	1.	1.1	1.21		1.46	1.61		1.95	2.14		2.59
Exact solution	1	1.11	1.22	1.35	1.49	1.65	1.82	2.01	2.23	2.46	2.72
Percentage Error	0	1	1		2	3		3	4		5

Table 13.2: Euler's Method for $y' = y$ compared to exact solution.

(13.9.2) Use two steps of Euler's method with $\Delta t = 0.2$ for the equation $y' = y - t$ with initial condition $y(0) = 1$.

(13.9.3) Use Euler's method to find $y(0.4)$ if $y' = 1 - y^2$ and $y(0) = 0.5$ Use a time step of 0.1.

(13.9.4) Find an approximate solution using 5 time steps over the interval $0 \leq t \leq 1$ for the equation $y' = -sin(y)$ with initial condition $y(0) = 0.1$. Compare this to the solution of $y' = -y$ which you should solve algebraically. We have seen that for small values of y that $sin(y) \approx y$. So we should expect these two equations to have nearly equal solutions.

(13.9.5) Describe how to modify Euler's method to deal with the initial condition of $y(5) = 3$ instead of $y(0) = 3$. Use this on the equation $y' = 1 + t$ using a time step of 0.1 and initial condition of $y(5) = 3$ and find $y(5.4)$. Now find the exact solution of this equation using calculus and find what the exact answer is. What is the percentage error in the value Euler's method gave at $t = 5.4$.

Chapter 14

Pollution of the marine environment. By Roger Nisbet.

Acute instances of environmental pollution[1], such as oil spills from tankers, make major news headlines, but these spectacular events are ultimately less important than the chronic exposure to low (sub-lethal) levels of contaminants that is experienced by organisms in almost any area impacted by human activity. One mechanism by which contaminants at sub-lethal levels have ecological impact is by affecting the rates at which living organisms assimilate and use energy from food. This lecture introduces some data on these effects, and shows how simple mathematical models may be used to predict the effects of contaminants on growth of marine animals.

14.1 A model of animal growth

The size of some animals is determined by the amount of food they eat. Animals convert food into body tissue, but lose weight through respiration. Development of a detailed model of animal growth would involve careful discussion of physiology, but a simple, yet remarkably powerful, model can be developed that describes growth in terms of a single differential equation.

Biologists use many measures of the size and weight of animals. In this lecture, we characterize an animal by its **carbon mass**, defined as the total mass of all the carbon atoms in the animal. Animals gain carbon from food and lose carbon (in the form of carbon dioxide) by excretion and respiration. If the symbol W represents the animal's carbon mass, I denotes the rate of input of carbon from food, E is the excretion rate, and R is the respiration rate, then

$$\frac{dW}{dt} = I - E - R.$$

Although carbon mass is a useful theoretical concept, it is not straightforward to measure; indeed measurement involves death of the organisms being studied. In its place, we often have measurements of the animal's **length**, which we represent by the symbol L. These two quantities are related if the animal retains a constant shape throughout its life, an adult being merely a scaled up version of a juvenile. Then the animal's volume is proportional to L^3. If furthermore, the animal's chemical composition remains approximately constant throughout life (no middle age spread), its carbon mass is proportional to its volume, and we can write

$$W = kL^3,$$

where k is a proportionality constant. We can differentiate this [using the product rule (or the chain rule)] and write

$$\frac{dW}{dt} = 3kL^2\frac{dL}{dt},$$

implying a relationship between the rate of change of length and the physiological rates I, E, and R, namely

$$\frac{dL}{dt} = \frac{1}{3kL^2}\left(I - E - R\right).$$

---

[1]From *Pollution of the Marine Environment* by Roger Nisbet. Reprinted by permission.

We now make a number of assumptions[2]:

- As feeding involves food crossing a surface, we assume that feeding rate is proportional to L^2, and write $I = \alpha L^2$, where α is a constant of proportionality.

- A fixed fraction, ε, of ingested food is excreted, implying $E = \varepsilon I$.

- The main contribution to respiration is "maintenance", i.e. the cost of keeping cells in working order. Maintenance rate (and hence respiration rate) is proportional to W, and hence to L^3. Thus we write $R = \mu L^3$, where μ is a constant of proportionality.

We can now substitute the expressions for I, E, and R into equation (14.1) to obtain

$$\frac{dL}{dt} = A - BL \quad \text{where} \quad A = \frac{\alpha(1-\varepsilon)}{3k} \quad \text{and} \quad B = \frac{\mu}{3k}.$$

You should check the steps for yourself.

You have learned how to solve a differential equation of this type. Suppose an organism is born at time $t = 0$. Then if we know its initial length, $L(0)$, its length at time (or age) t is (see (13.7) where this was discussed)

$$L(t) = L(0)exp(-Bt) + \frac{A}{B}\left[1 - exp(-Bt)\right].$$

Thus the complete pattern of growth is determined by three quantities A, B, and $L(0)$. If we assume that all individuals are born at approximately the same size, growth is controlled by the values taken by the parameters A and B.

Exercises

1. Sketch for yourself a plot of $L(t)$ against time. Show that growth of a new born individual requires $A/B > L(0)$. [Hint: for small values of t, $exp(-Bt) \approx 1 - Bt$.]

2. Show that the final size of the organism (as $t \to \infty$) is A/B [Hint: $exp(-Bt) \to 0$ as $t \to \infty$.].

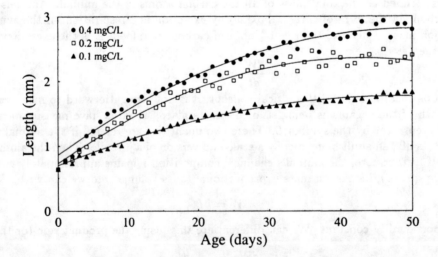

Figure 14.1: Growth curves for the waterflea *Daphnia pulex* at three concentrations of its algal food. Food density is measured as carbon mass (in milligrams) per unit volume (in Litres). Unpublished data of Dr. E. McCauley, University of Calgary.

---

[2]The biology behind these apparently simple assumptions involves many subtleties. An authoritative, though difficult, book with the details is S.A.L.M. Kooijman, "Dynamic Energy Budget Models in Biology". Cambridge University Press, 1993.

14.2 Feeding and growth

The rate of growth of many organisms is known to change in response to their environment. For humans, this effect is rather small, but for other organisms the effect is considerable. Fig. 14.1 shows growth of the waterflea *Daphnia* (a small freshwater animal) in laboratory environments where the experimenter controlled the availability of food. The figure shows that *Daphnia* grow faster, and to a larger size, when more food is available.

We can understand this from our simple model. It seems reasonable to expect that the parameter α, that relates feeding rate to size will be larger when there is more food available. Thus the parameter A is larger. If the other parameters do not change in response to food in the environment, then the final size A/B will increase. Also, our model predicts that the initial rate of growth will increase. To convince yourself of this, refer to the calculation you performed in exercise (1).

14.3 Effects of toxicants

Many toxicants cause a reduction in the rates at which organisms feed and an increase in respiration rate. Fig. 14.3 shows an example of this for the marine mussel *Mytilus edulis*. The changes in rates are well fitted by assuming that if the concentration of toxicant in the environment is c, then the values of the parameters α and μ become:

$$\alpha = \frac{\alpha_0}{1 + \frac{c}{K}} \quad \text{and} \quad \mu = \mu_0(1 + \frac{c}{K}),$$

where the symbols α_0 and μ_0 refer to the unpolluted environment and K is a new parameter. Try and create a "word definition" of the parameter K.

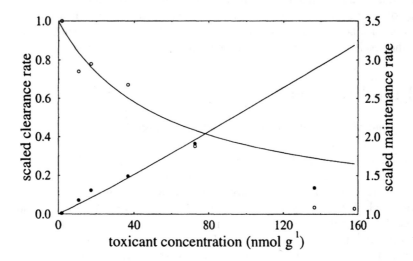

Figure 14.2: Variation in the assimilation and respiration rates of the mussel, *Mytilus edulis*, exposed to the organic toxicant polychlorophenol (PCP)(redrawn from data in a paper by J. Widdows and P. Donkin, *Comparative Biochemistry And Physiology C* **100**:69-75, 1991).

If toxicants affect the model parameters, they will also affect growth. This prediction was tested for a number of organisms/toxicant combinations by two UCSB ecologists, Drs. E.B.Muller and R.M. Nisbet. The continuous curves in Fig. 14.3 shows one sample set of results; for larvae of the mussel *Mytilus galloprovencialis*, growing in water contaminated by mercury. The curves certainly capture the decline in growth at higher levels of contamination, but the shape of the curves at these higher levels is wrong, with the data exhibiting a "flattening" of the growth curve not predicted by the model.

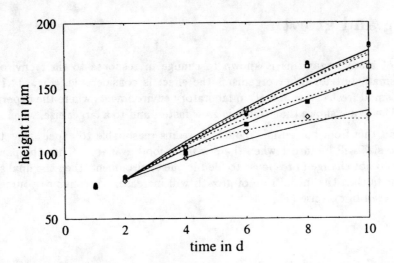

Figure 14.3: Growth of larvae of the mussel *Mytilus galloprovencialis* at 5 mercury concentrations (0, 5, 10, 20, and 40 nM). The solid lines are fits of the simple model that assumes constant toxicant concentration; the broken lines are fits of the more sophisticated model that allows for accumulation of toxicant. Data from R. Beiras and E. His, *Marine Ecology-Progress Series* **126**:185-189, 1995. Model fits by E.B. Muller.

Scientists are often as interested in deviations from the predictions of a mathematical model as in agreement between experiment and data (or at least, so they say when their models don't work). Certainly, in this case, the model failure at high levels of toxicant can be understood in terms of a mechanism *not* included in the model. In constructing the simple model, we ignored the fact that that physiological rates are influenced by the concentration of toxicant *within* the organism, and cavalierly identified the mercury concentrations in the environment with the variable *c*. This is sloppy, and a more careful analysis would require a representation of the rates of uptake and removal of toxicant by the organism. Evading this requirement, amounts to assuming that the internal toxicant is approximately in equilibrium with the ambient level in the environment. This is defensible if exchange of toxicant between organism and environment is rapid in comparison with growth, which is certainly not the case for a heavy metal such as mercury which accumulates within the organism. The broken curve in Fig. 14.3 is the result of fitting a modified model involving two differential equations, to the same data.

14.4   Model implications

Why should we care about such detailed analysis of simple experiments? You will of course form your own opinions, but here I highlight two.

First, the experiments and our model tell us that the sub-lethal effects of low levels of contaminants can be understood in terms of simple models describing inputs and outputs of carbon (or energy) within an organism. This is important, since many toxicants affect individual growth and reproduction, even when present at levels well below the limits at which they may be detected by chemical analysis. Some scientists therefore perform measurements that let them calculate directly the term $I - E - R$ in equation (14.1), a quantity known as *scope for growth*. [Why do you think it is given this name?] They then use scope for growth as an index of the amount of pollution in an area. The power of the insight obtained in this way is illustrated by a comprehensive study by a British biologist J. Widdows and several colleagues of pollution in the western sections of the North Sea. Mussels were collected from many sites and transported to Plymouth, where laboratory measurements were made of feeding and excretion rates (in standardized conditions), and respiration rates. The results, shown in Fig. 14.4, give powerful evidence of a south-north decrease in toxicity.

Second, toxicants reduce not only a animal's growth, but also *reproduction*, and hence may lead to changes in population size. These may be much more important ecologically than small changes in the size of individual organisms. For example, the number of eggs produced by a long-lived individual is often proportional to the square of its maximum size. According to our theory, a toxicant reduces feeding rate, with the observed feeding rate being a fraction x of the original value, where $x = 1/(1 + c/K)$. Respiration rate increases by a factor $1/x$. Thus the final length ($=A/B$) is reduced by a factor x^2, and the egg production rate of an organism at that final size is reduced to x^4 of its original value. Remember that for any number $x < 1$, x^2 and x^4 are both *less* than x. So, for example, we might not feel to concerned about contamination that reduces feeding rate by 10%, implying $x = 0.9$. However, the value of x^4 is then 0.65, i.e. the egg production is reduced by one third, which could lead to a considerable impact on population size.

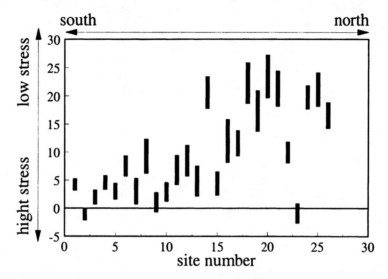

Figure 14.4: Variation in scope for growth (assimilation less respiration) for mussels collected from sites in the eastern United Kingdom. Redrawn from data in J. Widdows *et al.*, *Marine Ecology Progress Series* **127**: 131-148, 1995).

Chapter 15

Multi-Variable Calculus.

We have studied calculus involving functions of one variable. We now extend this to two or more variables. A circular cylinder with height h and base a circle of radius r has volume

$$V(r,h) = \pi r^2 h.$$

The volume depends on both r and h. We also express this by saying that V is a function of the two variables r and h. In general such a function is like a machine which takes **two** numbers as input and produces from them a single number as output.

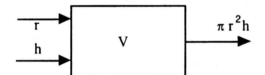

A function of 2 variables is a machine which takes two numbers input and produces from them a single number output.

Here are some more examples of functions of several variables. The pressure inside a car tire depends on the amount of air in the tire and also on the temperature of the tire. If x is the mass of air in the tire and T is the temperature then the pressure, p depends on both these quantities and is given by a formula

$$p(x,t) = K \times x \times t$$

where K is a constant that depends on how big the tire is. The volume V, of a rectangular box with width w, length L and height h is a function of 3 variables:

$$V(w,L,h) = w \times L \times h.$$

If you have money invested in two different stocks, the amount of interest you make depends on four things: the amount, x, you invest in fund number one and the interest rate, $A\%$, that the fund makes, and also the amount, y you invest in fund two and the interest rate, $B\%$, that the second fund makes. In one year the amount you make from fund one is $x \times A/100$ and from fund two is $y \times B/100$ so your income in one year, p, is a function of these four quantities:

$$p(x,y,A,B) = \frac{x \times A}{100} + \frac{y \times B}{100}.$$

Remember that for functions involving one variable, the derivative of the function tells us how fast the function increases. The rate of change. For example if $f(t)$ is the temperature (measured in degrees Celsius) at time t (measured in hours after midnight) the $f'(t)$ is how quickly the temperature is increasing in units of degrees Celsius per hour. Thus if you are told $f'(9) = 5$ you know that at 9a.m. the temperature was increasing at the rate of 5 degrees per hour.

When a function depends on more than one variable we still may want to know about how quickly it is changing. For example the area of a rectangle of width w and height h is

$$A(w, h) = w \times h.$$

How quickly does the area increase as the **width** is increased? How quickly does the area increase as the **height** is increased? These are **two** different rates of change. We have a function (area) that depends on **two** quantities (height, width) and we might wish to know how quickly the function changes as we change one, but not the other, of the two quantities involved. We will work out this example. Let us calculate how quickly the area of a rectangle increases as the width, w, increases. Look at the diagram.   To do this, we work out the area of the

	change in Area	rate of change of Area as x increases is
Area $A = hw$ h w	$\Delta A = h\Delta w$ Δw	$\dfrac{\Delta A}{\Delta w} = \dfrac{h\Delta w}{\Delta w} = h$

original rectangle of width w and height h. This is just $h \times w$. Next we increase the width by a small number Δw to be $w + \Delta w$. We calculate the area of the new bigger rectangle. It is $h \times (w + \Delta w)$. So the increase in area in going from the small rectangle to the big rectangle is

$$\Delta A = h \times (w + \Delta w) - h \times w = h \times \Delta w.$$

To find the **rate of increase** we divide the increase in area by the increase in w to get

$$\frac{\Delta A}{\Delta w} = \frac{h\Delta w}{\Delta w} = h.$$

So the rate of increase of area as width is increased, without changing height, is h. Let's plug in some numbers to make sure we really understand what this means. Suppose we start of with a rectangle of height $3cm$ and width $7cm$. Then we just worked out that the rate of change of area of this rectangle when you increase width is 3. Why should this be so?. If you make the rectangle a bit longer, by for example $0.1cm$, then the area increases by 3×0.1. This is because the height is 3 and so the extra area of the bigger rectangle is height times increase in width. Remember that the **units** of the rate of change, $f'(x)$, are

$$\frac{\text{units of f}}{\text{units of x}}.$$

In our situation, the units of area are cm^2 and the units of width are cm so the units of the rate of change of area as width changes are *centimeters squared per centimeter* written as cm^2/cm or even more simply as cm.

The mathematical notation for what we have just calculated is

$$\frac{\partial A}{\partial w}$$

and it is called a **partial derivative.** It is the partial derivative of area with respect to width. We discovered that

$$\frac{\partial}{\partial w}(h \times w) = h.$$

> The partial derivative $\dfrac{\partial f}{\partial x}$ is the rate of change of f as **x increases**.

We can also work out how quickly the area increases as the **height** of the rectangle increases. The answer is

$$\frac{\partial}{\partial h}(h \times w) = w.$$

There is one basic rule for partial derivatives that makes calculation very simple. Suppose that f is a function of two variables x and y. To find

$$\frac{\partial f}{\partial x}$$

Pretend y is a constant. Use the ordinary rules for taking derivatives using x.

For example

$$\frac{\partial}{\partial x}\left(4x^3 y^2 - 7xy^4 + 3\right) \;=\; 12x^2 y^2 - 7y^4.$$

The x's "got differentiated" the y's didn't. You just have to keep saying to yourself "when I work out the partial derivative with respect to x I pretend that y is constant." If you wish to calculate

$$\frac{\partial f}{\partial y}$$

then the roles of x and y are swapped. This time you pretend x is constant and differentiate as if y is the variable. Thus

$$\frac{\partial f}{\partial y}\left(4x^3 y^2 - 7xy^4 + 3\right) \;=\; 8x^3 y - 28xy^3.$$

This time the y's "got differentiated" and the x's didn't. Remember that

$$\frac{d}{dx}(e^x) \;=\; e^x.$$

From this you remember that

$$\frac{d}{dx}(e^{x+3}) \;=\; e^{x+3}$$

because $e^{x+3} \;=\; e^x e^3$ and e^3 is a **constant** so that the derivative of e^{x+3} is

$$\frac{d}{dx}(e^{x+3}) \;=\; \frac{d}{dx}\left(e^3 e^x\right) \;=\; e^3 \frac{d}{dx}\left(e^x\right) \;=\; e^3 e^x \;=\; e^{x+3}.$$

We can use this to find

$$\frac{\partial}{\partial x}\left(e^{x+2y}\right) \;=\; e^{x+2y} \qquad\qquad \frac{\partial}{\partial y}\left(e^{x+2y}\right) \;=\; 2e^{x+2y}.$$

The point is that you treat $2y$ like a constant when you work out $\partial/\partial x$ and do what was done before.

There is another notation for partial derivatives:

$$f_x \;=\; \frac{\partial f}{\partial x} \qquad f_y \;=\; \frac{\partial f}{\partial y}.$$

For example if $f(x,y) = 2x^3 + y^{-1}$ then $f_x(x,y) \;=\; 6x^2$ and $f_y(x,y) \;=\; -y^{-2}$. As in the single variable case, $f_x(2,3)$ is the partial derivative of f with respect to x at the point $x = 2$, $y = 3$. Thus

$$f_x(2,3) \;=\; 6(2^2) = 24 \qquad f_y(2,3) \;=\; -(3)^{-2} \;=\; -1/9.$$

Problems

(15.0.1) Calculate the following
(a) $\frac{\partial}{\partial x}\left(3x^2 y^4\right)$ (b) $\frac{\partial}{\partial y}\left(3x^2 y^4\right)$ (c) $\frac{\partial}{\partial x}\left(2x^2 y - 3xy^3 + 7x^4 y^5 + 3\right).$

(15.0.2) Find the partial derivatives with respect to x and y of
(a) $\frac{x}{y}$ (b) $3x + 7y + 2$ (c) $(2x - 3y)^2$ (d) $x^{-2}y^{-2}.$

(15.0.3) Find both partial derivatives of
(a) e^{2x-3y} (b) $sin(xy)$ (c) e^{2xy} (d) xe^y

(15.0.4) What is the rate of increase with respect to x of
(a) $e^y(x^2 + 2x + 1)$ (b) $e^x sin(3y)$ (c) $e^{-2y} sin(3x)$ (d) $x \cdot sin(2y + x)$.

(15.0.5) If $f(x, y) = 3x^2 + 4y^2 - 2$ find
(a) $f_x(1, 2)$ (b) $f_y(1, 2)$ (c)$f_x(0, 0)$.

(15.0.6) The pressure, $p(x, T)$, in a tire depends on the mass x of air in the tire and the temperature, T, of that air. What is the practical significance of $\frac{\partial p}{\partial x}$ and $\frac{\partial p}{\partial T}$.

(15.0.7) If \$$x$ is invested the amount of money after y years is $f(x, y)$. Explain the practical meaning of the two partial derivatives of f. In particular explain the practical meaning of $f_x(3000, 5) = 1.6$ and of $f_y(3000, 5) = 300$.

(15.0.8) A box has a square base of side length x meters and height h meters. The volume of the box is $V(h, x)$ and $S(h, x)$ is the combined surface area of all six sides of the box
(a) find both partial derivatives of V and of S.
(b) Find $S_x(2, 3)$.
(c) What are the units of your answer to (b) ?
(d) What is the practical significance of your answer to (b) ?

(15.0.9) The gravitational force of attraction between two masses m and M at a distance apart of R is

$$F = \frac{GmM}{R^2}.$$

Where G is the universal gravitational constant. Suppose that m is the mass of a rocket which decreases as it burn up its fuel, and that M is constant.
(a) What is the meaning of F_m and F_R. Calculate these partial derivatives.
(b) One of them is negative. What does this mean?

(15.0.10) A company sells cell-phones for \$$x$ each. It costs the company \$20 to manufacture one cell-phone. The number of cell-phones it sells is y.
(a) Express the profit $p(x, y)$ in terms of x and y.
(b) What do $\partial p/\partial x$ and $\partial p/\partial y$ mean in practical terms.

15.1   Planes

The simplest functions of one variable are ones whose graphs are straight lines: $f(x) = m \cdot x + b$. For two variables, the simplest functions are those with graphs which are flat **planes**:

$$g(x, y) = m \cdot x + n \cdot y + b$$

For example

$$z = 2x + 3y + 4$$

is the graph of a flat plane which meets the z-axis at the point $z = 4$ (just plug in $x = 0$ and $y = 0$). This plane is not horizontal, but tilted. To describe how a plane is tilted requires two numbers. To explain this, imagine taking a book. Hold it horizontally in front of you. Now raise the right side of the book a few inches until it slopes left-right with a slope of 2. Put the book horizontal again and then lift the side furthest from you until it has a slope of 3 in the near-far direction.

You can combine these two effects so the book has a slope of 2 in the left-right direction and a slope 3 in the near-far direction. Now it slopes in a diagonal way. In general, to describe how the book slopes takes two numbers.

Two ways to tilt a plane

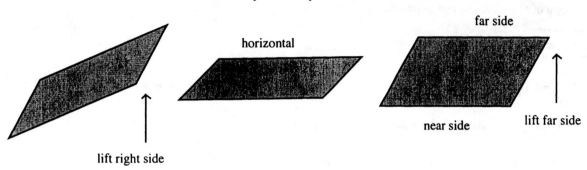

lift right side

One number is how much it slopes in the left-right direction (let us call this the x-axis) and the other number describes how much it slopes in the near-far direction (let us call this the y-axis.) Finally, raise the book, without changing how it slopes, 4 inches into the air. Congratulations! your book now has equation

$$z = 2x + 3y + 4.$$

It slopes upwards with a slope of 2 in the x-direction and slopes upwards with a slope of 3 in the y-direction. You can read this off from the equation in the same way you do for a line. The line $y = m \cdot x + c$ has slope m.

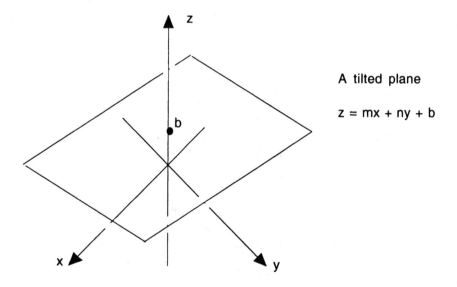

A tilted plane

z = mx + ny + b

Remember that this is because you can see from this equation that increasing x by 1 unit causes y to increase by m units, so the slope is indeed m. Well, if we look at the equation $z = 2x + 3y + 4$ we can see that increasing x by 1 unit causes z to increase by 2 units. Thus the coefficient of x tells us how quickly z increases as you move in the x-direction.

Remember that a **key property** of a straight line is that the slope is the same everywhere on the line. This distinguishes it from more complicated functions like x^2 where the slope varies depending on the x-value. The same holds true for flat planes. The slope in the x-direction of a flat plane (such as $z = 2x + 3y + 4$) does not change (it is always 2 in this example.) Similarly the slope in the y-direction does not change. This distinguishes flat planes from more complicated surfaces (like the bowl $z = x^2 + y^2$.)

Picturing planes described by equations is quite difficult. However the idea that the equation

$$g(x, y) = m \cdot x + n \cdot y + b$$

describes a plane is not so hard. Here b is the z-intercept. In other words it is simply the point on the z-axis where the plane cuts through. It is what you get by plugging $x = 0$ and $y = 0$ into the equation. The numbers

m and n describe how the plane slopes. The number m is the slope of the plane in x-direction. The number n is the slope of the plane in the y-direction.

> **Problems**

(15.1.1) Find the equation of the plane $z = mx + ny$ which contains the origin and the points $(1,0,2)$ and $(0,1,-4)$.

(15.1.2) Find the equation of the plane $z = mx + ny + b$ which contains the points $(0,0,2)$ and $(1,0,-3)$ and $(0,2,5)$.

(15.1.3) Write down the coordinates of three different points on the plane $z = 2x - 3y + 5$.

15.2 Partial Derivatives and Tangent Planes.

The derivative of a function $f(x)$ of a single variable x tells us the slope of the tangent line to the graph of $f(x)$. For example if $f'(5) = 1/2$ then the tangent line to the graph of $y = f(x)$ at the point on the graph with x-coordinate 5 has slope $1/2$. If we have a function $f(x,y)$ of two variables x and y, the partial derivative $\partial f/\partial x$ tells us something similar. To understand this we have to understand what the **graph** of a the function $g(x,y)$ is. It is NOT a curve drawn on a piece of paper. Instead it is a two-dimensional SURFACE in space. For example look at the graph shown of the function

$$g(x,y) = x^2 + y^2.$$

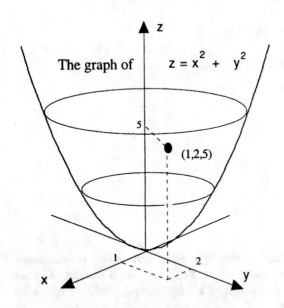

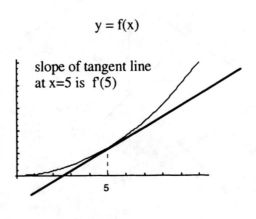

This is a bowl shaped surface. We must think about what a **graph** actually is. A graph shows the **values** a function takes for each possible input. For a function of one variable, the input is a single number, x, and so the graph shows a value for each x. The input for $g(x,y)$ is **two** numbers, x and y. So the graph must show a value $(x^2 + y^2)$ for each value of x and each value of y. So we imagine a horizontal plane with an x-axis and y-axis drawn on it. For every point in this xy-plane the function has a value, so we plot a point above (or below) the xy-plane showing this value. For example since $g(1,2) = 1^2 + 2^2 = 5$ we plot a point at height 5 above the xy-plane lying above the point with x-coordinate 1 and y-coordinate 2. The vertical direction has been indicated with a z-axis in the diagram. So we the diagram has a point on it with (x,y,z) coordinates $(1,2,5)$. This tells us that $g(1,2) = 5$.

The tangent line to the **curve** $y = f(x)$ at $x = 5$ has **slope** $f'(5)$. What is the **slope** of a surface supposed to be?. A surface has **different** slopes depending on which direction you move in. Imagine you are walking on a path which goes gently up the side of a steep mountain. The ground is a surface. If you look forwards along the path the ground slopes gently upwards. If you look sideways to the path, on one side the ground slopes steeply downwards and on the other side it slopes steeply upwards. The ground **has different slopes in different directions.** If we decide that the path is in the x-direction, then the partial derivative $\partial g/\partial x$ is how quickly the ground rises as we walk along the path. The y-axis points sideways from the path. The partial derivative $\partial g/\partial y$ is how quickly the ground rises when you move in that direction.

A partial derivative tells you the slope of a surface in one particular direction.

For example, if you look at the diagram you see the bottom of the bowl shape. It is a flat bottom, not pointed. It has slope zero in every direction at that point. The partial derivatives are

$$\frac{\partial g}{\partial x} = 2x \qquad \frac{\partial g}{\partial y} = 2y.$$

At the origin $x = 0 = y$ and so we see that at the origin both partial derivatives are indeed 0.

The simplest functions of one variable are $f(x) = m \cdot x + b$ whose graphs are lines. For two variables, the simplest functions are

$$g(x,y) = m \cdot x + n \cdot y + b$$

whose graphs are flat **planes.** For example

$$z = 2x + 3y + 4$$

is the graph of a flat plane which meets the z-axis at the point $z = 4$. The the coefficient of x tells us how quickly z increases as you move in the x-direction. You can also work this out by using partial derivatives, since

$$\frac{\partial}{\partial x}(2x + 3y + 4) = 2.$$

This again tells us that the surface slopes upwards with a slope of 2 in the x-direction.

Straight lines have a special role in calculus because they are so simple. In particular because they have constant slope. In a similar way flat planes have a special role in calculus of functions of 2 variables. A curve has a **tangent line.** A surface has a **tangent plane.** A tangent line to a curve has the same direction as the curve at a particular point. A tangent plane to a surface has the same slopes in **both x and y directions** as the surface has at a particular point. The derivative of $f(x)$ at $x = 5$ is $f'(5)$ and tells us the slope of the tangent line to the curve $y = f(x)$ at $x = 5$. The partial derivatives of $g(x,y)$ at a the point $(x,y) = (7,11)$ tells us the slopes (in the x-direction and y-direction) of the tangent plane to the surface at this point.

Problems

(15.2.1) The height above the xy plane, in feet, of a surface at a point y miles north and x miles east of the origin is given by the formula $h(x,y) = 20x + x^2 - 30y + y^2 + 50$.
(a) What is the height of the surface 4 miles north and 2 miles west of the origin.
(b) Does the surface rise or fall as you move north from that point? how rapidly? include units in your answer.

(15.2.2) A rectangle has width 5 cm and length 3 cm.
(a) What is the rate of increase of area as the width is increased?
(b) What is the rate of increase of area as the length is increased?
(c) Which dimension should be increased by 1 cm to cause the greatest increase in area.
(d) What is the rate of increase in the length of the perimeter as the width increases?

(15.2.3) A cylindrical can has height h and radius r both measured in cm. If the rate of increase of volume of the can with respect to height equals the rate of increase of volume with respect to radius, what is the relationship between height and radius?

(15.2.4) What is the equation of the tangent plane to $z = x^2 - y^2$ at the point $(x,y) = (2,1)$.

(15.2.5) What is the equation of the tangent plane to $z = 3x - 2y$ at $(x,y) = (4,3)$.

15.3   Tangent Plane Approximation.

The tangent line approximation for a function of one variable says

$$f(x + \Delta x) \approx f(x) + \Delta x f'(x).$$

For example if $f(x) = x^2$ then we can use this to find an approximate value for $(3.01)^2$ by using $x = 3$ and $\Delta x = 0.01$ Then

$$f(3.01) \approx f(3) + 0.01 \times f'(3) = 9 + 0.01 \times 6 = 9.06.$$

The actual value is 9.0601. This works for partial derivatives also

Tangent Plane Approximation $f(x + \Delta x, y + \Delta y) \approx f(x, y) + \Delta x \times f_x(x, y) + \Delta y \times f_y(x, y).$

This rather complicated formula can be thought of as saying:

$f(x + \Delta x, y + \Delta y) \approx f(x, y) + (\text{change due to } \Delta x) + (\text{change due to } \Delta y).$

For example if $f(x, y) = xy$ then we can find an approximate value for $f(3.01, 2.05)$ using $x = 3$, $y = 2$, $\Delta x = 0.01$ and $\Delta y = 0.05$.

$$f(3.01, 2.05) \approx f(3, 2) + 0.01 f_x(3, 2) + 0.05 f_y(3, 2).$$

Now $f_x(x, y) = y$ so $f_x(3, 2) = 2$. And $f_y(x, y) = x$ so $f_y(3, 2) = 3$. Also $f(2, 3) = 2 \times 3 = 6$. Thus

$$f(3.01, 2.05) \approx 6 + 0.01 \times 2 + 0.05 \times 3 = 6.17$$

The precise answer is 6.1705. This is useful when we do not have a formula for $f(x, y)$ but we do know the value of $f(x, y)$ at some particular point and also the rates of change at that point.

An important thing to realize about the tangent plane approximation is that the total change in f is **the sum of changes** due to the change in x and the change in y. To explain this further, if we set
$\Delta f = f(x + \Delta x, y + \Delta y) - f(x, y)$ then

$$\Delta f \approx \Delta x \times f_x(x, y) + \Delta y \times f_y(x, y).$$

There are two contributions to the change Δf in f. The first term

$$\Delta x \times f_x(x, y)$$

is just the change caused by changing x by Δx. The tangent line approximation for a function of the single variable x tells us that this causes a change in f of $\Delta x \times f_x(x, y)$. One just pretends that y is a constant while this change is made to x. In a similar way, the term $\Delta y \times f_y(x, y)$ is due to the change Δy. Finally the **combined effect** of these two changes, the one due to Δx and the one due to Δy is just the **sum** of the individual changes.

Problems

(15.3.1) Apply the tangent plane approximation to find $f(2.003, 1.04)$ where $f(x, y) = 3x^2 + y^2$.

(15.3.2) Apply the tangent plane approximation to find $h(4.001, 0.997)$ where $h(x, y) = x^3 + 2xy$.

(15.3.3) Suppose that $f(100, 50) = 230$ and $f_x(100, 50) = 2$ and $f_y(100, 50) = -3$.
(a) Approximately what is $f(101, 52)$?
(b) What value should y be in order that $f(101, y) = 230$?

(15.3.4) Use the tangent plane approximation to calculate approximately how much more area a rectangle that is 5.01 by 3.02 cm has than one which is 5 by 3. Draw a diagram showing the smaller rectangle inside the enlarged rectangle. On this diagram clearly indicate rectangles corresponding to the two terms in the tangent line approximation. What is the percentage error in using the tangent plane approximation to calculate the **increase in area.** In other words, if ΔA is the exact increase in area, and Q is the approximate increase what is the percentage error in using Q to approximate ΔA.

(15.3.5) A patient is to be given a mixture of two drugs Xeniccilin and Yasprin. Each of these drugs causes the body temperature of the patient to change. When x milligrams of Xeniccilin and y milligrams of Yasprin are given the temperature of the patient is $f(x,y)$ °F. Suppose that initially the patient is given 30 mg of Xeniccilin and 20 mg of Yasprin and that this combination does not affect the patient's body temperature.
(a) What is the practical significance of the facts that $f_x(30,20) = 0.3$ and $f_y(30,20) = -0.6$?
(b) If the dosage of Xeniccilin is increased by a small amount Δx milligrams, how much should the dosage of Yasprin be changed in order that the patient's temperature does not change ?

(15.3.6) The number of minutes taken for a chemical reaction is $f(t,x)$. It depends on the temperature t °C, and the quantity, x grams, of a catalyst present. When the temperature is 30°C and there are 5 grams of catalyst, the reaction takes 50 minutes. Increasing the temperature by 3°C reduces the time taken by 5 minutes. Increasing the amount of catalyst by 2 grams decreases the time taken by 3 minutes. Use this information to find the partial derivatives $f_x(30,5)$ and $f_t(30,5)$. Use the tangent plane approximation to find $f(33,4)$. Write an explanation of how you could have arrived at this answer using common sense instead of the tangent plane approximation.

(15.3.7) An internet advertiser gets revenue of $R(x,y)$ million dollars for x hours of advertising during one week which is watched by y million people.
(a) Last week $R(3,5) = 2$. What is the practical meaning of this ?
(b) $R_y(3,5) = 0.3$. What is the practical meaning of this ?
(c) $R_x(3,5) = 0.2$. What is the practical meaning of this ?
(d) Suppose the number of people watching this week will drop by one million. How many extra hours of advertising are necessary to achieve the same revenue as last week?

15.4 Max/Min with 2 variables.

At a **maximum** (or **minimum**) of the function $f(x)$ of a single variable, the derivative $f'(x) = 0$. See the diagram.

> At a maximum of a function $f(x,y)$ of two variables, **both** partial derivatives f_x and f_y are zero.

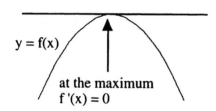

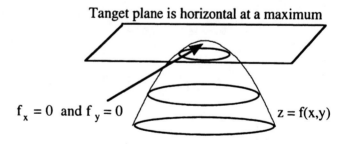

Tanget plane is horizontal at a maximum

This is because at a maximum, the tangent plane is horizontal. This means the slope of the tangent plane in both the x direction and y direction is zero. This in turn means the partial derivatives (which equal these slopes) are zero. This fact can be used to **find** local maxima and minima just as in the one variable case.

Example 15.4.1 *Find the minimum of $f(x,y) = 2x^2 + xy + y^2 + 2x - 3y + 4$.*

Solution: Differentiate to find
$$f_x = 4x + y + 2 \qquad f_y = x + 2y - 3.$$
At the maximum both $f_x = 0$ and $f_y = 0$. This gives
$$\begin{aligned} 4x + y &= -2 & (1) \\ x + 2y &= 3 & (2). \end{aligned}$$

From (2) we get
$$x = 3 - 2y.$$

Substitute into (1) gives

$$-2 \;\dot{=}\; 4(3 - 2y) + y \;=\; 12 - 7y.$$

This gives $y = 2$. Substitute $y = 2$ into (2) gives $x + 2(2) = 3$ so $x = -1$. So the minimum is at $(x, y) = (-1, 2)$. The minimum is $f(-1, 2) = 0$.

Problems

(15.4.1) Find the x and y values at the minimum of $f(x, y) = 2x^2 + xy + 2y^2 + 5y + 7$.

(15.4.2) Find the x and y values at the maximum of $-3x^2 + 2xy - 4y^2 + 2x + 1$.

(15.4.3) A rectangular box must have a volume of $2 \ m^3$. The material for the base and top costs \$5 per square meter. The material for the vertical sides costs \$20 per square meter.
(a) Express the total cost of the box in terms of the length and width of the base
(b) Find the dimensions of the box that costs least.

(15.4.4) The effect of applying x milligrams of drug A and y milligrams of drug B is measured by $f(x, y) = xy - 2x^2 - y^2 + 110x + 60y + 25$. What amounts of each drug should be used to get the biggest effect ?

Chapter 16

How to study Math and Science

From *How to Study Mathematics, Chemistry, Statistics, Physics* by Jason L. Frand. Copyright ©1979 by Jason L. Frand. Reprinted by permission.

16.1 Previewing

Before class briefly preview the text material that will be covered in the lecture.

1. Get an overview of the material by reading the introductory and summary passages, section headings and subheadings, and diagrams.
2. Look at the problems at the end of the chapter.
3. Make note of new terms and theorems.
4. Review (if necessary) old terms and definitions referred to in the new material.
5. Formulate possible questions for class.

Remember, the purpose of previewing is not to understand the material but to get a general idea of what the lecture will cover. This should not be a very time-consuming process.

16.2 Note-taking

When taking notes in class, listen actively; intend to learn from the lecture.

1. Write down the instructor's explanatory remarks about the problem.
- Note how one gets from one step of the problem to another.
- Note any particular conditions of the problem.
- Note why the approach to the problem is taken.
2. Try to anticipate the consequences of a theorem or the next step in a problem. During a proof, keep the conclusion in mind.
3. Note any concepts, rules, techniques, problems that the instructor emphasizes.
4. Question your instructor during class about any unclear concept or procedure.
5. If you miss something in the lecture or don't understand what's being presented, then write down what you can catch–especially key words. Be sure to skip several lines so you can fill in the missing material later.
6. As soon as possible after class, summarize, review, and edit your notes.
- Quickly read through your notes to get an overview of the material and to check for any errors or omissions.
- Fill in any information–especially explanatory remarks (see no. 1 above)–that you did not have time to write down or that the instructor did not provide.
- Use the margin or the back of the opposite page to summarize the material, list key terms or formulas, and rework examples. You can also use this space to take notes from the textbook.
- Note any relationship to previous material; i. e. , write down key similarities and differences between concepts in the new material and concepts in previously learned material.
7. Review your notes at regular intervals and review them with the intent to learn and retain.

16.3   Reading the text

If your class lectures provide a good overall structure of the course, you can use your text to clarify and supplement your lecture notes. In order to create a single study source, insert the notes you take from the text into your lecture notes themselves as well as in the margin or the back of the opposite page.

If your text provides the best overall structure of the material, then you can use your lecture notes as the supplementary source. In either case consider the following procedures:

1. Briefly preview the material. Get an overview of the content and look at the questions at the end of the chapter.
2. Read actively and read to understand thoroughly.
- Formulate questions before you read (from lecture notes or from previewing) and
- read to answer those questions.
- Know what every word and symbol means.
- Translate abstract formulas to verbal explanations.
- Analyze the example problems by asking yourself these questions:
What concepts, formulas, and rules were applied?
What methods were used to solve the problem? Why was this method used?
What was the first step?
Have any steps been combined?
What differences or similarities are there between the examples and homework problems?
Further analyze the example problems by using the following procedures:
Explain each step using your own words. Write these explanations on paper. Draw your own diagrams to illustrate and explain problems.
For practice, write down example problems from your book, close your book, and try to work the problems.
Check your work with the example to find what concepts, rules, or methods you are having trouble with.
Check to see how the material relates to previous material. Ask yourself these questions:
How was the material different from previous material? How was it the same?
What totally new concepts were introduced and how were they applied?
Where does this material "fit" within the overall structure of the course?
3. Stop periodically and recall the material that you have read.
4. Review prerequisite material, if necessary.

16.4   Problem Solving

Solving problems is usually the most important aspect of math or science courses. You must, therefore, spend much of your study time either working or studying problems. When working a problem, follow these steps:
1. Read through the problem at a moderate speed to get an overview of the problem.
2. Read through the problem again for the purpose of finding out what the problem is asking for (your unknown). Be able to state this in your own words.
3. If appropriate, draw a diagram and label the givens.
4. Read each phrase of the problem and write down (symbolically or otherwise) all information that is given.
5. Devise a tentative plan to solve the problem by using one or more of the following tactics:
- Form relationships among all facts given. (Write an equation that includes your unknown.)
- Think of every formula or definition that might be relevant to the problem.
- Work backwards; ask yourself, "What do I need to know in order to get the answer?"
- Relate the problem to a similar example from your textbook or notes.
- Solve a simpler case of the problem using extremely large or small numbers; then follow your example as if it is an example from the text.
- Break the problem into simpler problems. Work part of the problem and see if it relates to the whole.
- Guess an answer and then try to check it to see if it's correct. The method you use to check your answer may suggest a possible plan.
- If you are making no progress, take a break and return to the problem later.

6. Once you have a plan, carry it out. If it doesn't work, try another plan.

7. Check your solution.

- Check to see if the answer is in the proper form.
- Insert your answer back into the problem.
- Make sure your answer is "reasonable. "

During the problem solving process, it is often helpful to say out loud all of the things you are thinking. This verbalization process can help lead you to a solution.

16.5 Problem Analysis

After you have worked a problem, analyze it. This can help sharpen your understanding of the problem as well as aid you when working future problems.

1. Focus on the processes used (not the answer) and ask yourself these questions:

- What concept, formulas, and rules did I apply?
- What methods did I use?
- How did I begin?
- How does the solution compare with worked examples from the textbook or my notes?
- Can I do this problem another way? Can I simplify what I did?

2. Explain each step using your own words. Write these explanations on your paper.

16.6 Test preparation

If you have followed an approach to study as suggested in this chapter, your preparation for exams should not be overly difficult. Consider these procedures:

1. Quickly review your notes to determine what topics/problems have been emphasized.

2. Look over your notes and text. Make a concept list in which you list major concepts and formulas which will be covered.

3. Review and rework homework problems, noting why the procedures applied.

4. Note similarities and differences among problems. Do this for problems within the same chapter and for problems in different chapters.

5. Locate additional problems and use them to take a practice test. Test yourself under conditions that are as realistic as possible (e. g. , no notes, time restriction, random sequence of problems, etc.). Also try to predict test questions; make up your own problems and practice working them.

16.7 Test taking

1. Glance over the whole exam quickly, assessing questions as to their level of difficulty and point value. Also get a sense of how much time to spend on each question. Leave time at the end to check your work.

2. Begin to work the problems which seem easiest to you. Also give priority to those problems which are worth the most points.

3. Maximize partial credit possibilities by showing all your work.

4. If you have a lapse of memory on a certain problem, skip the problem and return to it later.

16.8 Test analysis

Analyzing returned tests can aid your studying for future tests. Ask yourself the following questions:

- Did most of the test come from the lecture, textbook, or homework?
- How were the problems different from those in my notes, text, and homework?
- Where was my greatest source of error (careless errors, lack of time, lack of understanding of material,

uncertainty of which method to choose, lack of prerequisite information, test anxiety, etc.)?
• How can I change my studying habits to adjust for the errors I am making?

IMPORTANT: The knowledge of most math/science courses is cumulative. Many concepts build on previous concepts, and a poor understanding of one concept will likely lead to a poor understanding of future concepts. Consequently, you should seek help early, if you encounter difficulty.

16.9   Problem solving tests

Preparing for Problem-Solving Tests:

1. Review notes and text – list the major concepts and formulas that have been covered.

2. Highlight those topics/problems that were emphasized. Note why they were emphasized.

3. The best way to prepare for problem-solving tests is to solve problems – lots of them. Work problems not previously assigned.

4. Analyze all problems you work:
• What concepts, formulas, and rules did I apply?
• What methods did I use?
• How did I begin?
• Have I seen this problem before?
• Is it similar or dissimilar to other problems I've done?
• How does my solution compare with the examples from the book and class?
• Could this problem be worked another way? Can I simplify what I did?
5. In your own words, next to each problem-solving step, explain what you did and why.

6. Look for fundamental problem types. Usually a course has approximately 5 fundamental groups of problems – make sure you can recognize what they are.

7. Try working problems out of sequence. For example, work a problem from Chapter 7, then one from Chapter 5, then one from Chapter 10, etc. This randomness will allow you to see how different problems relate to each other and will simulate taking tests.

8. Work with a time limit – aim to solve as many problems as you will have on the test within the test time limit (eg , 5 problems in 50 minutes).

9. Make up a practice test. You might cut/paste/xerox a test from your homework problems.

16.9.1   Taking Problem-Solving Tests

1. Before starting the test, turn it over and jot down all the formulas, relationships, definitions, etc. , that you are trying to keep current in memory.

2. Look the whole test over, skimming the questions and developing a general plan for your work. If any thoughts come to you immediately as you look at a problem, note these down in the margin.

3. Plan your time. Allow more time for high point value problems; reserve time at the end of the period to review your work and for emergencies.

4. Start with the easier problems, the ones for which you can specify a solution method quickly. This will reduce anxiety and facilitate clear thinking.
5. For the more difficult problems:

- Make absolutely sure that you understand the posed problem; mark key words, identify the givens and unknowns in your own words, sketch a diagram or picture of the problem, anticipate the form and characteristics of the solution (e. g. , it has to be an integer, the solution is an algebraic expression, etc.).
- Make a note, in symbols, diagrams, graphs or tables, of all the information given. For complex problems, list all the formulas you consider might be relevant to the solution; then decide which you will need to begin with.

6. If you still have no solution method:
- If possible, write out an equation to express the relationships among all the givens and unknowns, accounting for all the data and facts of the problem.
- Think back to similar practice problems to select a solution method.
- Work backwards: Ask yourself, "What do I need to know in order to get the answer?"
- Solve a simpler form of the problem if dealing with complex configurations OR substitute simple numbers for unknowns to reduce the amount of abstract thinking required.
- Break the problem into a series of smaller problems and work each part, thus building up to a solution.
- Guess an answer and check it. This process may suggest a solution method.

7. If all else fails, mark it to come back to later and work another problem. You may find clues in subsequent problems.

8. For all problems, easy and difficult: Once you have the solution method, follow it carefully. Check each step for consistency in notation. Document all your work so that it may be read easily. Write legibly. Evaluate your solutions. Check your answer against the original problem to make sure it fits.

9. Try all test problems. If your mind goes blank, relax for a moment and contemplate the problem OR mark it to come back to later.

10. If you run out of time and still have some problems left, try to gain at least partial credit by setting the problem up in a solution plan (even if you can't follow through on the calculations).

16.9.2 Analyzing Returned Problem-Solving Tests

1. Read the comments and suggestions.
2. Locate the source of the test: did the problems come from the lectures, textbook, or homework?
3. Note any transformations – how were the problems changed from those in the notes, text, and homework?
4. Determine the source of your errors.
- Were your errors due to carelessness? For example, did you fail to carry a negative sign from one step to another?
- Did you misread questions? For example, did you fail to account for all the given data in your solution method?
- Did you consistently miss the same kind of problem?
- Could you produce the formulas, or did you remember them incorrectly?
- Were you unable to finish the test because you ran out of time?
- Were you unable to solve problems because you had not practiced doing similar ones?

5. Did you have a difficult time during the test because you were too anxious to focus on the question?

16.10   Reading In The Sciences

What is different about reading in the sciences?

1. Readings contain a large number of facts and details (specifics). This kind of reading requires an overall understanding of the information presented in order to process details.

2. Placement of main ideas and details is usually straight forward (headings and subheadings), but grasping all of the details can be awesome. Information may be dense.

3. Organizational pattern is usually "relationships ", ie. , ideas and details building upon previous information. A solid background of the basics in the discipline is necessary to understand and comprehend the information.

4. Terminology is subject specific and must be understood to comprehend information presented. Often important terms are boldfaced or italicized. A review of terminology before reading increases understanding and comprehension.

5. Diagrams, figures, charts, and graphs are numerous. Time is needed to study these both before reading and as one reads as these visual aids help integrate information. Visualizations aid in the understanding and retention of information.

6. Knowledge and comprehension of the readings and lectures must be taken to an "application " level, ie. , the ideas learned should be applied to new or different situations other than those presented in the texts or lectures.

7. Research is an extension of information. In order to read research and understand it, the basics of the discipline must be understood. Think of research as taking an idea, analyzing, synthesizing and extending it.

8. Difficulty of material often necessitates more than one reading for thorough understanding of information.

16.10.1   How does one read the sciences?

BEFORE THE LECTURE

1. Preview the chapter. - Briefly look over titles, introductions, subheadings, first few sentences beneath subheadings, figures, diagrams, italicized or boldfaced words and terms, and summaries. As you preview, ask yourself:

- What is this about?
- What do I know about this . . . and don't know or don't remember?
- Where does the author begin and where is he going?
- What is the organizational pattern (relationships, chronological, topics?)
- How does this fit into what we are learning in this course?
- How difficult or how easy is this?
- Is there terminology that is unfamiliar or that I will need to review?
- How important is this information? Are there parts I could skim and get the main ideas?
- Where can I make logical breaks in the reading to divide up my study time?
- In what order might I read the information in the chapter? Would it be easier and more motivating to read the most interesting section first?

2. Skim the chapter - in more detail, but don't try to read it thoroughly yet. Read first and last sentences of paragraphs. Pull out some major ideas and details. Examine charts and figures. Try to understand the more important and frequently repeated terminology. Think about the over-all organization of ideas.

3. Don't panic or become overwhelmed with the readings. They may be dense, but not unconquerable. By previewing and skimming the materials before the lecture, you can then use the lecture to clarify the level and depth of comprehension you'll need to achieve when you actually read the chapter more thoroughly.

4. If the material is quite difficult and detailed, and if you have little recent background in the discipline, it might be useful to review the fundamentals and basic terminology in an introductory text in the field.

IN CLASS

1. Be prepared to anticipate information acquired from pre-skimming your textbook, and listen for clues during the lecture that will help you focus on an appropriate level of comprehension when you read the related chapter(s) after the lecture.

2. Take lecture notes on the right side of your notebook page, and leave the facing left page free to add related notes summarized from your textbook.

AFTER CLASS

1. Review and edit your notes taken from the lecture. Begin thinking about what additional information you'll need to add from the text.

2. Read the related textbook material that you have previously skimmed.

- Re-preview and and break the reading into logical sub-sections to be tackled one at a time.
- Plan far enough ahead of time that you'll be able to take a break and move away from the material at the end of each sub-section if you feel overwhelmed. Often time is needed to allow the mind to gradually absorb complex ideas.
- Read carefully and methodically, referring to figures and diagrams as appropriate.
- "Self-pacing" by moving a card or pencil finger along as you read may help keep your attention focused on the task.
- After reading a sub-section, stop and recall what you've read: tell it to yourself in your own words; take relevant notes along side the related lecture notes in your notebook , and/or make marginal notes in the textbook and highlight key details. (However, just highlighting in dense texts may not be the best form of recall since nearly all of many paragraphs may need to be highlighted and review would be difficult.)

- Draw your own diagrams or charts to summarize and translate information.
- Review your notes and the reading periodically. Information needs to be reviewed and used periodically for it to be stored in long term memory.

3. Reflect upon the information in various ways - e. g. , How are these functions related to each other? How do they affect each other? Apply ideas learned to other or new situations: What would happen to the body if one of these areas/functions/organs were damaged or destroyed? , etc.

4. Anticipate and practice responding to the kinds of test questions which might be asked.

16.11   Tactics for solving new problems

TACTIC	ACTIVITY	OUTCOME
clarifying	re-read the problem as if you were editing it: analyze givens, make unstated assumptions explicit, clarify goals	a re-statement of problems, its givens, assumptions, goals in your own words
visualizing	draw a figure and label givens, close your eyes to form a mental picture and then imagine what the experiment would look like if you set up the equipment	a figure, diagram, model which should help you see relationships between givens and unknowns
analogy	recall or use the text or notes to find a similar problem, method, result, useful theorem technique	a model to follow in solving your problem
subgoal	break problem into simpler problems; do only part of problem	partial solutions leading towards goal
algebraic	introduce variables for unknown; write equations, relations	a symbolic representation of the problem
brainstorming	think of every formula or definition related to the concept or terms	a list of formulas, conversion factors, or definitions to be used
questioning	assume you are going to ask the instructor for help: what would you ask? Identify what you need to know to solve the problem.	a list of questions whose answers lead to a solution
unit analysis	compare units in answer you want to compute with units in given information, look for conversion factors involving these units	a series of relationships involving units which can be multiplied or divided to get desired goal
identifying	identify concept behind problem, type of problem, section of book from which taken	once you know concept behind problem you can use brainstorming, analogy or other tactics
team	work with classmate, friend	discussion of ideas which can lead to broader understanding
trial-error	hit-miss attempts; try special cases	corrective feedback, better

		understanding of problem; may lead to induction
induction	try cases searching for a pattern	generalizations and insights about problem
work backwards	begin with answer if given, or approximate an answer, and try to figure out how it was obtained	the process for solving the problem
look back	check and verify your work; is solution reasonable	verification of solution
incubate	if making no progress after 30 minutes, stop working on problem, sleep on it, or leave it for a few hours	opportunity for insights and Ideas to develop
go for help	after trying all other tactics, ask for hints or explanations	obtaining insights and strategies required for solving the problem

Math 34A

entrance quiz

Print Name

Put answers in Boxes on this page

TaSDi

Score ⟋10

---

(1) Solve the equation for x

$$ax + 7 = 3x + 5$$

x=

(3) Solve for x

$$2/(2x-k) = 3/(4x+3)$$

x=

(5) A Lillith Fair concert sold 20000 tickets. 5000 of the tickets sold for $40 each. The rest of the tickets sold for $20 each. It cost $100000 to put on the concert. The rest of the money was shared equally between 5 charities. How much did each charity receive?

(7) 3 robots can dig a ditch 30 feet long in 2 days. How long would it take 4 robots to dig a ditch that is 100 feet long ?

(9) Car A left Santa Barbara at noon traveling along a route of 120 miles to Paso Robles. Car B left Paso Robles 1 hour after Car A left Santa Barbara and traveled at the same speed as Car A. They meet at 2pm. How fast do the cars travel?

mph

(2) Multipily out and simplify

$$(2+w)(w^2 -3w + 2)$$

=

(4) What is 3/4 of 80% of 2/3 give the answer as a fraction

(6) calculate

$$10 - \cfrac{6}{\left[\left(\frac{1}{3} - \frac{1}{4} \right)^{-1} \Big/ 2 \right]} =$$

(8) What is the area of the shaded region

area =

(10) solve for x and y

$$3x - y = 13$$
$$x + y = 7$$

x=

y=

---

1 point for correct answer 0 points if there is **any** mistake

(1) $x = 2/(3-a)$ (2) $w^3 - w^2 - 4w + 4$ (3) $-3 - 3K/2$ (4) 2/5 (5) $80000 (6) 3/2 (7) 5 days (8) 33 square units (9) 40mph (10) x=5 y=2

STAPLE   PRINT
Version   NAME   TaSDi

[Practice]

Math 34A Winter 99   Midterm 1   Quality   SCORE
Prof Cooper   Bonus   /2   /16
No Calculators

Put final answers in boxes on this page. When a question says **SHOW WORK** put **high quality** work in the blue book. Points are awarded for this **Number your solutions in the blue book** .
For the other questions, you are encouraged to put work in blue book, this may affect your quality bonus.
At the end of the exam STAPLE this page to the INSIDE front blue cover of the blue book, so that this side faces the white writing pages of the blue book. If you used a 3"x5" card, this MUST also be stapled here, behind this page.

Bring Photo ID + Blue Book + Stapler + ruler + sharp pencil.

(1)[/2] Solve for x the equation

$$\frac{3a+2x}{4-Kx} = 5$$ x=

(2)[/2] Multiply out and combine like terms

$$(a+b)(2a+b)(a-2b)$$

(3)[/2] Substitute   x=a+1
into

$$\frac{x^2-5x+2}{x+9}$$

and simplify as much as possible.

(4)[/2] Solve for x and y the equations

$$3x + 2y = 3a$$
$$x + y = 2a-b$$

x=   y=

(5)[/4] Car A leaves Sacramento at noon travelling at 60 mph on a road   560 miles long to Los Angeles. Car B leaves Los Angeles at   2 pm travelling at constant speed along the same road to Sacramento. They meet at   6 pm.   What was the speed of Car B ?
SHOW WORK   mph

(6)[/4] A farmer wants to make a field in the shape of a rectangle. It is divided in half by a fence into two smaller rectangles of equal area. The field is also surrounded by a fence. The area of the entire large field is to be 3000 square meters. Express the total length of all the fence required in terms of the length of the fence that divides the big field into two halves.
SHOW WORK

answer

fence	fence
fence	fence
fence	fence

You will not be allowed to leave early because this disrupts other students.

STAPLE
Version A

PRINT NAME

TaSDi

Math 34A   Fall 99   Midterm 1   Quality Bonus [/2]   SCORE [/16]
Prof Cooper
No Calculators

Put final answers in boxes on this page. When a question says **SHOW WORK** put **high quality** work in the blue book. Points are awarded for this **Number your solutions in the blue book** .
For the other questions, you are encouraged to put work in blue book, this may affect your quality bonus.
At the end of the exam STAPLE this page to the INSIDE front blue cover of the blue book, so that this side faces the white writing pages of the blue book. If you used a 3"x5" card, this MUST also be stapled here, behind this page.

(1)[/2] Solve for y the equation

$$y - \frac{y^2 + 3y}{y + 2} = n$$

y =

(2)[/2] Multiply out and simplify

$$(1+x+2y)(x - 2y) - (x+y)^2$$

(3)[/2] FIRST substitute u = k+1 into

$$(u - 1)^2 + 2u - 2$$

THEN substitute k = 2w into the result from the first step. Simplify the result as much as possible.
FINALLY write the result of this simplification here:

(4)[/2] Solve for x and y the equations

$$2x + k = ay$$
$$x - 2y = 0$$

[you may assume a ≠ 4]

x=

y=

(5)[/4] Car A and Car B travel along the same route. They start at the same time. They finish the journey at the same time. The total journey takes 8 hours. Car A travels at 60mph for the first 3 hours and at 40mph for the rest of the time. Car B travels twice as fast for the first 2 hours as it does for the remaining time. How fast does car B travel during the **first 2 hours** of the journey?

SHOW WORK

[you may leave the answer as a fraction if you want]

mph

(6)[/4] A box has vertical sides. The top and bottom of the box are squares. The vertical sides are rectangles. The volume of the box is 20 cubic meters. Express the **total surface area** of the box in terms of the **length of one side of the base** of the box.

SHOW WORK   draw a labelled diagram

area=

(1) y = -2n/(n+1) (2) x - 2y -2xy -5y² (3) 4w² + 4w (4) 2k/(a-4) (5) 76 mph (6) 2x² + 80/x

STAPLE
Version
A

PRINT
NAME

TaSDi

Math 34A   Winter 99   Midterm 1   Quality $\boxed{2}$ SCORE $\boxed{16}$
Prof Cooper   Bonus
No Calculators
Put final answers in boxes on this page. When a question says **SHOW WORK** put **high quality**
work in the blue book. Points are awarded for this. **Number your solutions in the blue book** .
For the other questions, you are encouraged to put work in blue book, this may affect your quality bonus.
At the end of the exam STAPLE this page to the INSIDE front blue cover of the blue book, so that
this side faces the white writing pages of the blue book. If you used a 3"x5" card, this MUST also be
stapled here, behind this page.

Bring Photo ID + Blue Book + Stapler + ruler + sharp pencil.

(1)[/2] Solve for x the equation

$$\frac{a}{3+x} - \frac{5}{k+2x} = 0 \qquad x =$$

(2)[/2] Multiply out and simplify

$$y^3\left(\frac{2}{y} - 3y^{-1}\right)^2 \qquad =$$

(3)[/2] Substitute $x = 1 + \sqrt{a}$
into
$$x^2 - 2x + 5$$

and simplify as much as possible.

(4)[/2] Solve for x and y the equations

$$ax + y = 3$$
$$x + y = c$$

x=   y=

(5)[/4] Car A and Car B leave Los Angeles at the
same time and travel along the same route. Both cars
arrive at the end of the route after 6 hours. Car A
goes at 40mph for the first 4 hours then at 60mph for
the rest of the time. Car B goes at 50mph for the first
2 hours and travels at a (different) constant speed for
the remaining time. What is the speed with which
car B travels the second part of the route?
SHOW WORK

[you may leave the answer
as a fraction if you want]   mph

(6)[/4] A farmer wants to make a field
in the shape of a rectangle using 1000 meters of
fence. Express the area of the field in terms of the
width of the field.

SHOW   WORK   draw a labelled diagram

area=

Math 34A W99 Midterm 1 Version $\boxed{A}$ Solutions   Prof Cooper

(1) $a / (3+x) = 5 / (k+2x)$
multiplying both sides by $(3+x)(k+2x)$:
     $a(k+2x) = 5(3+x)$
so $ka+2ax = 15 + 5x$
so $2ax-5x = 15 - ka$
factoring $(2a-5)x = 15-ka$

thus $\boxed{x = (15-ka)/(2a-5)}$

(2) $(2/y - 3y^{-1}) = (2y^{-1} - 3y^{-1}) = -y^{-1}$
thus

$y^3(2/y - 3y^{-1})^2 = y^3(-y^{-1})^2 = y^3 \, y^{-2} = \boxed{y}$

(3) $(1 + \sqrt{a})^2 \; -2 (1 + \sqrt{a}) + 5$

$= (1 + 2\sqrt{a} + a) -2(1 + \sqrt{a}) +5$

$= \boxed{a + 4}$

(4) $ax + y = 3$ (i)
       $x + y = c$ (ii)

subtract (ii) from (i) get $ax-x = 3-c$
thus $x(a-1) = 3-c$ thus $\boxed{x = (3-c) / (a-1)}$ **1 pt**

substitute into (ii) for x gives:

$(3-c) / (a-1) + y = c$

    thus $\boxed{y = c - (3-c) / (a-1)}$ **1 pt**

(5)
Plan: first find **length of route** .

Car A travels
(4hrs at 40mph)= 160 miles
then
(2hrs at 60mph) = 120 miles
so total length of route is
120+160=280 miles.   **2 pt**

Car B goes
(2 hrs at 50mph) = 100 miles.

This leaves car B 4 hrs to travel
(280-100) = 180 miles.   **1 pt**

Thus speed of car B is

$\boxed{180/4 = 45 \text{ mph.}}$ **1 pt**

(6)

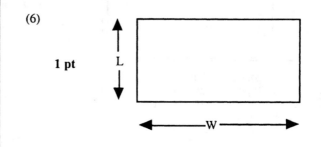

1 pt

W = width of field in meters
L = length of field in meters   **1 pt**

total length of fence = 2W + 2L = 1000.
Thus L = 500-W.   **1 pt**

Area = WL = $\boxed{W(500-W)}$ **1 pt**

STAPLE
Version
[Practice]

PRINT
NAME

TaSDi

Math 34A Winter 99   Midterm 2   Quality   /3   SCORE   /24
Prof Cooper   **Bonus**
No Calculators
Put final answers in boxes on this page. When a question says **SHOW WORK** put **high quality**
work in the blue book. Points are awarded for this. **Number your solutions in the blue book** .
For the other questions, you are encouraged to put work in blue book, this may affect your quality bonus.
At the end of the exam STAPLE this page to the INSIDE front blue cover of the blue book, so that
this side faces the white writing pages of the blue book. If you used a 3"x5" card, this MUST also be
stapled here, behind this page.

Bring Photo ID + Blue Book + Stapler + ruler + sharp pencil.

(1)[/6] Use the graph given to find

(a) Log(.047) =

(b) antilog(3.85) =

(c) Log($5600^{0.1}$)=

use a ruler and a sharp pencil to get an accurate answer
show construction lines

$y = 10^x$

(2)[/3] Solve the following equation.
Leave Logs in your answer.

$3^{4+7x} = 8$ x=

(3)[/3] The half life of a certain element
is 100 years. Initially there are 48 grams.
How much remains after 400 years ?

[you may leave logs
in your answer, but
logs are not necessary]   grams

(4)[/4] Line A goes through the points
(3,1) and (5,9). Line B goes through the
point (2,7) and has twice the slope of line A.
What is the equation of line B? **SHOW**
WORK. Draw a diagram. Give the answer
in the form
y = mx+b   y =

(5)[/4] Solution A contains 5% salt.
Solution B contains 20% salt.
How much solution A do you combine with
3 liters of solution B, to obtain a result
containing 10% salt? **SHOW WORK.**
check your answer

liters

(6)[/4]   A square has 7 times the area of a
circle of radius R. Express the length of the
perimeter of the square in terms of R.
SHOW WORK   draw labelled diagrams.

perimeter =

Bring:
Blue Book
Sharp pencil
Ruler
Stapler

(1)(a) -1.33 ± .01 (b) 7000∓200 (c) 0.375 ± .001 (2) ([log(8)/log(3)] - 4)/7 (3) 3 gms (4) y=8x-9
(5) 6 liters (6) $4\sqrt{7\pi}$ R

STAPLE
Version
A

PRINT
NAME

TaSDi

Math 34A Fall   99   Midterm 2   Quality
Prof Cooper   **Bonus** /2
No Calculators

SCORE /24

Put final answers in boxes on this page. When a question says **SHOW WORK** put **high quality** work in the blue book. Points are awarded for this. **Number your solutions in the blue book** . For the other questions, you are encouraged to put work in blue book, this may affect your quality bonus. At the end of the exam STAPLE this page to the INSIDE front blue cover of the blue book, so that this side faces the white writing pages of the blue book. If you used a 3"x5" card, this MUST also be stapled here, behind this page.

(1)[/6] Use the graph given to find

(a) antilog(-1.41) =

(b) Log(870)=

(c) Log($29^{.2}$)=

use a ruler and a sharp pencil to get an accurate answer
show construction lines

$y = 10^x$

(2)[/3] Solve the following equation.
Leave Logs in your answer.

$7^{3x} = 2^{x+4}$ x=

SHOW WORK

(3)[/3] At the start of an epidemic among lobsters the death rate is 300 lobsters per day. The death rate doubles every 3 days. What is the death rate 12 days after the start of the epidemic ? **SHOW WORK**
[you may leave logs in your answer but logs are not necessary]   lobsters/day

(4)[/4] Line A goes through the points (1,2) and (3,7). Line B has twice the slope of line A and hits the y-axis at y=3.

(a) What is the equation of line B? Give the answer in the form y = mx+b

SHOW WORK.
Draw diagrams.   y =

(b) What is the x coordinate
of the point where line A
meets the line x+y=2.   x =

(5)[/4] I have some cans of paint. **SHOW WORK**
Can A contains 10% blue and 90% red.

(a) how many liters of paint from can A should I mix with k liters of blue paint to get paint which is 60% blue   liters

(b) how many liters of paint from can A should I mix with 2 liters of blue paint to get paint which is w% blue   liters

(6)[/4] A rectangular field has a brick wall along one side and wooden fences along the other 3 sides. The brick wall costs $50 per meter. The fence costs $20 per meter. The area of the field is 800 square meters. Express the total cost of the brick wall plus fences in terms of the length of the brick wall.

cost=

fence
fence   fence
wall

SHOW WORK

(1) (a) .039±.002 (b) 2.94±.02 (c) .29±.01 (2) 4Log(2)/[-Log(2)+3Log(7)] (3) 4800 lobsters/day
(4)(a) y=5x+3 (b) x=-5/7 (5)(a) 4k/5 liters (b) (200-2w)/(w-10) (6) x=wall length; cost=70x+32000/x

STAPLE
Version
A

PRINT
NAME

TaSDi

Math 34A Winter 99   Midterm 2   Quality
Prof Cooper   **Bonus**   /2
No Calculators

SCORE   /24

Put final answers in boxes on this page. When a question says **SHOW WORK**   put **high quality**
work in the blue book. Points are awarded for this. **Number your solutions in the blue book**   .
For the other questions, you are encouraged to put work in blue book, this may affect your quality bonus.
At the end of the exam STAPLE this page to the INSIDE front blue cover of the blue book, so that
this side faces the white writing pages of the blue book. If you used a 3"x5" card, this MUST also be
stapled here, behind this page.

Bring Photo ID + Blue Book + Stapler + ruler + sharp pencil.

(1)[/6] Use the graph given to find

(a) Log(5900) =

(b) antilog(-2.25) =

(c) Log($\sqrt{240}$)=

use a ruler and a sharp pencil to get an accurate answer
show construction lines

$y = 10^x$

(2)[/3] Solve the following equation.
 Leave Logs in your answer.

$5^{2+3x} - 2^x = 0$ x=

SHOW WORK

(3)[/3] Bacteria double in mass every 10 hours.
Initially there are 30mg of bacteria. After how many
hours will there be 240mg.   **SHOW WORK**
[you may leave logs in
your answer but logs
are not necessary]

hours

(4)[/4] Line A goes through the point
(3,1) and has slope 2.
(a) What is the equation of line A? Give the
answer in the form y = mx+b

y =

(b) Line B is y = 3 + x. What are the coordinates
of the point where line A meets line B.

x =   y =

SHOW WORK.   Draw a diagram.

(5)[/4] I have two cans of paint.
Can A contains 10% blue and 90% yellow.
Can B contains 80% blue and 20% yellow.
(a) If I mix 2 liters from can A with x liters from can
B, what is the percentage of blue in the result.

%

(b) how much paint from can B should I mix with 2
liters from can A to get paint which is 60% blue

SHOW WORK   liters

SHOW WORK

(6)[/4] A rectangular field is surrounded by a
fence and divided in half by another fence. The total
length of all the fence used is 7000m. Express the
area of the entire field in terms of the length of the
dividing fence (shown in heavy ink)

area=

fence	fence
fence	fence

fence   fence   fence

(1) (a) Log(5900)

= Log(1000) + Log(5.9) $5.9 \to$

= 3 + 0.77 from graph

= $\boxed{3.77}$ allowed error $\pm$ .01   0.77

(b) antilog(-2.25)

= $10^{-2.25}$ $5.6 \leftarrow$

= $10^{-3} 10^{.75}$ .75

= .001 x 5.6 from graph

= $\boxed{.0056}$ allowed error $\pm$ .0001

(c) Log($\sqrt{240}$)

= Log($240^{0.5}$) $2.4 \to$

= (0.5)Log(240)   0.38

= (0.5)[Log(100)+Log(2.4)]

= (0.5)[2 + .38] from graph

= $\boxed{1.19}$ allowed error $\pm$ .01

2 pts per part.
-1 pt if error more than allowed
must show work and
construction lines for quality

5(a) Plan: work out how much blue paint is in (2 liters of A plus x liters of B). Then divide by total volume of paint to get % of blue in mixture.

(amount of blue in 2 liters of A plus x liters of B)

= (10% of 2) + (80% of x)   liters

= 0.2 + 0.8x   **1pt**

(total volume of paint) = 2+x   liters
so :
(percentage of blue paint in mixture)

= 100 (0.2 + 0.8x)/(2+x)

= $\boxed{(20+80x)/(2+x)}$ **1pt**

(b) To get 60% blue need

60 = (20+80x)/(2+x)   **1pt**

Solve for x:

60(2+x) = (20+80x)
120 + 60x = 20 + 80x
Thus 20x = 100 so $\boxed{x = 5}$ **1pt**

---

(2) $5^{2+3x} = 2^x$

Take logs of both sides

Log(5^{2+3x}) = Log (2^x)   **1pt**

use 4'th law Log(a^P) = pLog(a)

(2+3x)Log(5) = xLog(2)   **1pt**

Thus

2Log(5) = xLog(2) - 3xLog(5)
       = x[Log(2)-3Log(5)]

dividing:

$$x = \frac{2Log(5)}{Log(2)-3Log(5)}$$ **1pt**

(4)(a) use point-slope formula for line:
y - 1 = m(x-3)

m=slope=2
so equation is

y - 1 = 2(x-3) **1pt**
thus
$\boxed{y = 2x - 5}$

slope=2

1pt

(6)   **1pt**
x = length of center fence in m
y = width of **entire** field in m

total length of fence
= 7000
= 3x + 2y   (*) **1pt**

Area of field = xy
need to express y in terms of x. From (*) get

y = (7000-3x)/2 **1pt**

Thus Area = $\boxed{(7000-3x)(x/2)}$ **1pt**

---

(3) Initially have 30mg. This increases to 240mg.
Ratio = 240/30 = 8.

Thus end up with 8 times initial amount.

$8 = 2^3$
so it takes
3 doubling periods .

This is 3 x 10 = $\boxed{30 \text{ hours.}}$

partial credit at graders discretion

common mistake: leaving out parentheses here. Graders: be **harsh** on this !

(b) y = 2x -5 meets y = 3+x when **both** these equations are satisfied at the same time (since the x coordinate of the point where the lines meet is on **both** lines) Thus need to solve

2x - 5 = 3 + x   **1pt**
hence
x = 3+5 = 8
Now we know x, plug this into equation for either line ato find
y = 3+8 = 11

Thus $\boxed{x=8 \quad y = 11}$ **1pt**

Graders: if the student does not say **in English** what the variables are, deduct 1 point !

STAPLE   PRINT
Version   NAME   TaSDi
[Practice]

Math 34A   Winter 99   Midterm 3   Quality
Prof Cooper   Bonus   /3   SCORE   /32
No Calculators
Put final answers in boxes on this page. When a question says**SHOW WORK** put**high quality**
work in the blue book. Points are awarded for this**Number your solutions in the blue book**.
For the other questions, you are encouraged to put work in blue book, this may affect your quality bonus.
At the end of the exam STAPLE this page to the INSIDE front blue cover of the blue book, so that
this side faces the white writing pages of the blue book. If you used a 3"x5" card, his MUST also be
stapled here, behind this page.

Bring Photo ID + Blue Book + Stapler + ruler + sharp pencil.

(1)[/6] Use the graph given to find
(use a ruler and a sharp pencil to get an accurate answer
show construction lines)

(a) Log(1/460) =

(b) antilog(-4.6) =

(c) the derivative
    of 10^x
    at x=0.3 is

$$y = 10^x$$

(2)[/6] Find the derivatives of

(a) $2 / x^3$

(b) $(2x+1)(x+a)$
[a is constant]

(c) $5 e^{4x}$

(3)[/ 4] $f(x) = 2x^3 - 3x^2 - 12x + 1$

(a) Where is f(x) increasing?

(b) where is f(x) concave up?

(4) [/4] A water tank has a circular base
of radius 2 meters. The height of water in the
tank after t hours is $h(t) = 4 - t^2$

(a) when is the tank empty

(b) how quickly is water
leaving the tank after 1 hour m^3 /hr
SHOW WORK

(5) [/4] Find the equation of the
tangent line to the graph $y = x^2$ at x=3
Give the answer in the form y = mx+b

y =

SHOW WORK Draw Labelled Diagram

(6) [/4] A volcano erupts at noon.
The temperature of some lava which is cooling
is p(t) degrees celsius t hours after noon
If p(2)=500 and p'(2) = -40,
what approximately was the
temperature at 2:15pm.
SHOW WORK

(7) [/4] The temperature in degrees celsius
at a height of x meters above sea level is
h(x). Write a**brief gramatical clear English**
sentence explaining the meaning of
(a) h(2000) = 15
(b) h'(2000) = -0.03   (c) $h^{-1}(-5) = 5000$

STAPLE
Version
A

PRINT
NAME

TaSDi

Math 34A   F 99   Midterm 3   Quality   /2   SCORE   /30
Prof Cooper   **No Calculators**   Bonus

Put final answers in boxes on this page. Put **high quality** work in the blue book for **ALL QUESTIONS**.
Points are awarded for this. **Number your solutions in the blue book** . At the end of the exam STAPLE this
page to the INSIDE front blue cover of the blue book, so that this side faces the white writing pages of the blue
book. If you used a 3"x5" card, this MUST also be stapled here, behind this page.

---

(1)[/6] Use the graph given to find

(use a ruler and a sharp pencil to get an accurate answer
show construction lines)

(a) Log(0.01/27) =
[do **NOT** do any
division!]

(b) antilog(-3.63) =

(c) the slope of
the graph at x = 0.65

$y = 10^x$

(2)[/6] Find the derivative with
respect to x of the following

(a) $(2x-c)^2$
[c is constant]

(b) $2/x$

(c) $k^2 + k\,e^{kx}$
[k is constant]

(3) [/6] The height above the ground of an
object t seconds after it is launched vertically
upwards is h(t) = $100t - 5\,t^2$ meters

(a) what is the velocity of
the object after 3 seconds
[upwards is positive]   m/s

(b) what is the acceleration
after 3 seconds ? m/s^2

(c) how many seconds after
launch did the object hit the   secs
ground ?

(5) [/6] A water tank has a square base and
vertical rectangular sides. The length of a side of the
base is 3 meters. The height of water in the tank after t
hours is h(t) meters.

(a) what is the volume of
water in the tank after m^3
t hours
[draw a labelled diagram]

(b) After 5 hours the height of water in the tank is 1
meter. Also h'(5) = 3. What approximately is the
volume of water in the tank at 5:15pm ?

m^3

[hint: first work out the height at 5:15]

(4) [/6] Consider the graph of $y = x^2 + 4x$

(a) what is the slope of
the graph at x=2   slope=

(b) what is the equation of
the tangent line at x=2.
[give the answer in the form   y =
y = mx+b]
[draw a labelled diagram]

(c) what is the x-coordinate
of the point on the graph where   x =
the tangent line has slope zero.

---

(4) (a) 8 (b) y=8x-4 (c) x=-2 (5)(a) 9h(t) (b) height 7/4 meters, volume 63/4 cubic meters
(1) -3.43±.02 (b) .00023±.00002 (c) 10±3 2(a) 8x-4c (b) -2x^{-2} (c) k^2 k e^{kx} 3(a) 70m/s (b) -10 (c) 20 secs

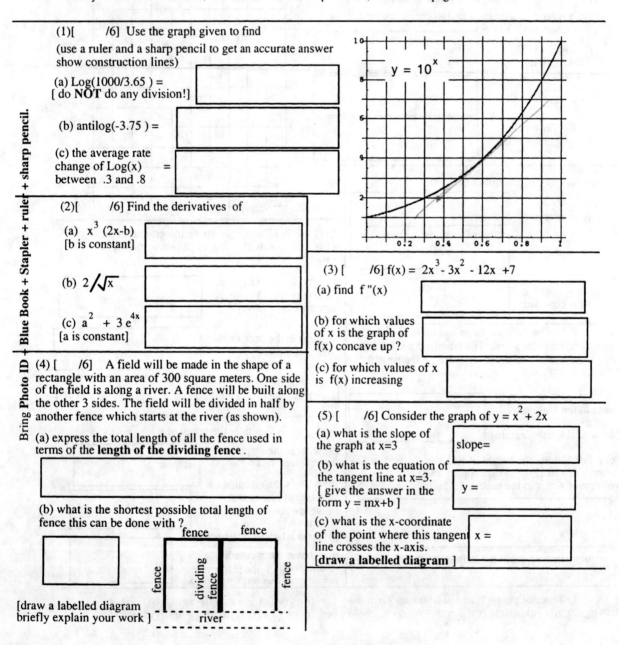

STAPLE   PRINT
Version   NAME   TaSDi
A

Math 34A   Winter 99   Midterm 3   Quality
Prof Cooper   **No Calculators**   Bonus   /2   SCORE   /30

Put final answers in boxes on this page. Put **high quality** work in the blue book for **ALL QUESTIONS**.
Points are awarded for this. **Number your solutions in the blue book**. At the end of the exam STAPLE this
page to the INSIDE front blue cover of the blue book, so that this side faces the white writing pages of the blue
book. If you used a 3"x5" card, this MUST also be stapled here, behind this page.

Bring Photo ID + Blue Book + Stapler + ruler + sharp pencil.

(1)[/6] Use the graph given to find

(use a ruler and a sharp pencil to get an accurate answer
show construction lines)

(a) Log(1000/3.65) =
[do **NOT** do any division!]

(b) antilog(-3.75) =

(c) the average rate
change of Log(x)   =
between .3 and .8

$y = 10^x$

(2)[/6] Find the derivatives of

(a) $x^3 (2x-b)$
[b is constant]

(b) $2/\sqrt{x}$

(c) $a^2 + 3 e^{4x}$
[a is constant]

(3) [/6] f(x) = $2x^3 - 3x^2 - 12x + 7$

(a) find f "(x)

(b) for which values
of x is the graph of
f(x) concave up ?

(c) for which values of x
is f(x) increasing

(4) [/6]   A field will be made in the shape of a
rectangle with an area of 300 square meters. One side
of the field is along a river. A fence will be built along
the other 3 sides. The field will be divided in half by
another fence which starts at the river (as shown).

(a) express the total length of all the fence used in
terms of the **length of the dividing fence** .

(b) what is the shortest possible total length of
fence this can be done with ?

[draw a labelled diagram
briefly explain your work]

fence   fence
fence   dividing fence   fence
river

(5) [/6] Consider the graph of y = $x^2 + 2x$

(a) what is the slope of
the graph at x=3   slope=

(b) what is the equation of
the tangent line at x=3.
[give the answer in the   y =
form y = mx+b]

(c) what is the x-coordinate
of the point where this tangent   x =
line crosses the x-axis.
[draw a labelled diagram]

Math 34A W99 Midterm 3   Version A   Prof Cooper

2 pts per part.

(1)(a) Log(1000/3.65)
=Log(1000)-Log(3.65)
[using Log(b/a)=Log(b)-Log(a)]
= 3 - 0.56 from graph

= $\boxed{2.44}$ allowed error ± .02

(b) antilog(-3.75)

= $10^{-3.75}$

= $10^{-4}10^{.25}$

=.0001 x 1.8   from graph
= $\boxed{.00018}$ allowed error ± .00002

(c) (av. rate of change)
= (Log(.8)-Log(.3))/(.8-.3)
= ([Log(8)-1]-[Log(3)-1]) / (.5)
= (Log(8)-Log(3))/(.5)
=(.9-.48)/(.5)
=(.42)/(.5)

= $\boxed{.84}$ allowed error ± .04

2 pts per part.
-1 pt if error more than allowed
must show work and construction
lines for full credit

(2)(a) $x^3(2x-b)$

= $2x^4 - bx^3$ **-1pt per error**

taking the derivative gives
$\frac{d}{dx}\left(2x^4 - bx^3\right)$

= $\boxed{8x^3 - 3bx^2}$

(b) this equals $2x^{-1/2}$

$\frac{d}{dx}\left(2x^{-1/2}\right)$

= $\boxed{-x^{-3/2}}$

(c) a^2 is a constant so has
derivative 0. Thus

$\frac{d}{dx}\left(a^2 + 3e^{4x}\right)$

= $\boxed{12e^{4x}}$

(3)(a)

$f(x) = 2x^3 - 3x^2 - 12x + 7$

$f'(x) = 6x^2 - 6x - 12$ **1pt**

$\boxed{f''(x) = 12x - 6}$ **1pt**

(b) graph is concave up when
$f''(x) > 0$ **1pt**
this is when
$12x-6 > 0$
ie when
$12x > 6$
ie when $\boxed{x > 1/2}$ **1pt**

(c) f is increasing when
$f'(x) > 0$
Factoring gives
$f'(x) = 6(x-2)(x+1)$
this is positive when
$(x-2)(x+1) > 0$.   **1pt**
This is positive if
EITHER both terms are positive
OR both terms are negative.

The first case is x-2>0 and x+1>0
this happens when x>2

The second case is when
x-2<0 and x+1<0
this happens when x<-1.

$\boxed{\text{Thus f is increasing when } x > 2 \text{ OR } x < -1}$ **1pt**

Graders: if the student does not say in English
what the variables are, deduct 1 point !

(4)
x = length of dividing fence in m
y = length in m of side parallel to river.   **1pt**

area of entire field
= 300 square meters
= xy   **1pt**

thus   y = 300/x

f = total length of all fence
  = 3x + y   **1pt**
goal express f in terms of x
substitute for y using above

$f(x) = 3x + 300x^{-1}$
Minimum when f '(x)=0. **1pt**

$f'(x) = 3 - 300x^{-2} = 0$
gives $x^2 = 100$ so x = 10   **1pt**

When x = 10
y=300/x=300/10=30
so total length of fence = $\boxed{60\ m}$ **1pt**

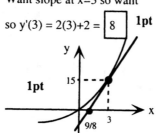

(5) (a) slope = derivative at x=3.
y '(x) = 2x+2 ← - - **1pt**
this is the slope at x.
Want slope at x=3 so want

so y'(3) = 2(3)+2 = $\boxed{8}$ **1pt**

(b) tangent line has same slope as
graph so has slope 8. Thus equation of
tangent line is y=8x+b. To find b
use that tangent line goes thru point on
graph where x=3. y-coord of this point
is $3^2 + 2(3) = 15$. Thus need 15=8(3)+b
so b=-9. **1pt**
Tangent line is $\boxed{y = 8x-9}$ **1pt**

(c) point where tangent line
meets x-axis is where y=0 on
tangent line. Then 0=8x-9 so
$\boxed{x = 9/8}$ **1pt**

STAPLE
Version
Practice

PRINT
NAME

TaSDi

Math 34A   Winter 99   Final
Prof Cooper

Quality
Bonus   /4

SCORE   /62

Final: Fri Mar19 1pm-4pm Chem 1179

Put final answers in boxes on this page. Put **high quality** work in the blue book for all answers. Points might be awarded for this. **Number your solutions in the blue book** . At the end of the exam STAPLE this page to the INSIDE front blue cover of the blue book, so that this side faces the white writing pages of the blue book. If you used a 3"x5" card, this MUST also be stapled here, behind this page. **Put away your photo ID after we check it.**

Bring **Photo ID** + Blue Book + Stapler + ruler + sharp pencil.

(1)[/6] Use the graph given to find

(a) $4^{1.3}$ =

(b) $\int_{.2}^{.5} 3(10^x)\, dx$ =

(c) $f(x) = 10^x$

  find $f'(0.45)$

$y = 10^x$

(2)[/6] Find the integrals

(a) $\int_a^b (8x - 3)\, dx$

(b) $\int (20\ x^4)\, dx$

(c) $\int_1^3 \left(\frac{b}{a\ x}\right)^{-2} dx$

(3) [/5] $f(x) = 3x^2 + 4$

(a) Sketch clearly the graph of y=f(x)
(b) Find the average value of   f(x)
     between   x=0   and x= 2 .
(c) Clearly mark the point on your graph where
     this average value occurs.

average=

(4) [/6] Find where the function
$f(x) = x(2x^2 - 9x + 12)$

(a) is increasing when

(b) concave up when

(c) has horizontal
     tangent when   x=

(5) [/4]   The table shows the total number of children in a school infected by measles during an epidemic in June. What was the highest rate of new infections? During which period did the **rate of increase** start to decline?

highest rate   14-15

period   15 - 17

day of June	5	9	12	14	5	7	21
no. children	1	2	4	2	17	23	24

$2\frac{1}{4}$ $2/3$ $\frac{4}{5}$ $\frac{5}{3}$

(6) [/4] A circular swimming pool has radius 10 m. Initially it is empty. Water is pumped in at a rate of   30000   liters per hour. Express the depth of water in the pool in terms of the number of hours the water is pumped. [there are 1000 liters in one cubic meter]

A

(7) [/6] Find the following

(a) $\dfrac{d}{dx}\left((x+2a)^2\right)$

Given $h(x) = 3x^{-1} + kx + \pi$ find

(b) $h'(-1) =$

(c) $h''(2) =$

(8) [/6] The height (in feet) of water in the harbour t hours after 6am is given by the formula

$h(t) = t^2 - 6t + 20$

[this formula is valid when $0 < t < 6$]

(a) How high is the water at 10am

(b) What time is low tide? [=lowest water level]

am

(c) How quickly is the water level changing at 10am? rising or falling?

(9) [/4] A farmer grows apples. On August 1 she has 80 tons of apples. Every day the crop increases by 2 tons. The price of apples on August 1 is $100 per ton. The price falls by $2 per ton each day.
(a) what function do you maximize

$f(x)=$

(b) On **which day of August** should she harvest and sell the apples to get the most money?

Aug

(10) [/4] The speed of a car in miles per hour is shown on the graph.
(a) How long does it take to travel 200 miles.
(b) Briefly explain your method.

mph

80

40

hours

2   4   6

hours

(11) [/ 5] Sketch the graph $y = x^2 - 1$ and include on the diagram the tangent line to the graph at x=2 and the tangent line to the graph at x=-1. Find the y coordinate of the point where these two tangent lines cross.

$y=$

(12) [/6] A rectangular field will have a fence on three sides and a wall on the fourth side. The fence costs $5 per meter. The wall costs $15 per meter. The area of the field is to be 800 square meters.

(a) Express the total cost in terms of the length of the wall

$

(b) what length should the wall be for the lowest cost

m

draw a labelled diagram

STAPLE
Version

PRINT
NAME

TaSDi

A

Math 34A   Winter 99   Final
Prof Cooper   Mar 19

Quality
Bonus   4

SCORE   72

Put final answers in boxes on this page. Put **high quality** work in the blue book **for all answers** .
Points might be awarded for this. **Number your solutions in the blue book**   . At the end of the exam
STAPLE this page to the INSIDE front blue cover of the blue book, so that this side faces the white
writing pages of the blue book. If you used a 3"x5" card, this MUST also be stapled here, behind this
page. **Put away your photo ID after we check it.**

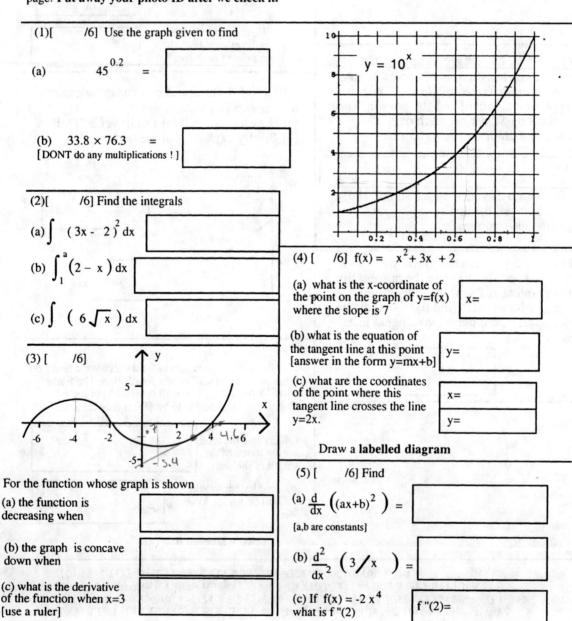

(1)[/6] Use the graph given to find

(a) $45^{0.2}$ =

(b) 33.8×76.3 =
[DONT do any multiplications !]

$y = 10^x$

(2)[/6] Find the integrals

(a) $\int (3x - 2)^2 \, dx$

(b) $\int_1^a (2 - x) \, dx$

(c) $\int (6\sqrt{x}) \, dx$

(3) [/6]

For the function whose graph is shown

(a) the function is
decreasing when

(b) the graph is concave
down when

(c) what is the derivative
of the function when x=3
[use a ruler]

(4) [/6] $f(x) = x^2 + 3x + 2$

(a) what is the x-coordinate of
the point on the graph of y=f(x)
where the slope is 7

x=

(b) what is the equation of
the tangent line at this point
[answer in the form y=mx+b]

y=

(c) what are the coordinates
of the point where this
tangent line crosses the line
y=2x.

x=

y=

Draw a **labelled diagram**

(5) [/6] Find

(a) $\frac{d}{dx}\left((ax+b)^2\right)$ =

[a,b are constants]

(b) $\frac{d^2}{dx^2}\left(3/x\right)$ =

(c) If $f(x) = -2x^4$
what is f "(2)

f "(2)=

A

(6) [/6] The number of billions of barrels of oil remaining in the world x years after 1950 is given by the function f(x). Write brief clear sentences explaining the meaning of the following statements
(a) f(20) = 3000
(b) f '(20) = -30   (c) f $^{-1}$(2500) = 28

(8) [/6] The height above the ground in meters of a rocket t seconds after launch is

h(t) = 10t + 2t^2

(a) what is the velocity of the rocket after t seconds   m/s

(b) how many seconds after launch until the velocity is 50 m/s   secs

(c) what is the acceleration of the rocket 3 seconds after launch   m/s^2

(9) [/6] An artist sells copies of a picture. All the copies will be sold for the same price. If the price was $70 she would sell 100 pictures. Every $2 reduction in price means 5 more pictures are sold. It costs $10 to make a picture.

(a) express the total profit in terms of the number of $2 price reductions.

(b) What price should she charge for each picture to make the greatest total profit?   price=

(11) [/6] A rectangular field is to be made with a wooden fence round all four sides and with a brick wall dividing the field into two smaller equal rectangles. The fence costs $10 per meter. The wall costs $20 per meter. The total area of the field must be 200 square meters.

(a) Express the **total cost** of fence and wall in terms of the length of the wall

(b) What length should the wall be to minimize the **total cost**

fence	fence
fence wall fence	
fence	fence

(7) [/6] The table shows the height of water in a reservoir. Each one inch depth of water in the reservoir contains 1 million gallons of water.

day of January	1	4	10	16	22	25	31
depth in inches	200	194	226	220	211	201	192

(a) draw a neat graph showing the information in the table

(b) during which period was the level of the reservoir falling most quickly

(c) on average how many gallons of water was the reservoir losing per day during Jan 10 to 22   million gallons per day

(10) [/6] The graph shows the rate that rain fell in centimeters per hour during a storm which lasts 5 hours.

(a) How many centimeters of rain fell during the entire storm   cm

(b) How many hours after the start of the storm was it until 7 centimeters had fallen   hours

(12) [/6] A box with a square base and top has rectangular sides. The volume of the box is 10 cubic meters. The cost of the material for the base is $7 per square meter. The cost of the material for the top and sides is $3 per square meter. Express the total cost of the box in terms of the width of the base. Draw a **labelled diagram** .

cost=

Math 34A   W99   Final   Version   [A]   Prof Cooper

(1)(a) $\text{Log}(45^{0.2})$
$=0.2\text{Log}(45)$
[using $\text{Log}(a^p) = p\text{Log}(b)$]

$=0.2[1+\text{Log}(4.5)]$
$=0.2[1+0.65]$ from graph
$=0.33$

So answer is
antilog(0.33)= [2.1]
allowed error ± .2
[from graph]

(b) $\text{Log}(ab)=\text{Log}(a) + \text{Log}(b)$
$\text{Log}(33.8)=1+\text{Log}(3.38)$
$= 1+0.53$
[from graph]
$=1.53$

$\text{Log}(76.3)$
$=1+\text{Log}(7.63)$
$=1+0.88$
[from graph]
$=1.88$

Thus $\text{Log}(\text{answer})$
$=1.53+1.88$
$=3.41$

So answer
$= \text{antilog}(3.41)$
$=1000\text{antilog}(0.41)$

$= \boxed{2600}$
allowed error ± 200
[from graph]

3 pts per part.
-1 pt if error more than allowed
must show work and construction
lines for full credit

(6)(a) There were 3000 billion barrels of
oil left in 1970. (b) In 1970 the amount
of oil left was decreasing at a rate of 30
billion barrels per year. (c) In 1978 there
were 2500 billion barrels of oil left.
[**2pts** per part. 0 points if the sentence does
not make sense! graders judgement is final]

inches   (a)
230
220
210
3pt
200
190
                                          days
1   4   10   16   22 25   31

(2)(a) $(3x-2)^2 =9x^2-12x +4$

$\int (3x - 2)^2 dx$

$= \boxed{3x^3- 6x^2 + 4x + C}$

(b) $\int_1^a (2-x)\, dx = \left[2x -x^2/2\right]_1^a$

$= \boxed{(2a - a^2/2) - 3/2}$

(c) $\int 6 x^{0.5} dx = \boxed{4 x^{1.5} + C}$

2 pts per part.
-1 pt per error

(5) (a) $(ax+b)^2= a^2 x^2+ 2abx + b^2$

differentiating gives

$\boxed{2a^2 x +2ab}$ **2pt**

(b) derivative of $3x^{-1}$ is $-3 x^2$

so second derivative is $\boxed{6x^{-3}}$ **2pt**

(c) $f'(x) = -8 x^3$
so $f''(x) = -24 x^2$

thus $f''(2) = -24(2)^2= \boxed{-96}$ **2pt**

(9) x = number of $2 price cuts **1pt**

price = 70-2x
number pictures sold = 100+5x **1pt**
profit on one picture
= price-cost = 70-2x - 10 = 60-2x

p = total profit
= (100+5x)(60-2x) **1pt**

For maximum profit **1pt**
p'(x)=0 now
$p=6000 + 100x - 10x^2$
so p'(x)=100-20x
this is 0 when x=5 **1pt**

price of one picture when
x=5 is 70-5(2)= [$60] **1pt**

(3) (a) [-3.7 < x < 0.9]

(b) [x < -1.9]

allowed error ± .3 for
(a) and (b)

2 pts per part.

(c) [1.7]

anything between 1.3
and 2.1 OK

(4) (a) f '(x) = 2x+3 **1pt**
so slope is 7 when
2x+3=7 ie when [x=2] **1pt**

(b) tangent line has slope
7 so equation is y=7x+b

when x=2 get f(2)=12 **1pt**
so 12=7(2)+b thus b=-2

[y=7x-2] **1pt**

(c) this meets y=2x **1pt**
when 7x-2=2x
thus 5x=2 so [x=2/5] **1pt**
and
y=2x so [y= 4/5]

8(a) [h'(t) = 10 + 4t] **2pt**

(b) velocity is 50 m/s
when 10+40t = 50 ie
when [t=10] **2pt**

(c) acceleration =

h''(t) = [4 m/s²] **2pt**

2pt

(7) (b) [Jan 22-25]

(c) (226-211)/(22-10)

[15/12 = 5/4] **1pt**

Math 34A W99   Final Version [A]   page2   Prof Cooper

(10)(a)

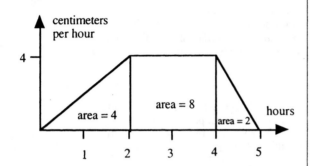

Total area under graph = 4+8+2 = 14   **1pt**

Graph is rate of increase of height of water. By fundamental theorem of calculus this area represents the change in height of water between y=0 and t=5. Thus a total of 14cm falls.   **1pt**

(b)

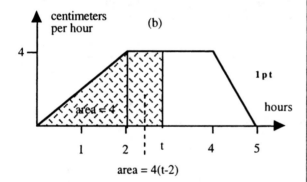

area = 4(t-2)

1pt

amount of rain which falls up to time t is represented by area shown = 4 + 4(t-2) = 4t-4. This is 7 when 4t-4=7 ie when 4t = 11 thus t = 11/4   **1pt**   **1pt**

11(a)

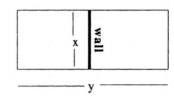

x = length of wall in meters   **1pt**
y = width of field in meters

area of field
=200   **1pt**
=xy

thus y = 200x^{-1}

C = total cost = (cost of wall) + (cost of fence)

cost of wall = 20x
length of fence = 2y + 2x
cost of fence = 10(2x+2y) = 20x+20y   **1pt**

so C
= 20x + (20x+20y)
= 40x + 20y
= 40x + 20(200x^{-1})
= 40x + 4000x^{-1} **1pt**

(b) for a minimum C'(x)=0

C'(x) = 40 - 4000x^{-2} **1pt**

This is zero when
40 = 4000x^{-2}

ie when 40x^2 = 4000
thus x^2= 4000/40 = 100

so x = 10   **1pt**

(12)

1pt

1pt

x = width of base in meters
y = height of box in meters

Volume of box = 10
= (area of base)xheight

= x^{2}y   **1pt**

1pt

Total cost
= cost base + (cost top and sides)   **1pt**

= 7x^2 + 3(x^2 + 4xy)
=10x^2 + 12xy   **1pt**

See that xy = 10/x so

total cost = 10x^2 + 120/x   **1pt**

STAPLE
Version
Practice

PRINT
NAME

TaSDi

Math 34B   Winter 99   midterm 1   Quality
Prof Cooper   Fri Jan 22   Bonus   /4
No Calculators

SCORE   /25

Put final answers in boxes on this page.**SHOW WORK** for **ALL QUESTIONS**. Put **high quality**
work in the blue book. Points are awarded for this**Number your solutions in the blue book** .
At the end of the exam STAPLE this page to the INSIDE front blue cover of the blue book, so that
this side faces the white writing pages of the blue book. If you used a 3"x5" card, this MUST also be
stapled here, behind this page.

Bring Photo ID + Blue Book + Stapler + ruler + sharp pencil.

(1) [/ 5] How many liters of water
containing 3 grams of salt per liter must
be combined with 4 liters of water which
contains y grams [y is less than 2] of salt per
liter to yield a solution with 2 grams of salt
per liter.

(2) [/5] Adult movie tickets cost $6. Child
tickets cost $3. If 700 people go to a movie
and the total ticket sales are $3300 how many
children and how many adults where there?

no. children

no. adults

(3)[/6] Find

(a) $\dfrac{d}{dx}$ $(4x-3)^2$

(b) $\displaystyle\int_1^2 x^2 \, dx$

(c) $\displaystyle\int_a^3 a/x^2 \, dx$

(4) [/ 5] The height after t seconds of a ball
thrown vertically upwards is $20t-5t^2$ meters

(a) what is the velocity after
t seconds

(b) what is the maximum height

(5) [/4] What is the equation of
the tangent line to $y = -x^2$ at the point
x=3. Draw a **clear** labelled diagram.
Give the answer in the form y = mx + b.

STAPLE
Version
A

PRINT
NAME

TaSDi

Math 34B   Winter 99   midterm 1   Quality
Prof Cooper   Fri Jan 22   Bonus   /2
No Calculators

SCORE   /25

Put final answers in boxes on this page.**SHOW WORK** for **ALL QUESTIONS**. Put **high quality**
work in the blue book. Points are awarded for this.**Number your solutions in the blue book** .
At the end of the exam STAPLE this page to the INSIDE front blue cover of the blue book, so that
this side faces the white writing pages of the blue book. If you used a 3"x5" card, this MUST also be
stapled here, behind this page.

Bring Photo ID + Blue Book + Stapler + ruler + sharp pencil.

(1)[/6] Find

(a) $\dfrac{d}{dx}$ $(6x^2 + (7/x))$

(b) $\displaystyle\int_0^4 (a+x)\ dx$

(c) $\displaystyle\int (x^2 - 3x)\ dx$

SHOW WORK

(3) [/ 4] When a radio is sold for a price of
$x, the number of radios sold is $(2000 - 100x)$.
(a) In total, how much money is obtained by
selling the radios for this price.

(b) what price should the radios be sold for in
order to maximize this total amount of money.

x=

SHOW WORK

(5) [/5] A rectangular box has a square
base. The volume of the box is 200 cm^3 .
Express the total surface area of the sides, top
and bottom of the box in terms of the length of
one side of the square base.

SHOW WORK   draw a labelled diagram
briefly explain steps

(2) [/4] (a) What is the x-coordinate
of the point on the graph $y = x^2$ at which
the slope of the graph is 10.

x=

(b) What is the equation of the tangent line to this
graph at that point.
Give the answer in the form y = mx + b.
 Draw a **clear** labelled diagram.

y =

SHOW WORK

(4) [/ 6] A pool initially contains 100 liters
of pure water. Then liquid which is 4% detergent
(and the rest water) is pumped into the pool at a
rate of 20 liters per hour.

(a) how much liquid is in the pool after t hours

(b) how much detergent is in the pool after t hours

(c) After how many hours of pumping will the
concentration of detergent in the pool be 1% .
[you may leave the answer as a fraction]

Math 34B W99 Midterm 1   Version  A   Prof Cooper

(1) (a) $12x - 7x^{-2}$

                              2 pts each

(b) $4a + 8$

(c) $(x^3/3) - (3x^2/2) + C$

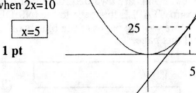

(2) (a) $y' = 2x$ so slope of graph is 10 when $2x=10$ ie when

$x=5$

1 pt

5   **1 pt**

___

(3)

(a) $p = x(2000-100x)$ **2 pts**

(b) for a maximum $p'=0$.
$p' = 2000-200x$
this is 0 when   **1 pt**

$x=10$ **1 pt**

(b) the slope of the tangent line is 10. The point on the graph where $x=5$ has $y = 5^2 = 25$. The equation of the tangent line is $y = 10x+b$. Substitute $x=5$ and $y=25$ to find b: $25 = 10(5)+b$ thus $b = -25$. Hence equation of tangent line is

$y = 10x-25$ **2 pts**

___

(5)

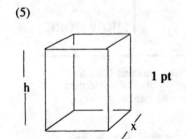

1 pt

(4) (a) $100+20t$ liters

                              2 pts each

(b) 4% of (20t) = $4t/5$ liters

(c) there is 1% detergent when $4t/5$ equals 1% of $(100+20t)$

ie when $4t/5 = (1/100)(100+20t) = 1 + t/5$.

thus $3t/5 =1$ so $t=5/3$

x = length in cm of side of base of box
h =height in cm of box   **1 pt**

volume = $x^2h = 200$ **1 pt**

Area = (top+base) + (4 sides)

      $=2x^2$ $+ 4xh$ **1 pt**

subsitute for h using $h = 200/x^2$

Area = $2x^2 + 800/x$ **1 pt**

STAPLE
Version

Practise

PRINT
NAME

TaSDi

Math 34B Midterm 2 Winter 99   Quality
Prof Cooper   Fri Feb 12   Bonus   /4
No Calculators

SCORE   /25

Put final answers in boxes on this page. **SHOW WORK** for **ALL QUESTIONS**. Put **high quality**
work in the blue book. Points are awarded for this **Number your solutions in the blue book** .
At the end of the exam **STAPLE** this page to the INSIDE front blue cover of the blue book, so that
this side faces the white writing pages of the blue book. If you used a 3"x5" card, this MUST also be
stapled here, behind this page.

Bring Photo ID + Blue Book + Stapler + ruler + sharp pencil.

(1) [/ 3] Consider $5\sin(4t + 3) + 9$
where t is the time in seconds

(a) what is the period

(b) what is the
frequency

(c) what is the
amplitude

(2) [/5] Find the local max and min of   .

$x + 9x^{-1}$ Clearly show how you apply the
second derivative test.

local max is   x=   y=

local min is   x=   y=

(3)[/6] Find the derivatives with
    respect to x of
(a) $7\sin(kx)+4x+2$

(b) $7e^{2x} \sin(4x)$

(c) $f(x)[g(x)+h(x)]$

(4) [/ 6] Two planes A and B can
each fly through **still air** at 500 mph. The
distance from Paris to Los Angeles is 5000
miles. Wind blows from Los Angeles
towards Paris at a speed of 100 mph. Plane
A leaves Paris at noon flying towards Los
Angeles. Plane B departs Los Angeles 1
hour later to fly to Paris. How far from Los
Angeles are the two planes when they pass
each other ?

[hint: what are the speeds of the two planes
relative to the ground]

(5) [/5] A metal can in the shape of
circular cylinder has a base and top. The
volume of the can is 400 cm.^{3} The
diameter of the base is D cm. Express
the total surface area of the can in terms
of the diameter of the base. Draw a
labelled diagram.
Briefly Explain.

[1](a) $\pi/2$ secs (b) $2/\pi$ Hz (c) 5 [2] loc max x=-3 y=-3 loc min x=3 y=6 [3] (a) $7k\cos(kx)+4$
(b) $e^{2x}[14\sin(4x)+28\cos(4x)]$ (c) $f'(x)[g(x)+h(x)] + f(x)[g'(x)+h'(x)]$ [4] 2760 miles [5] $\pi D^2/2 + 1600/D$

STAPLE   PRINT
Version   NAME   TaSDi

A

Math 34B Midterm 2 Winter 99   Quality   SCORE
Prof Cooper   Fri Feb 12   Bonus   / 2   / 25
                    No Calculators

Put final answers in boxes on this page. **SHOW WORK** for **ALL QUESTIONS**. Put **high quality**
work in the blue book. Points are awarded for this **Number your solutions in the blue book** .
At the end of the exam **STAPLE** this page to the INSIDE front blue cover of the blue book, so that
this side faces the white writing pages of the blue book. If you used a 3"x5" card, this MUST also be
stapled here, behind this page.

Bring Photo ID + Blue Book + Stapler + ruler + sharp pencil.

(1) [/ 4] The height of water in a harbor varies
like a sine wave with a period of 11 hours. The
maximum height is 30 feet and the minimum height
is 20 feet. T is the time measured in hours. The
height at time T=0 is 25 feet. Write down a function
which gives the **height of the water** at time T.

(2) [/5] Find the local max and min of the function
below. **Clearly** show how you apply the second
derivative test.

$$x^3 + 6x^2 + 9x + 2$$

local max is | x=   y=

local min is | x=   y=

(3)[/6] Find the derivatives with respect to x of

(a) $[f(x)]^2$

(b) $(3 + (1/x))\sin(4x)$

(c) $e^{2x}\ln(x)$

(4) [/ 5] A car travels along a route 350 miles long
in 7 hours. The speed of the car is 60 mph for the first
2 hours. It travels 80 miles in the next 2 hours. It
travels at a (different) constant speed for the remaining
part of the journey.

(a) what is the speed of the car after 5 hours.

mph

(b) how far has the car travelled after 5 hours

miles

SHOW WORK.

(5)[/5] A park is in the shape of a **square**. In
the middle of the park is a smaller **square** play
area. The park is surrounded by a fence. The play
area is also surrounded by a fence. The part of the
park outside the play area is grass. The area of the
grass is 2000 square meters. Express the total
length of all the fences added together in terms of
the **length of one side of the play area** .

SHOW WORK. draw a labelled diagram

length of   =
fence

grass

play
area

fence

fence

Math 34B   W99   Midterm 2   Version A   Prof Cooper

(any integer multiple of π is OK)

(1) $25 + 5\sin(2\pi T/11 + 0)$

1pt 1pt 1pt   1pt

Water varies between 20 and 30 feet
so amplitude = (30-20)/2 = 5

Water varies 5ft above and 5ft below
the level of 25ft.

(2) $f(x) = x^3 + 6x^2 + 9x + 2$

critical points are when $f'(x)=0$

$f'(x) = 3x^2 + 12x + 9$

1pt

This is zero when
$3(x+1)(x+3)=0$
ie when
$x=-1$ or $x=-3$ 1pt

$f''(x) = 6x+12$ 1pt

second derivative test at x=-3

$f''(-3) = 6(-3)+12 < 0$ ⌢

so local max at $\boxed{x=-3}$ is
$y = f(-3)$
    $= (-27)+6(9)+9(-3)+2$
$\boxed{y = 2}$ 1pt

second derivative test at
x=-1

$f''(-1) = 6(-1)+12 > 0$ ⌣

so local min at $\boxed{x=-1}$ is
$y = f(-1)$
    $= (-1)+6(1)+9(-1)+2$
$\boxed{y = -2}$ 1pt

(3) (a) product rule gives $2f(x)f'(x)$

(b) $(3 + (1/x))4\cos(4x) - (1/x^2)\sin(4x)$ **2 pts per part**

(c) $e^{2x}[2\ln(x) + (1/x)]$

(4)   350 miles in 7 hours

120 miles	80 miles	
60mph 2 hours	40 mph 2 hours	350-(120+80) miles = 150 miles

in 7-(2+2)=3 hours

**partial credit at
graders discretion**

speed
= 150/3 = 50mph

car travels 120+80= 200 miles in first 4 hours
this leaves 350-200=150 miles for remaining time.
Remaining time is 7-(2+2)=3 hours.
(a) Thus speed during last 3 hours is 150/3 = $\boxed{50\text{mph}}$

(b) car travels 200 miles in first 4 hours. Then 50 miles
during 5'th hour. Thus total distance in 5 hours is
200+50= $\boxed{250\text{ miles}}$

(5)

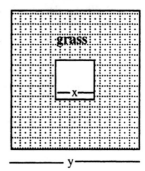

x = length of one side of play area
y = length of one side of park   1pt

Length of fence
= (fence round play area)
+ (fence round park)
= 4x+4y   1pt

Need to express y in terms of x

area of grass
= (area big square) - (area of play area)

$= y^2 - x^2$ 1pt
$= 2000$

Thus $y^2 = 2000 + x^2$

so $y = \sqrt{2000 + x^2}$ 1pt

$\boxed{\text{Length of fence} = 4x + 4\sqrt{2000 + x^2}}$ 1pt

STAPLE   PRINT
Version   NAME

Practice

TaSDi

Math 34B Midterm 3 Winter 99   Quality
Prof Cooper   Fri Mar 5   Bonus   /2
                  No Calculators

SCORE   /25

Put final answers in boxes on this page. **SHOW WORK** for **ALL QUESTIONS**. Put **high quality** work in the blue book. Points are awarded for this. **Number your solutions in the blue book**.
At the end of the exam **STAPLE** this page to the INSIDE front blue cover of the blue book, so that this side faces the white writing pages of the blue book. If you used a 3"x5" card, this MUST also be stapled here, behind this page.

Bring Photo ID + Blue Book + Stapler + ruler + sharp pencil.

(1) [/ 3] Find the solution of the differential equation $y' = 2(3-y)$ satisfying the initial condition $y(0) = 7$

(3)[/ 4] (a) What is the general solution of $y''(t) = 12t - 2$

(b) What is the particular solution for which $y(0) = 3$ and $y(1) = 5$

(5) [/ 6] (a) The mass of a colony of bacteria is growing at a rate proportional to its mass. Initially the mass is 10 grams. Initially the mass is increasing at a rate of 5 grams per day. What is the differential equation which describes this?

(b) What is the mass after t days ?

(2) [/6] (a) **Carefully** sketch the slope field for the differential equation $y' = -1 - y$ for the range $-2 < y < 0$ and $0 < t < 2$ using a grid of .5 in both the y-direction and the t-direction.

(b) Sketch the solution curve for which $y(1) = -1.8$

(c) Write a brief clear sentence describing the long term behaviour of this particular solution.

(4) [/ 6] The number of **pounds of salmon caught each day** is growing at the rate of $200+20t$ pounds per day, t days after fishing quotas are elliminated.
(a) Express this as a differential equation. **Clearly define** all the variables in your equation

(b) If initially 100000 pounds per day are caught, how much is caught on the 10'th day?

(c) When will the mass be 60 grams? [you may leave ln(...) in your answer]

(6)$\ln 2 = t$ (c) $\ e^{5t} \ 10 = (t)m$ (b) $(t)m(2/1) = (t)'m$ (a) (5) 000301 (b) $t02+002 = (t)'y$ (4) $\ 3 + t + {}_2t - {}_3t2 = y$ (b)

(1) $y(t) = 3 + 4e^{-2t}$ (2) (c) y increases towards -1 as t increases but never gets there. (3) (a) $y = 2t^3 - t^2 + At + B$

STAPLE
Version

[A]

PRINT
NAME

TaSDi

Math 34B   Midterm 3   Winter 99
Prof Cooper   **No Calculators**

Quality
Bonus   /2

SCORE   /30

Put final answers in boxes on this page. **SHOW WORK** for **ALL QUESTIONS** . Put **high quality** work in the blue book. Points are awarded for this. **Number your solutions in the blue book** .
At the end of the exam **STAPLE** this page to the INSIDE front blue cover of the blue book, so that this side faces the white writing pages of the blue book. If you used a 3"x5" card, this MUST also be stapled here, behind this page.

Bring Photo ID + Blue Book + Stapler + ruler + sharp pencil.

(1)[/6] (a) What is the general solution of
$y''(t) = 18t - 4$

(b) What is the particular solution for which
$y'(0) = 3$ and
$y(1) = 5$

(c) what is the general solution of
$y'''(t) = 0$

(3))[/6] The mass of a colony of bacteria is growing at a rate which is 0.3 times the mass of the colony. Time is measured in days and mass in grams.

(a) What is the differential equation which describes this

(b) If the mass after 2 days is 20 grams, what was the initial mass?

(c) How many days does it take for the mass to double?

(2) [/6] (a) **Carefully** sketch the slope field for the differential equation $y' = 2+y$ for the range $-4 < y < 0$ and $0 < t < 4$ using a grid of 1 in both the y-direction and the t-direction.

(b) Sketch the solution curve for which $y(2)=-2.5$.

(c) write a brief clear sentence describing the long term behaviour of this particular solution.

(4))[/6] A lava flow is cooling down at a rate given by Newton's law of cooling. The initial temperature of the lava is $500\,^{\circ}$C. The temperature of the surroundings is $20\,^{\circ}$C. The initial rate at which the lava cools is $70\,^{\circ}$C per day.

(a) what is the differential equation which describes the temperature of the lava.

(b) what will the temperature of the lava be after 5 days ?

(a) what is the speed of Car B

mph

(b) How far (measured round the track) from the starting point on the race track is the flag.

miles

(5) [/6] A race track is in the shape of a square each side of which is 3 miles long. Car A travels at a constant speed of 50mph. Car B travels at a constant speed which is more than 50mph. They start at the same corner going in the same direction round the race track. Car B immediately gets ahead of car A. The next time that both cars are at the same place at the same time is 2 hours later and a flag is posted there.

PLEASE: Check your algebra .

Math 34B W99 Midterm 3   Version A   Prof Cooper

(1)(a) y "= 18t - 4
intergrating gives

y ' = 9t^2- 4t + A
integrate again:

$$y = 3t^3 - 2t^2 + At + B$$

(b) y '(0) = 3 = 9(0)^2 -4(0) + A
thus A=3

y(1) = 5 = 3(1)^3- 2(1)^2+ 3(1) + B
        = 4+B
thus B=1

hence particular solution is

$$y= 3t^3 - 2t^2 + 3t + 1$$

y '" = 0   **2pts per part**
integrate to get   **-1pt per error**

y " = A
integrate to get

y ' = At + B
integrate to get

$$y = (1/2) At^2 + Bt + C$$

[it is also correct to leave out the 1/2]

─────────────────────────

(3) m(t) = mass in grams of   **1pt**
colonony after t hours

(a) $m'(t) = 0.3m(t)$ **1pt**

(b) general solution m(t) = Ae$^{0.3t}$
                                1pt

initial mass = m(0) = Ae^0 = A
goal: find A

after 2 days m(2) = 20 = Ae$^{0.3(2)}$

thus A = $20e^{-0.6}$ grams   **1pt**

(c) particular solution is m(t) = 20e$^{-0.6}$ e$^{0.3t}$

mass is double initial mass when   **1pt**
m(t)=2m(0)=2A **goal**: solve this for t.

Ae$^{0.3t}$ = 2A   thus 0.3t = ln(2)   so t = ln(2)/0.3
                                                1pt

─────────────────────────

(2)   (a) + (b)

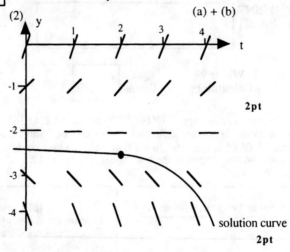

                                    solution curve
                                         2pt

Equation is homogeneous

y	-4	-3	-2	-1	0
2+y	-2	-1	0	1	2

1pt

(c) the particular solution goes to minus infinity as t increases. **1pt**

─────────────────────────
                          1pt

(4) Newton's law cooling   y '(t) = k(M - y) where
y(t)=temperature in ^{o} C after t days   M = temperature of
surroundings k=constant. **General solution**   y(t) = M +Ae^{-kt} **1pt**
So **initial temperature** is y(0) = M + A.

Told M=20 and   initial temperature is
y(0) = 500 = M+A thus A = 480   so y(t) = 20 + 480e^{-kt} **1pt**
goal : find k.
Told initial rate of decrease of temp = -70 = y '(0) = -480ke^0
thus k = 70/480. **1pt**

(a) $y ' = (7/48)(20 - y)$ **1pt**

(b) temp after 5 days = y(5) = $20 + 480e^{-5(7/48)}$ **1pt**

─────────────────────────

(5) length of race track = 4x3 = 12 miles **1pt**
car A goes 2x50 = 100 miles in 2 hours
100 = (8x12) + 4 so car A goes 8 times round
track and is 4 miles past start when B catches up
again
            1pt

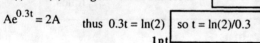

                                                    1pt

                                            3miles

                                            start

Car B must travel an extra time round race track to catch A
up again. This is an extra 12 miles. So car B goes extra 12
miles in 2 hours. So car B has an extra (12/2)=6mph speed.
Thus speed of car B = 56 mph   **1pt**
            1pt

STAPLE
Version
[Practice]

PRINT
NAME

TaSDi

Math 34B Winter 99   Final
Prof Cooper   **4pm-7pm**
            Wens Mar 24

Quality
Bonus   /4

SCORE   /72

No Calculators

Put final answers in boxes on this page. **SHOW WORK for ALL QUESTIONS**. Put **high quality** work in the blue book. Points are awarded for this **Number your solutions in the blue book** .
At the end of the exam STAPLE this page to the INSIDE front blue cover of the blue book, so that this side faces the white writing pages of the blue book. If you used a 3"x5" card, his MUST also be stapled here, behind this page. **Put away your photo ID after we check it.**

Bring Photo ID + Blue Book + Stapler + ruler + sharp pencil.

(1)[/ 6] (a) What is the general solution of $y' = 4y$

(b) what is the particular solution for which $y'(0) = 6$

(c) For the particular solution found in (b), what is y when t=2?

(2) [__ / 6] Plane A flies at a constant speed from New York to Los Angeles along a route which is 2000 miles. Plane B flies in the opposite direction at a constant speed which is 100 mph faster than plane A. Plane B takes off one hour after plane A. They land at the same moment. How far are they from Los Angeles when they pass?

miles

(3)[/6] Find the integrals

(a) $\int_1^2 (ax-b)^2 \, dx$

(b) $\int 7e^{3t} \, dt$

(c) $\int_a^b c \, dx$

(4) [/6] Beaker A contains 1 liter which is 10% oil and the rest is vinegar, thoroughly mixed up. Beaker B contains 2 liters which is 50% oil and 50% vinegar, completely mixed up. Half of the contents of B are poured into A, then completely mixed up. How much oil should now be added to A to produce a mixture which is 60% oil?

liters

(5) [/6] Sketch the graph of
$y = 3-2\sin(3t + (\pi/2))$
for t between π and π . Label the axes.

(6) [/ 6] Find the equation of the tangent line to
$y = x^2 e^{3x}$ at x=2

(7) [/6] $f(x) = \sqrt{x+9}$
(a) find the linear approximation to f(x) which agrees with f(x) at x=0 and at x=27

(b) sketch the graph of f(x) and this linear approximation
(c) what is the percentage error of this approximation at x=16.

(d) Write a brief sentence explaining whether you would expect the percentage error to be bigger or smaller at x=1

(8) [/6] $f(x,y) = 3x^2y + 5xy - e^y$
Find the following partial derivatives

(a) f_x

(b) f_y

(c) f_{xy}

(9) [/6] (a) Sketch the slope field for y' = y-t
for y and t in the range from -2 to 2 using a grid
spacing of 1 in each direction.
(b) Sketch the solution curve for which y(0)=0.5.
(c) Write a brief clear sentence describing the
long term behaviour of this solution.

(11) [/6] The population of a country is
growing at a rate given by the logistic
equation. The maximum population the
country can sustain is 10 million. In 1950 the
population was 5 million, and growing at a
rate of 1 million per year.

(a) write down the differential equation that
describes this

(b) what is the solution of this equation
that fits the given information.

(c) sketch this solution

(10) [/6] A cup of tea is cooling at a rate given
by Newton's law of cooling. The temperature of the
tea when it was first made was 100° C. The room
temperature was 20°C. After 4 minutes the
temperature of the tea is 90° C

(a) write down the differential equation
 which governs this

(b) What is the temperature after 8 minutes?

(12) [/6] A two stage rocket is moving
vertically upwards. At the moment when the
second stage ignites, the rocket is 10000 meters
above the ground moving upwards with a
velocity vertically upwards of 200 m/s. The
acceleration produced by the second stage is
10+6t m/s^2 t seconds after the second stage
ignites. What is the height of the rocket 10
seconds after the second stage ignites?

STAPLE
Version
A

PRINT
NAME

TaSDi

Math 34B Winter 99   Final
Prof Cooper   **4pm-7pm**
              Wens Mar 24

Quality
Bonus   /4

SCORE   /72

No Calculators

Put final answers in boxes on this page.**SHOW WORK** for **ALL QUESTIONS**. Put **high quality**
work in the blue book. Points are awarded for this **Number your solutions in the blue book** .
At the end of the exam STAPLE this page to the INSIDE front blue cover of the blue book, so that
this side faces the white writing pages of the blue book. If you used a 3"x5" card, his MUST also be
stapled here, behind this page.**Put away your photo ID after we check it.**

---

(1) [/6] Find the integrals

(a) $\int 3x(x+2)\ dx$

(b) $\int_0^3 (3t + 2e^{4t})\ dt$

(c) $\int_1^2 (5/x^2)\ dx$

(2) [/6] $f(x,y) = 2x\,y^2 + 5x - 7y + 3$
Find the following partial derivatives

(a) f_x

(b) f_y

(c) f_{yx}

---

(3) [/6]   A plane flies along a route 2000
miles long. It flies at 200mph for the first and last
half hour of the flight. It flies at a higher constant
speed for the rest of the flight. The plane is 1000
miles from the starting point
after 2 hours of flying. How
many hours is the entire flight?

hours

(4) [/6] A rectangular water tank has a square
base. The length of a side of the base of the tank is 3
meters. Initially the water tank is empty. Water enters
the tank at a rate of 2000 liters per hour. [there are
1000 liters in a cubic meter]

(a) what is the height
of water in the tank
after t hours

meters

(b) How quickly is the
water level in the tank
rising

meters/hour

---

(5) [/6] Due to over-fishing, the number of fish
in a lake dropped to 2000, then fishing stopped.
t years after fishing had stopped the number of fish
in the lake was increasing at a rate of 300+200t fish
per year.

(a) How many fish are in the lake after t years?

(b) How many years after
fishing stopped was it until
the number of fish in the
lake reach 3800

years

(6) [/6] Find the local maximum and minimum
of the function below. Show **clearly** how you use
the second derivative test.

$$f(x) = -x^3 + 12x + 5$$

local minimum is   when   x=

local maximum is   when   x=

A

(7) [/6] A colony of bacteria is growing at a rate equal to 0.3 times its mass. Here time is measured in hours and mass is measured in grams. There was 5 grams of bacteria after 3 hours.

(a) Write down the differential equation that the mass of bacteria satisfies.

(b) What was the mass of bacteria after t hours.

(c) How quickly was the mass increasing after 6 hours

(8) [$_o$ /6] A steel bar is initially at a temperature of 550°C and cools down according to Netwon's law of cooling. The temperature of the surroundings is 50°C. Initially the steel bar is cooling at a rate of 20°C per minute.

(a) write down the differential equation that governs the temperature of the steel bar.

(b) What was the temperature after 20 minutes

(c) carefully draw a sketch graph showing the temperature over a long period.

(9) [/6] Rabbits are introduced onto an island. The population of rabbits is described by the logisitc equation. Initially there were 200 rabbits and this number was increasing at a rate of 20 per month. The maximum number of rabbits the island can support is 2000.

(a) What is the differential equation which the number of rabbits satisfies?

(b) What was the rate of increase of the number of rabbits when the rabbit population was 1000

(c) approximately how many rabbits are there after 1000 years

(10) [/6] (a) Sketch the slope field for y ' = y+t for y and t in the range from -2 to 2 using a grid spacing of 1 in each direction.
(b) sketch the solution curve for which y(-1)=1 and label it.
(c) sketch the solution curve for which y(-1)=-1 and label it.
(d) what is the long term behaviour of the solution curve in (b)

(11) [/6]

(a) what is the amplitude of the sine wave shown   meters

(b) What is the frequency of the sine wave shown   hertz

(c) what is the formula of the sine wave shown

(12) [/6] A rectangle has an initial width of 30cm and a length of 20cm. The area is increasing at a rate of 40 cm^2 per minute. The width is increasing at a rate of 3 cm per minute.
What is the rate of change of the length ?   cm/minute

Math 34B Final Version $\boxed{A}$ Solutions   Prof Cooper   Winter 99

(1) (a) $\boxed{x^3 + 3x^2 + C}$

(b) $\left[2t^2/2 + (1/2)e^{4t}\right]_0^3$

$= (27/2 + (1/2)e^{12}) - (0+1/2)$

$= \boxed{13 + e^{12}/2}$

(c) $\left[-5x\right]_1^2 = (-5/2) - (-5/1) = \boxed{5/2}$

-1pt per error. 2pts per part

(2)(a) $2y^2 + 5$ (b) $4xy - 7$ (c) $4y$

-1pt per error. 2pts per part

(3) v = speed of plane in
middle part of flight   1pt

plane takes 2 hours to go 1000 miles. Does first 1/2 hour at 200 mph so travels 100 miles in first half hour. Thus travels remaining 900 miles in 3/2 hours. Thus $(3/2)v = 900$ so $v = 600$ mph. 2pt

- - - - - - - - 2000 miles - - - - - - - - -

1000 miles _____

200mph	900 miles	200mph
1/2 hr	v mph	1/2 hr
100 miles		100 miles

1pt

Middle part of journey is 2000-200=1800 miles. Speed of this part is 600 mph so time taken for middle of journey is 1800/600 = 3 1pt hours. Total time for flight is $(1/2)+(1/2)+3 = \boxed{4\ \text{hours}}$ 1pt

(4)

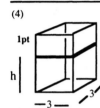

t = time in hours
h = height of water in meters after t hours 1pt

volume of water which enters tanks in t hours is 2000t liters = 2t cubic meters. 1pt
Volume of water in tank
= (area base)×height
= $9h = 2t$ 1pt   thus $h = \boxed{2t/9}$ 1pt

and rate water rises is $\boxed{2/9}$ meters per hour 1pt

(5) f(t) = number of fish in lake after t years. Told $f'(t) = 300+200t$ 1pt integrate to get
$f(t) = 300t + 100t^2 + C$ 1pt

initially $f(0) = 2000 = C$

thus $f(t) = \boxed{300t + 100t^2 + 2000}$ 1pt

There are 3800 fish when

$100t^2 + 300t + 2000 = 3800$ 1pt

simplify: $t^2 + 3t - 18 = 0$
factor $(t-3)(t+6)=0$

so $\boxed{t = \quad 3 \quad \text{years}}$ 1pt

(6) $f'(x) = -3x^2 + 12$ 1pt

$f''(x) = -6x$ 1pt

for max/min $f'(x)=0$ so
$3x^2 = 12$ hence $x^2 = 4$ so
$x = 2$ or $x = -2$. 1pt

when x=-2 $f''(-2) = -6(-2)=12 > 0$
thus have a local minimum 1pt

$\boxed{f(-2) = 8-24+5 = -11}$ 1pt $\boxed{x = -2}$

when x=2 $f''(2) = -6(2)=-12 < 0$
thus have a local maximum

$\boxed{f(2) = -8+24+5 = 21}$ 1pt $\boxed{x = 2}$

(7) m(t) = mass in grams after t hours

(a) $m'(t) = 0.3\,m$ 1pt

General solution $m(t) = Ae^{.3t}$ 1pt

told $m(3) = 5 = Ae^9$ so $A = 5e^{-.9}$

(b) thus $m(t) = \boxed{5\,e^{.3t-.9}}$ 1pt

(c) $m'(6) = 0.3m(6) = \boxed{1.5\,e^{.9} \ \text{gms/hr}}$ 1pt

(8) y(t) = temperature of steel after t minutes. Newtons law gives $y'(t) = k(M-y)$ where M=temp of surroundings and k is a constant. Told M = 50.
General solution $y(t) = 50 + Ae^{-kt}$ 1pt

initally $y(0) = 550 = 50 + A$
thus A=500 so
$y(t) = 50 + 500e^{-kt}$

Initally $y'(0)=-20 = k(50-550)$
so $k=20/500=1/25$. 1pt

(a) $\boxed{y'(t) = (1/25)(50-y)}$ 1pt

(b) $\boxed{y(20) = 50 + 500e^{-20/25}}$ 1pt

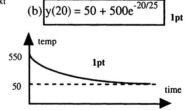

(9) y(t) = number of rabbits after t months 1pt
Logistic equation:
$y'(t) = ky(1-y/2000)$ 1pt

Initially $y'(0) = 20 = k(200)(1-200/2000)$
so $1 = 9k$ so $k = 1/9$ 1pt

(a) thus $y'(t) = \boxed{(1/9)y(1-y/2000)}$ 1pt

(b) when y=1000 equation says
$y' = (1/9)(1000)(1-1000/2000) = \boxed{500/9}$ 1pt

(c) After 1000 years population is very close to maximum possible thus $\boxed{2000}$ 1pt

(10)

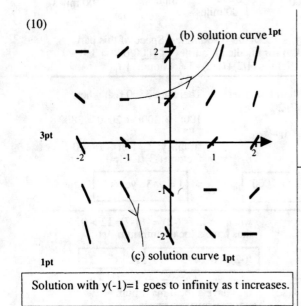

3pt

1pt

Solution with y(-1)=1 goes to infinity as t increases.

(11) (a) amplitude (17-3)/2 = $\boxed{7}$ meters   **1pt**

allowed error 2 meters either way

(b) frequency = $\boxed{1/28 \text{ Hertz}}$ **2pt**

(c) $10 + 7\sin(2\pi t/28)$ **3pt**

(12)

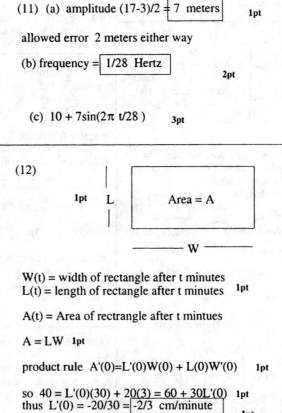

W(t) = width of rectangle after t minutes
L(t) = length of rectangle after t minutes   **1pt**

A(t) = Area of rectrangle after t mintues

A = LW   **1pt**

product rule   A'(0)=L'(0)W(0) + L(0)W'(0)   **1pt**

so 40 = L'(0)(30) + 20(3) = 60 + 30L'(0)   **1pt**
thus L'(0) = -20/30 = $\boxed{-2/3 \text{ cm/minute}}$
                                                            1pt

x	0.00	0.01	0.02	0.03	0.04	0.05	0.06	0.07	0.08	0.09
1.0	.0000	.0043	.0086	.0128	.0170	.0212	.0253	.0294	.0334	.0374
1.1	.0414	.0453	.0492	.0531	.0569	.0607	.0645	.0682	.0719	.0755
1.2	.0792	.0828	.0864	.0899	.0934	.0969	.1004	.1038	.1072	.1106
1.3	.1139	.1173	.1206	.1239	.1271	.1303	.1335	.1367	.1399	.1430
1.4	.1461	.1492	.1523	.1553	.1584	.1614	.1644	.1673	.1703	.1732
1.5	.1761	.1790	.1818	.1847	.1875	.1903	.1931	.1959	.1987	.2014
1.6	.2041	.2068	.2095	.2122	.2148	.2175	.2201	.2227	.2253	.2279
1.7	.2304	.2330	.2355	.2380	.2405	.2430	.2455	.2480	.2504	.2529
1.8	.2553	.2577	.2601	.2625	.2648	.2672	.2695	.2718	.2742	.2765
1.9	.2788	.2810	.2833	.2856	.2878	.2900	.2923	.2945	.2967	.2989
2.0	.3010	.3032	.3054	.3075	.3096	.3118	.3139	.3160	.3181	.3201
2.1	.3222	.3243	.3263	.3284	.3304	.3324	.3345	.3365	.3385	.3404
2.2	.3424	.3444	.3464	.3483	.3502	.3522	.3541	.3560	.3579	.3598
2.3	.3617	.3636	.3655	.3674	.3692	.3711	.3729	.3747	.3766	.3784
2.4	.3802	.3820	.3838	.3856	.3874	.3892	.3909	.3927	.3945	.3962
2.5	.3979	.3997	.4014	.4031	.4048	.4065	.4082	.4099	.4116	.4133
2.6	.4150	.4166	.4183	.4200	.4216	.4232	.4249	.4265	.4281	.4298
2.7	.4314	.4330	.4346	.4362	.4378	.4393	.4409	.4425	.4440	.4456
2.8	.4472	.4487	.4502	.4518	.4533	.4548	.4564	.4579	.4594	.4609
2.9	.4624	.4639	.4654	.4669	.4683	.4698	.4713	.4728	.4742	.4757
3.0	.4771	.4786	.4800	.4814	.4829	.4843	.4857	.4871	.4886	.4900
3.1	.4914	.4928	.4942	.4955	.4969	.4983	.4997	.5011	.5024	.5038
3.2	.5051	.5065	.5079	.5092	.5105	.5119	.5132	.5145	.5159	.5172
3.3	.5185	.5198	.5211	.5224	.5237	.5250	.5263	.5276	.5289	.5302
3.4	.5315	.5328	.5340	.5353	.5366	.5378	.5391	.5403	.5416	.5428
3.5	.5441	.5453	.5465	.5478	.5490	.5502	.5514	.5527	.5539	.5551
3.6	.5563	.5575	.5587	.5599	.5611	.5623	.5635	.5647	.5658	.5670
3.7	.5682	.5694	.5705	.5717	.5729	.5740	.5752	.5763	.5775	.5786
3.8	.5798	.5809	.5821	.5832	.5843	.5855	.5866	.5877	.5888	.5899
3.9	.5911	.5922	.5933	.5944	.5955	.5966	.5977	.5988	.5999	.6010
4.0	.6021	.6031	.6042	.6053	.6064	.6075	.6085	.6096	.6107	.6117
4.1	.6128	.6138	.6149	.6160	.6170	.6180	.6191	.6201	.6212	.6222
4.2	.6232	.6243	.6253	.6263	.6274	.6284	.6294	.6304	.6314	.6325
4.3	.6335	.6345	.6355	.6365	.6375	.6385	.6395	.6405	.6415	.6425
4.4	.6435	.6444	.6454	.6464	.6474	.6484	.6493	.6503	.6513	.6522
4.5	.6532	.6542	.6551	.6561	.6571	.6580	.6590	.6599	.6609	.6618
4.6	.6628	.6637	.6646	.6656	.6665	.6675	.6684	.6693	.6702	.6712
4.7	.6721	.6730	.6739	.6749	.6758	.6767	.6776	.6785	.6794	.6803
4.8	.6812	.6821	.6830	.6839	.6848	.6857	.6866	.6875	.6884	.6893
4.9	.6902	.6911	.6920	.6928	.6937	.6946	.6955	.6964	.6972	.6981
5.0	.6990	.6998	.7007	.7016	.7024	.7033	.7042	.7050	.7059	.7067
5.1	.7076	.7084	.7093	.7101	.7110	.7118	.7126	.7135	.7143	.7152
5.2	.7160	.7168	.7177	.7185	.7193	.7202	.7210	.7218	.7226	.7235
5.3	.7243	.7251	.7259	.7267	.7275	.7284	.7292	.7300	.7308	.7316
5.4	.7324	.7332	.7340	.7348	.7356	.7364	.7372	.7380	.7388	.7396
5.5	.7404	.7412	.7419	.7427	.7435	.7443	.7451	.7459	.7466	.7474
5.6	.7482	.7490	.7497	.7505	.7513	.7520	.7528	.7536	.7543	.7551
5.7	.7559	.7566	.7574	.7582	.7589	.7597	.7604	.7612	.7619	.7627
5.8	.7634	.7642	.7649	.7657	.7664	.7672	.7679	.7686	.7694	.7701
5.9	.7709	.7716	.7723	.7731	.7738	.7745	.7752	.7760	.7767	.7774
6.0	.7782	.7789	.7796	.7803	.7810	.7818	.7825	.7832	.7839	.7846
6.1	.7853	.7860	.7868	.7875	.7882	.7889	.7896	.7903	.7910	.7917
6.2	.7924	.7931	.7938	.7945	.7952	.7959	.7966	.7973	.7980	.7987
6.3	.7993	.8000	.8007	.8014	.8021	.8028	.8035	.8041	.8048	.8055
6.4	.8062	.8069	.8075	.8082	.8089	.8096	.8102	.8109	.8116	.8122
6.5	.8129	.8136	.8142	.8149	.8156	.8162	.8169	.8176	.8182	.8189
6.6	.8195	.8202	.8209	.8215	.8222	.8228	.8235	.8241	.8248	.8254
6.7	.8261	.8267	.8274	.8280	.8287	.8293	.8299	.8306	.8312	.8319
6.8	.8325	.8331	.8338	.8344	.8351	.8357	.8363	.8370	.8376	.8382
6.9	.8388	.8395	.8401	.8407	.8414	.8420	.8426	.8432	.8439	.8445
7.0	.8451	.8457	.8463	.8470	.8476	.8482	.8488	.8494	.8500	.8506
7.1	.8513	.8519	.8525	.8531	.8537	.8543	.8549	.8555	.8561	.8567
7.2	.8573	.8579	.8585	.8591	.8597	.8603	.8609	.8615	.8621	.8627
7.3	.8633	.8639	.8645	.8651	.8657	.8663	.8669	.8675	.8681	.8686
7.4	.8692	.8698	.8704	.8710	.8716	.8722	.8727	.8733	.8739	.8745
7.5	.8751	.8756	.8762	.8768	.8774	.8779	.8785	.8791	.8797	.8802
7.6	.8808	.8814	.8820	.8825	.8831	.8837	.8842	.8848	.8854	.8859
7.7	.8865	.8871	.8876	.8882	.8887	.8893	.8899	.8904	.8910	.8915
7.8	.8921	.8927	.8932	.8938	.8943	.8949	.8954	.8960	.8965	.8971
7.9	.8976	.8982	.8987	.8993	.8998	.9004	.9009	.9015	.9020	.9025
8.0	.9031	.9036	.9042	.9047	.9053	.9058	.9063	.9069	.9074	.9079
8.1	.9085	.9090	.9096	.9101	.9106	.9112	.9117	.9122	.9128	.9133
8.2	.9138	.9143	.9149	.9154	.9159	.9165	.9170	.9175	.9180	.9186
8.3	.9191	.9196	.9201	.9206	.9212	.9217	.9222	.9227	.9232	.9238
8.4	.9243	.9248	.9253	.9258	.9263	.9269	.9274	.9279	.9284	.9289
8.5	.9294	.9299	.9304	.9309	.9315	.9320	.9325	.9330	.9335	.9340
8.6	.9345	.9350	.9355	.9360	.9365	.9370	.9375	.9380	.9385	.9390
8.7	.9395	.9400	.9405	.9410	.9415	.9420	.9425	.9430	.9435	.9440
8.8	.9445	.9450	.9455	.9460	.9465	.9469	.9474	.9479	.9484	.9489
8.9	.9494	.9499	.9504	.9509	.9513	.9518	.9523	.9528	.9533	.9538
9.0	.9542	.9547	.9552	.9557	.9562	.9566	.9571	.9576	.9581	.9586
9.1	.9590	.9595	.9600	.9605	.9609	.9614	.9619	.9624	.9628	.9633
9.2	.9638	.9643	.9647	.9652	.9657	.9661	.9666	.9671	.9675	.9680
9.3	.9685	.9689	.9694	.9699	.9703	.9708	.9713	.9717	.9722	.9727
9.4	.9731	.9736	.9741	.9745	.9750	.9754	.9759	.9763	.9768	.9773
9.5	.9777	.9782	.9786	.9791	.9795	.9800	.9805	.9809	.9814	.9818
9.6	.9823	.9827	.9832	.9836	.9841	.9845	.9850	.9854	.9859	.9863
9.7	.9868	.9872	.9877	.9881	.9886	.9890	.9894	.9899	.9903	.9908
9.8	.9912	.9917	.9921	.9926	.9930	.9934	.9939	.9943	.9948	.9952
9.9	.9956	.9961	.9965	.9969	.9974	.9978	.9983	.9987	.9991	.9996

Logs to Base 10   Log(x)

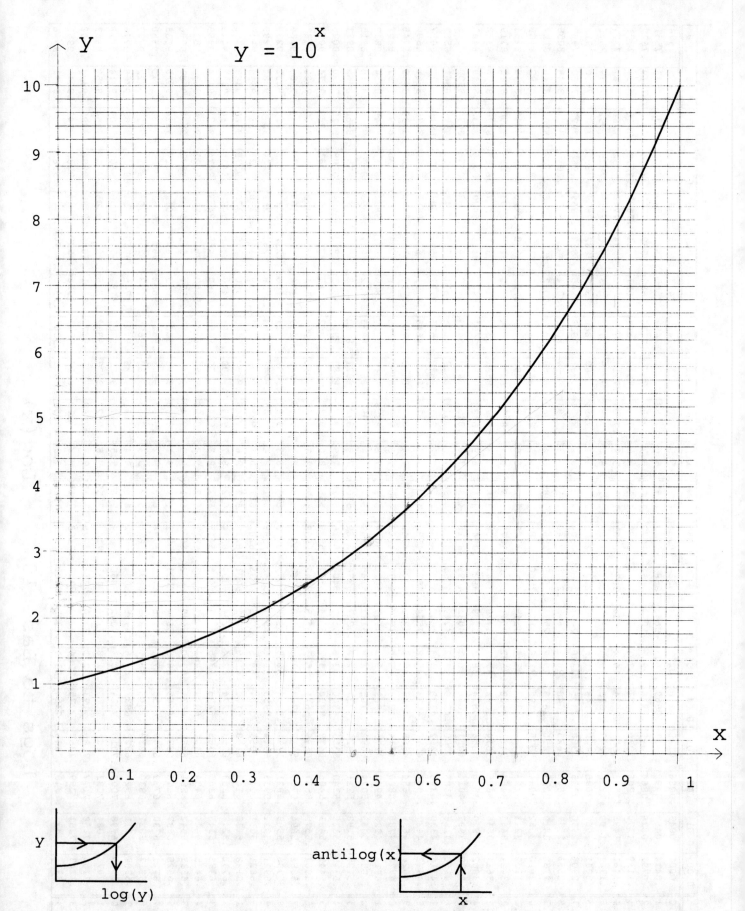

x	0.00	0.01	0.02	0.03	0.04	0.05	0.06	0.07	0.08	0.09
1.0	0.000	0.010	0.020	0.030	0.039	0.049	0.058	0.068	0.077	0.086
1.1	0.095	0.104	0.113	0.122	0.131	0.140	0.148	0.157	0.166	0.174
1.2	0.182	0.191	0.199	0.207	0.215	0.223	0.231	0.239	0.247	0.255
1.3	0.262	0.270	0.278	0.285	0.293	0.300	0.307	0.315	0.322	0.329
1.4	0.336	0.344	0.351	0.358	0.365	0.372	0.378	0.385	0.392	0.399
1.5	0.405	0.412	0.419	0.425	0.432	0.438	0.445	0.451	0.457	0.464
1.6	0.470	0.476	0.482	0.489	0.495	0.501	0.507	0.513	0.519	0.525
1.7	0.531	0.536	0.542	0.548	0.554	0.560	0.565	0.571	0.577	0.582
1.8	0.588	0.593	0.599	0.604	0.610	0.615	0.621	0.626	0.631	0.637
1.9	0.642	0.647	0.652	0.658	0.663	0.668	0.673	0.678	0.683	0.688
2.0	0.693	0.698	0.703	0.708	0.713	0.718	0.723	0.728	0.732	0.737
2.1	0.742	0.747	0.751	0.756	0.761	0.765	0.770	0.775	0.779	0.784
2.2	0.788	0.793	0.798	0.802	0.806	0.811	0.815	0.820	0.824	0.829
2.3	0.833	0.837	0.842	0.846	0.850	0.854	0.859	0.863	0.867	0.871
2.4	0.875	0.880	0.884	0.888	0.892	0.896	0.900	0.904	0.908	0.912
2.5	0.916	0.920	0.924	0.928	0.932	0.936	0.940	0.944	0.948	0.952
2.6	0.956	0.959	0.963	0.967	0.971	0.975	0.978	0.982	0.986	0.990
2.7	0.993	0.997	1.001	1.004	1.008	1.012	1.015	1.019	1.022	1.026
2.8	1.030	1.033	1.037	1.040	1.044	1.047	1.051	1.054	1.058	1.061
2.9	1.065	1.068	1.072	1.075	1.078	1.082	1.085	1.089	1.092	1.095
3.0	1.099	1.102	1.105	1.109	1.112	1.115	1.118	1.122	1.125	1.128
3.1	1.131	1.135	1.138	1.141	1.144	1.147	1.151	1.154	1.157	1.160
3.2	1.163	1.166	1.169	1.172	1.176	1.179	1.182	1.185	1.188	1.191
3.3	1.194	1.197	1.200	1.203	1.206	1.209	1.212	1.215	1.218	1.221
3.4	1.224	1.227	1.230	1.233	1.235	1.238	1.241	1.244	1.247	1.250
3.5	1.253	1.256	1.258	1.261	1.264	1.267	1.270	1.273	1.275	1.278
3.6	1.281	1.284	1.286	1.289	1.292	1.295	1.297	1.300	1.303	1.306
3.7	1.308	1.311	1.314	1.316	1.319	1.322	1.324	1.327	1.330	1.332
3.8	1.335	1.338	1.340	1.343	1.345	1.348	1.351	1.353	1.356	1.358
3.9	1.361	1.364	1.366	1.369	1.371	1.374	1.376	1.379	1.381	1.384
4.0	1.386	1.389	1.391	1.394	1.396	1.399	1.401	1.404	1.406	1.409
4.1	1.411	1.413	1.416	1.418	1.421	1.423	1.426	1.428	1.430	1.433
4.2	1.435	1.437	1.440	1.442	1.445	1.447	1.449	1.452	1.454	1.456
4.3	1.459	1.461	1.463	1.466	1.468	1.470	1.472	1.475	1.477	1.479
4.4	1.482	1.484	1.486	1.488	1.491	1.493	1.495	1.497	1.500	1.502
4.5	1.504	1.506	1.509	1.511	1.513	1.515	1.517	1.520	1.522	1.524
4.6	1.526	1.528	1.530	1.533	1.535	1.537	1.539	1.541	1.543	1.545
4.7	1.548	1.550	1.552	1.554	1.556	1.558	1.560	1.562	1.564	1.567
4.8	1.569	1.571	1.573	1.575	1.577	1.579	1.581	1.583	1.585	1.587
4.9	1.589	1.591	1.593	1.595	1.597	1.599	1.601	1.603	1.605	1.607
5.0	1.609	1.611	1.613	1.615	1.617	1.619	1.621	1.623	1.625	1.627
5.1	1.629	1.631	1.633	1.635	1.637	1.639	1.641	1.643	1.645	1.647
5.2	1.649	1.651	1.652	1.654	1.656	1.658	1.660	1.662	1.664	1.666
5.3	1.668	1.670	1.671	1.673	1.675	1.677	1.679	1.681	1.683	1.685
5.4	1.686	1.688	1.690	1.692	1.694	1.696	1.697	1.699	1.701	1.703

x	0.00	0.01	0.02	0.03	0.04	0.05	0.06	0.07	0.08	0.09
5.5	1.705	1.707	1.708	1.710	1.712	1.714	1.716	1.717	1.719	1.721
5.6	1.723	1.725	1.726	1.728	1.730	1.732	1.733	1.735	1.737	1.739
5.7	1.740	1.742	1.744	1.746	1.747	1.749	1.751	1.753	1.754	1.756
5.8	1.758	1.760	1.761	1.763	1.765	1.766	1.768	1.770	1.772	1.773
5.9	1.775	1.777	1.778	1.780	1.782	1.783	1.785	1.787	1.788	1.790
6.0	1.792	1.793	1.795	1.797	1.798	1.800	1.802	1.803	1.805	1.807
6.1	1.808	1.810	1.812	1.813	1.815	1.816	1.818	1.820	1.821	1.823
6.2	1.825	1.826	1.828	1.829	1.831	1.833	1.834	1.836	1.837	1.839
6.3	1.841	1.842	1.844	1.845	1.847	1.848	1.850	1.852	1.853	1.855
6.4	1.856	1.858	1.859	1.861	1.863	1.864	1.866	1.867	1.869	1.870
6.5	1.872	1.873	1.875	1.876	1.878	1.879	1.881	1.883	1.884	1.886
6.6	1.887	1.889	1.890	1.892	1.893	1.895	1.896	1.898	1.899	1.901
6.7	1.902	1.904	1.905	1.907	1.908	1.910	1.911	1.913	1.914	1.915
6.8	1.917	1.918	1.920	1.921	1.923	1.924	1.926	1.927	1.929	1.930
6.9	1.932	1.933	1.934	1.936	1.937	1.939	1.940	1.942	1.943	1.944
7.0	1.946	1.947	1.949	1.950	1.952	1.953	1.954	1.956	1.957	1.959
7.1	1.960	1.962	1.963	1.964	1.966	1.967	1.969	1.970	1.971	1.973
7.2	1.974	1.975	1.977	1.978	1.980	1.981	1.982	1.984	1.985	1.987
7.3	1.988	1.989	1.991	1.992	1.993	1.995	1.996	1.997	1.999	2.000
7.4	2.001	2.003	2.004	2.006	2.007	2.008	2.010	2.011	2.012	2.014
7.5	2.015	2.016	2.018	2.019	2.020	2.022	2.023	2.024	2.026	2.027
7.6	2.028	2.029	2.031	2.032	2.033	2.035	2.036	2.037	2.039	2.040
7.7	2.041	2.043	2.044	2.045	2.046	2.048	2.049	2.050	2.052	2.053
7.8	2.054	2.055	2.057	2.058	2.059	2.061	2.062	2.063	2.064	2.066
7.9	2.067	2.068	2.069	2.071	2.072	2.073	2.074	2.076	2.077	2.078
8.0	2.079	2.081	2.082	2.083	2.084	2.086	2.087	2.088	2.089	2.091
8.1	2.092	2.093	2.094	2.096	2.097	2.098	2.099	2.100	2.102	2.103
8.2	2.104	2.105	2.107	2.108	2.109	2.110	2.111	2.113	2.114	2.115
8.3	2.116	2.117	2.119	2.120	2.121	2.122	2.123	2.125	2.126	2.127
8.4	2.128	2.129	2.131	2.132	2.133	2.134	2.135	2.137	2.138	2.139
8.5	2.140	2.141	2.142	2.144	2.145	2.146	2.147	2.148	2.149	2.151
8.6	2.152	2.153	2.154	2.155	2.156	2.158	2.159	2.160	2.161	2.162
8.7	2.163	2.164	2.166	2.167	2.168	2.169	2.170	2.171	2.172	2.174
8.8	2.175	2.176	2.177	2.178	2.179	2.180	2.182	2.183	2.184	2.185
8.9	2.186	2.187	2.188	2.189	2.191	2.192	2.193	2.194	2.195	2.196
9.0	2.197	2.198	2.199	2.201	2.202	2.203	2.204	2.205	2.206	2.207
9.1	2.208	2.209	2.210	2.212	2.213	2.214	2.215	2.216	2.217	2.218
9.2	2.219	2.220	2.221	2.222	2.224	2.225	2.226	2.227	2.228	2.229
9.3	2.230	2.231	2.232	2.233	2.234	2.235	2.236	2.238	2.239	2.240
9.4	2.241	2.242	2.243	2.244	2.245	2.246	2.247	2.248	2.249	2.250
9.5	2.251	2.252	2.253	2.254	2.255	2.257	2.258	2.259	2.260	2.261
9.6	2.262	2.263	2.264	2.265	2.266	2.267	2.268	2.269	2.270	2.271
9.7	2.272	2.273	2.274	2.275	2.276	2.277	2.278	2.279	2.280	2.281
9.8	2.282	2.283	2.284	2.285	2.286	2.287	2.288	2.289	2.291	2.292
9.9	2.293	2.294	2.295	2.296	2.297	2.298	2.299	2.300	2.301	2.302

NATURAL LOGS   Ln(10) = 2.302

e^x for $0 \le x \le 9$

e^x	0	0.01	0.02	0.03	0.04	0.05	0.06	0.07	0.08	0.09
0	1	1.0101	1.0202	1.0305	1.0408	1.0513	1.0618	1.0725	1.0833	1.0942
0.1	1.1052	1.1163	1.1275	1.1388	1.1503	1.1618	1.1735	1.1853	1.1972	1.2092
0.2	1.2214	1.2337	1.2461	1.2586	1.2712	1.284	1.2969	1.31	1.3231	1.3364
0.3	1.3499	1.3634	1.3771	1.391	1.4049	1.4191	1.4333	1.4477	1.4623	1.477
0.4	1.4918	1.5068	1.522	1.5373	1.5527	1.5683	1.5841	1.6	1.6161	1.6323
0.5	1.6487	1.6653	1.682	1.6989	1.716	1.7333	1.7507	1.7683	1.786	1.804
0.6	1.8221	1.8404	1.8589	1.8776	1.8965	1.9155	1.9348	1.9542	1.9739	1.9937
0.7	2.0138	2.034	2.0544	2.0751	2.0959	2.117	2.1383	2.1598	2.1815	2.2034
0.8	2.2255	2.2479	2.2705	2.2933	2.3164	2.3396	2.3632	2.3869	2.4109	2.4351
0.9	2.4596	2.4843	2.5093	2.5345	2.56	2.5857	2.6117	2.6379	2.6645	2.6912
1	2.7183	2.7456	2.7732	2.8011	2.8292	2.8577	2.8864	2.9154	2.9447	2.9743
1.1	3.0042	3.0344	3.0649	3.0957	3.1268	3.1582	3.1899	3.222	3.2544	3.2871
1.2	3.3201	3.3535	3.3872	3.4212	3.4556	3.4903	3.5254	3.5609	3.5966	3.6328
1.3	3.6693	3.7062	3.7434	3.781	3.819	3.8574	3.8962	3.9354	3.9749	4.0149
1.4	4.0552	4.096	4.1371	4.1787	4.2207	4.2631	4.306	4.3492	4.3929	4.4371
1.5	4.4817	4.5267	4.5722	4.6182	4.6646	4.7115	4.7588	4.8066	4.855	4.9037
1.6	4.953	5.0028	5.0531	5.1039	5.1552	5.207	5.2593	5.3122	5.3656	5.4195
1.7	5.4739	5.529	5.5845	5.6407	5.6973	5.7546	5.8124	5.8709	5.9299	5.9895
1.8	6.0496	6.1104	6.1719	6.2339	6.2965	6.3598	6.4237	6.4883	6.5535	6.6194
1.9	6.6859	6.7531	6.821	6.8895	6.9588	7.0287	7.0993	7.1707	7.2427	7.3155
2	7.3891	7.4633	7.5383	7.6141	7.6906	7.7679	7.846	7.9248	8.0045	8.0849
2.1	8.1662	8.2482	8.3311	8.4149	8.4994	8.5849	8.6711	8.7583	8.8463	8.9352
2.2	9.025	9.1157	9.2073	9.2999	9.3933	9.4877	9.5831	9.6794	9.7767	9.8749
2.3	9.9742	10.074	10.176	10.278	10.381	10.486	10.591	10.697	10.805	10.913
2.4	11.023	11.134	11.246	11.359	11.473	11.588	11.705	11.822	11.941	12.061
2.5	12.182	12.305	12.429	12.554	12.68	12.807	12.936	13.066	13.197	13.33
2.6	13.464	13.599	13.736	13.874	14.013	14.154	14.296	14.44	14.585	14.732
2.7	14.88	15.029	15.18	15.333	15.487	15.643	15.8	15.959	16.119	16.281
2.8	16.445	16.61	16.777	16.945	17.116	17.288	17.462	17.637	17.814	17.993
2.9	18.174	18.357	18.541	18.728	18.916	19.106	19.298	19.492	19.688	19.886
3	20.086	20.287	20.491	20.697	20.905	21.115	21.328	21.542	21.758	21.977
3.1	22.198	22.421	22.646	22.874	23.104	23.336	23.571	23.807	24.047	24.288
3.2	24.533	24.779	25.028	25.28	25.534	25.79	26.05	26.311	26.576	26.843
3.3	27.113	27.385	27.66	27.938	28.219	28.503	28.789	29.079	29.371	29.666
3.4	29.964	30.265	30.569	30.877	31.187	31.5	31.817	32.137	32.46	32.786
3.5	33.115	33.448	33.784	34.124	34.467	34.813	35.163	35.517	35.874	36.234
3.6	36.598	36.966	37.338	37.713	38.092	38.475	38.861	39.252	39.646	40.045
3.7	40.447	40.854	41.264	41.679	42.098	42.521	42.948	43.38	43.816	44.256
3.8	44.701	45.15	45.604	46.063	46.525	46.993	47.465	47.942	48.424	48.911
3.9	49.402	49.899	50.4	50.907	51.419	51.935	52.457	52.985	53.517	54.055
4	54.598	55.147	55.701	56.261	56.826	57.397	57.974	58.557	59.145	59.74
4.1	60.34	60.947	61.559	62.178	62.803	63.434	64.072	64.715	65.366	66.023
4.2	66.686	67.357	68.033	68.717	69.408	70.105	70.81	71.522	72.24	72.966
4.3	73.7	74.44	75.189	75.944	76.708	77.478	78.257	79.044	79.838	80.64
4.4	81.451	82.269	83.096	83.931	84.775	85.627	86.488	87.357	88.235	89.121

e^x	0	0.01	0.02	0.03	0.04	0.05	0.06	0.07	0.08	0.09
4.5	90.02	90.92	91.84	92.76	93.69	94.63	95.58	96.54	97.51	98.49
4.6	99.48	100.5	101.5	102.5	103.5	104.6	105.6	106.7	107.8	108.9
4.7	109.9	111.1	112.2	113.3	114.4	115.6	116.7	117.9	119.1	120.3
4.8	121.5	122.7	124	125.2	126.5	127.7	129	130.3	131.6	133
4.9	134.3	135.6	137	138.4	139.8	141.2	142.6	144	145.5	146.9
5	148.4	149.9	151.4	152.9	154.5	156	157.6	159.2	160.8	162.4
5.1	164	165.7	167.3	169	170.7	172.4	174.2	175.9	177.7	179.5
5.2	181.3	183.1	184.9	186.8	188.7	190.6	192.5	194.4	196.4	198.3
5.3	200.3	202.4	204.4	206.4	208.5	210.6	212.7	214.9	217	219.2
5.4	221.4	223.6	225.9	228.1	230.4	232.7	235	237.5	239.8	242.3
5.5	244.7	247.2	249.6	252.1	254.7	257.2	259.8	262.4	265.1	267.7
5.6	270.4	273.1	275.9	278.7	281.5	284.3	287.1	290	292.9	295.9
5.7	298.9	301.9	304.9	308	311	314.2	317.3	320.5	323.8	327
5.8	330.3	333.6	337	340.4	343.8	347.2	350.7	354.2	357.8	361.4
5.9	365	368.7	372.4	376.2	379.9	383.8	387.6	391.5	395.4	399.4
6	403.4	407.5	411.6	415.7	419.9	424.1	428.4	432.7	437	441.4
6.1	445.9	450.3	454.9	459.4	464.1	468.7	473.4	478.2	483	487.9
6.2	492.7	497.7	502.7	507.8	512.9	518	523.2	528.5	533.8	539.1
6.3	544.6	550	555.6	561.2	566.8	572.5	578.3	584.1	589.9	595.9
6.4	601.8	607.9	614	620.2	626.4	632.7	639.1	645.5	652	658.5
6.5	665.1	671.8	678.6	685.4	692.3	699.2	706.3	713.4	720.5	727.8
6.6	735.1	742.5	749.9	757.5	765.1	772.8	780.6	788.5	796.3	804.3
6.7	812.4	820.6	828.8	837.1	845.6	854.1	862.6	871.3	880.1	888.9
6.8	897.8	906.9	916	925.2	934.5	943.9	953.4	962.9	972.6	982.4
6.9	992.3	1002	1012	1022	1033	1043	1054	1064	1075	1086
7	1097	1108	1119	1130	1141	1153	1164	1176	1188	1200
7.1	1212	1224	1236	1249	1261	1274	1287	1300	1313	1326
7.2	1339	1353	1366	1380	1394	1408	1422	1437	1451	1466
7.3	1480	1495	1510	1525	1541	1556	1572	1588	1604	1620
7.4	1636	1652	1669	1686	1703	1720	1737	1755	1772	1790
7.5	1808	1826	1845	1863	1882	1901	1920	1939	1959	1978
7.6	1998	2018	2039	2059	2080	2101	2122	2143	2165	2186
7.7	2208	2231	2253	2276	2298	2322	2345	2368	2392	2416
7.8	2441	2465	2490	2515	2540	2566	2592	2618	2644	2670
7.9	2697	2724	2752	2779	2807	2836	2864	2893	2922	2951
8	2981	3011	3041	3072	3103	3134	3165	3197	3229	3262
8.1	3294	3328	3361	3395	3429	3463	3498	3533	3569	3605
8.2	3641	3678	3715	3752	3790	3828	3866	3905	3944	3984
8.3	4024	4064	4105	4146	4188	4230	4273	4316	4359	4403
8.4	4447	4492	4537	4583	4629	4675	4722	4770	4817	4866
8.5	4915	4964	5014	5064	5115	5167	5219	5271	5324	5378
8.6	5432	5486	5541	5597	5653	5710	5768	5825	5884	5943
8.7	6003	6063	6124	6186	6248	6311	6374	6438	6503	6568
8.8	6634	6701	6768	6836	6905	6974	7044	7115	7187	7259
8.9	7332	7406	7480	7555	7631	7708	7785	7864	7943	8022

e^{-x} for $0 \leq x \leq 9$

e^{-x}	0	0.01	0.02	0.03	0.04	0.05	0.06	0.07	0.08	0.09
0.0	1.000	0.990	0.980	0.970	0.961	0.951	0.942	0.932	0.923	0.914
0.1	0.905	0.896	0.887	0.878	0.869	0.861	0.852	0.844	0.835	0.827
0.2	0.819	0.811	0.803	0.795	0.787	0.779	0.771	0.763	0.756	0.748
0.3	0.741	0.733	0.726	0.719	0.712	0.705	0.698	0.691	0.684	0.677
0.4	0.670	0.664	0.657	0.651	0.644	0.638	0.631	0.625	0.619	0.613
0.5	0.607	0.600	0.595	0.589	0.583	0.577	0.571	0.566	0.560	0.554
0.6	0.549	0.543	0.538	0.533	0.527	0.522	0.517	0.512	0.507	0.502
0.7	0.497	0.492	0.487	0.482	0.477	0.472	0.468	0.463	0.458	0.454
0.8	0.449	0.445	0.440	0.436	0.432	0.427	0.423	0.419	0.415	0.411
0.9	0.407	0.403	0.399	0.395	0.391	0.387	0.383	0.379	0.375	0.372
1.0	0.368	0.364	0.361	0.357	0.353	0.350	0.346	0.343	0.340	0.336
1.1	0.333	0.330	0.326	0.323	0.320	0.317	0.313	0.310	0.307	0.304
1.2	0.301	0.298	0.295	0.292	0.289	0.287	0.284	0.281	0.278	0.275
1.3	0.273	0.270	0.267	0.264	0.262	0.259	0.257	0.254	0.252	0.249
1.4	0.247	0.244	0.242	0.239	0.237	0.235	0.232	0.230	0.228	0.225
1.5	0.223	0.221	0.219	0.217	0.214	0.212	0.210	0.208	0.206	0.204
1.6	0.202	0.200	0.198	0.196	0.194	0.192	0.190	0.188	0.186	0.185
1.7	0.183	0.181	0.179	0.177	0.176	0.174	0.172	0.170	0.169	0.167
1.8	0.165	0.164	0.162	0.160	0.159	0.157	0.156	0.154	0.153	0.151
1.9	0.150	0.148	0.147	0.145	0.144	0.142	0.141	0.139	0.138	0.137
2.0	0.135	0.134	0.133	0.131	0.130	0.129	0.127	0.126	0.125	0.124
2.1	0.122	0.121	0.120	0.119	0.118	0.116	0.115	0.114	0.113	0.112
2.2	0.111	0.110	0.109	0.108	0.106	0.105	0.104	0.103	0.102	0.101
2.3	0.100	0.099	0.098	0.097	0.096	0.095	0.094	0.093	0.093	0.092
2.4	0.091	0.090	0.089	0.088	0.087	0.086	0.085	0.085	0.084	0.083
2.5	0.082	0.081	0.080	0.080	0.079	0.078	0.077	0.077	0.076	0.075
2.6	0.074	0.074	0.073	0.072	0.071	0.071	0.070	0.069	0.069	0.068
2.7	0.067	0.067	0.066	0.065	0.065	0.064	0.063	0.063	0.062	0.061
2.8	0.061	0.060	0.060	0.059	0.058	0.058	0.057	0.057	0.056	0.056
2.9	0.055	0.054	0.054	0.053	0.053	0.052	0.052	0.051	0.051	0.050
3.0	0.050	0.049	0.049	0.048	0.048	0.047	0.047	0.046	0.046	0.046
3.1	0.045	0.045	0.044	0.044	0.043	0.043	0.042	0.042	0.042	0.041
3.2	0.041	0.040	0.040	0.040	0.039	0.039	0.038	0.038	0.038	0.037
3.3	0.037	0.037	0.036	0.036	0.035	0.035	0.035	0.034	0.034	0.034
3.4	0.033	0.033	0.033	0.032	0.032	0.032	0.031	0.031	0.031	0.031
3.5	0.030	0.030	0.030	0.029	0.029	0.029	0.028	0.028	0.028	0.028
3.6	0.027	0.027	0.027	0.027	0.026	0.026	0.026	0.025	0.025	0.025
3.7	0.025	0.024	0.024	0.024	0.024	0.024	0.023	0.023	0.023	0.023
3.8	0.022	0.022	0.022	0.022	0.021	0.021	0.021	0.021	0.021	0.020
3.9	0.020	0.020	0.020	0.020	0.019	0.019	0.019	0.019	0.019	0.018
4.0	0.018	0.018	0.018	0.018	0.018	0.017	0.017	0.017	0.017	0.017
4.1	0.017	0.016	0.016	0.016	0.016	0.016	0.016	0.015	0.015	0.015
4.2	0.015	0.015	0.015	0.015	0.014	0.014	0.014	0.014	0.014	0.014
4.3	0.014	0.013	0.013	0.013	0.013	0.013	0.013	0.013	0.013	0.012
4.4	0.012	0.012	0.012	0.012	0.012	0.012	0.012	0.011	0.011	0.011
4.5	0.011	0.011	0.011	0.011	0.012	0.012	0.012	0.012	0.012	0.012
4.6	0.010	0.010	0.010	0.010	0.010	0.011	0.011	0.011	0.011	0.011
4.7	0.009	0.009	0.009	0.009	0.009	0.010	0.010	0.010	0.010	0.010
4.8	0.008	0.008	0.008	0.008	0.008	0.009	0.009	0.009	0.009	0.009
4.9	0.007	0.008	0.008	0.008	0.008	0.008	0.008	0.008	0.008	0.008
5.0	0.007	0.007	0.007	0.007	0.007	0.007	0.007	0.007	0.007	0.007
5.1	0.006	0.006	0.006	0.006	0.006	0.006	0.006	0.007	0.007	0.007
5.2	0.006	0.006	0.006	0.006	0.006	0.006	0.006	0.006	0.006	0.006
5.3	0.005	0.005	0.005	0.005	0.005	0.005	0.005	0.006	0.006	0.006
5.4	0.005	0.005	0.005	0.005	0.005	0.005	0.005	0.005	0.005	0.005
5.5	0.004	0.004	0.004	0.004	0.004	0.004	0.004	0.005	0.005	0.005
5.6	0.004	0.004	0.004	0.004	0.004	0.004	0.004	0.004	0.004	0.004
5.7	0.003	0.003	0.003	0.003	0.004	0.004	0.004	0.004	0.004	0.004
5.8	0.003	0.003	0.003	0.003	0.003	0.003	0.003	0.003	0.003	0.003
5.9	0.003	0.003	0.003	0.003	0.003	0.003	0.003	0.003	0.003	0.003
6.0	0.002	0.002	0.003	0.003	0.003	0.003	0.003	0.003	0.003	0.003
6.1	0.002	0.002	0.002	0.002	0.002	0.002	0.002	0.002	0.002	0.002
6.2	0.002	0.002	0.002	0.002	0.002	0.002	0.002	0.002	0.002	0.002
6.3	0.002	0.002	0.002	0.002	0.002	0.002	0.002	0.002	0.002	0.002
6.4	0.002	0.002	0.002	0.002	0.002	0.002	0.002	0.002	0.002	0.002
6.5	0.002	0.002	0.002	0.002	0.002	0.002	0.002	0.002	0.001	0.001
6.6	0.001	0.001	0.001	0.001	0.001	0.001	0.001	0.001	0.001	0.001
6.7	0.001	0.001	0.001	0.001	0.001	0.001	0.001	0.001	0.001	0.001
6.8	0.001	0.001	0.001	0.001	0.001	0.001	0.001	0.001	0.001	0.001
6.9	0.001	0.001	0.001	0.001	0.001	0.001	0.001	0.001	0.001	0.001
7										
7.1										
7.2										
7.3										
7.4										
7.5										
7.6										
7.7										
7.8										
7.9										
8										
8.1										
8.2										
8.3										
8.4										
8.5										
8.6										
8.7										
8.8										
8.9										

Square roots for 1 ≤ x ≤ 10

x	0	0.01	0.02	0.03	0.04	0.05	0.06	0.07	0.08	0.09
1	1.000	1.005	1.010	1.015	1.020	1.025	1.030	1.034	1.039	1.044
1.1	1.049	1.054	1.058	1.063	1.068	1.072	1.077	1.082	1.086	1.091
1.2	1.095	1.100	1.105	1.109	1.114	1.118	1.122	1.127	1.131	1.136
1.3	1.140	1.145	1.149	1.153	1.158	1.162	1.166	1.170	1.175	1.179
1.4	1.183	1.187	1.192	1.196	1.200	1.204	1.208	1.212	1.217	1.221
1.5	1.225	1.229	1.233	1.237	1.241	1.245	1.249	1.253	1.257	1.261
1.6	1.265	1.269	1.273	1.277	1.281	1.285	1.288	1.292	1.296	1.300
1.7	1.304	1.308	1.311	1.315	1.319	1.323	1.327	1.330	1.334	1.338
1.8	1.342	1.345	1.349	1.353	1.356	1.360	1.364	1.367	1.371	1.375
1.9	1.378	1.382	1.386	1.389	1.393	1.396	1.400	1.404	1.407	1.411
2	1.414	1.418	1.421	1.425	1.428	1.432	1.435	1.439	1.442	1.446
2.1	1.449	1.453	1.456	1.459	1.463	1.466	1.470	1.473	1.476	1.480
2.2	1.483	1.487	1.490	1.493	1.497	1.500	1.503	1.507	1.510	1.513
2.3	1.517	1.520	1.523	1.526	1.530	1.533	1.536	1.539	1.543	1.546
2.4	1.549	1.552	1.556	1.559	1.562	1.565	1.568	1.572	1.575	1.578
2.5	1.581	1.584	1.587	1.591	1.594	1.597	1.600	1.603	1.606	1.609
2.6	1.612	1.616	1.619	1.622	1.625	1.628	1.631	1.634	1.637	1.640
2.7	1.643	1.646	1.649	1.652	1.655	1.658	1.661	1.664	1.667	1.670
2.8	1.673	1.676	1.679	1.682	1.685	1.688	1.691	1.694	1.697	1.700
2.9	1.703	1.706	1.709	1.712	1.715	1.718	1.720	1.723	1.726	1.729
3	1.732	1.735	1.738	1.741	1.744	1.746	1.749	1.752	1.755	1.758
3.1	1.761	1.764	1.766	1.769	1.772	1.775	1.778	1.780	1.783	1.786
3.2	1.789	1.792	1.794	1.797	1.800	1.803	1.806	1.808	1.811	1.814
3.3	1.817	1.819	1.822	1.825	1.828	1.830	1.833	1.836	1.838	1.841
3.4	1.844	1.847	1.849	1.852	1.855	1.857	1.860	1.863	1.865	1.868
3.5	1.871	1.873	1.876	1.879	1.881	1.884	1.887	1.889	1.892	1.895
3.6	1.897	1.900	1.903	1.905	1.908	1.910	1.913	1.916	1.918	1.921
3.7	1.924	1.926	1.929	1.931	1.934	1.936	1.939	1.942	1.944	1.947
3.8	1.949	1.952	1.954	1.957	1.960	1.962	1.965	1.967	1.970	1.972
3.9	1.975	1.977	1.980	1.982	1.985	1.987	1.990	1.992	1.995	1.997
4	2.000	2.002	2.005	2.007	2.010	2.012	2.015	2.017	2.020	2.022
4.1	2.025	2.027	2.030	2.032	2.035	2.037	2.040	2.042	2.045	2.047
4.2	2.049	2.052	2.054	2.057	2.059	2.062	2.064	2.066	2.069	2.071
4.3	2.074	2.076	2.078	2.081	2.083	2.086	2.088	2.090	2.093	2.095
4.4	2.098	2.100	2.102	2.105	2.107	2.110	2.112	2.114	2.117	2.119
4.5	2.121	2.124	2.126	2.128	2.131	2.133	2.135	2.138	2.140	2.142
4.6	2.145	2.147	2.149	2.152	2.154	2.156	2.159	2.161	2.163	2.166
4.7	2.168	2.170	2.173	2.175	2.177	2.179	2.182	2.184	2.186	2.189
4.8	2.191	2.193	2.195	2.198	2.200	2.202	2.205	2.207	2.209	2.211
4.9	2.214	2.216	2.218	2.220	2.223	2.225	2.227	2.229	2.232	2.234
5	2.236	2.238	2.241	2.243	2.245	2.247	2.249	2.252	2.254	2.256
5.1	2.258	2.261	2.263	2.265	2.267	2.269	2.272	2.274	2.276	2.278
5.2	2.280	2.283	2.285	2.287	2.289	2.291	2.293	2.296	2.298	2.300
5.3	2.302	2.304	2.307	2.309	2.311	2.313	2.315	2.317	2.319	2.322
5.4	2.324	2.326	2.328	2.330	2.332	2.335	2.337	2.339	2.341	2.343

x	0	0.01	0.02	0.03	0.04	0.05	0.06	0.07	0.08	0.09
5.5	2.345	2.347	2.349	2.352	2.354	2.356	2.358	2.360	2.362	2.364
5.6	2.366	2.369	2.371	2.373	2.375	2.377	2.379	2.381	2.383	2.385
5.7	2.387	2.390	2.392	2.394	2.396	2.398	2.400	2.402	2.404	2.406
5.8	2.408	2.410	2.412	2.415	2.417	2.419	2.421	2.423	2.425	2.427
5.9	2.429	2.431	2.433	2.435	2.437	2.439	2.441	2.443	2.445	2.447
6	2.449	2.452	2.454	2.456	2.458	2.460	2.462	2.464	2.466	2.468
6.1	2.470	2.472	2.474	2.476	2.478	2.480	2.482	2.484	2.486	2.488
6.2	2.490	2.492	2.494	2.496	2.498	2.500	2.502	2.504	2.506	2.508
6.3	2.510	2.512	2.514	2.516	2.518	2.520	2.522	2.524	2.526	2.528
6.4	2.530	2.532	2.534	2.536	2.538	2.540	2.542	2.544	2.546	2.548
6.5	2.550	2.551	2.553	2.555	2.557	2.559	2.561	2.563	2.565	2.567
6.6	2.569	2.571	2.573	2.575	2.577	2.579	2.581	2.583	2.585	2.587
6.7	2.588	2.590	2.592	2.594	2.596	2.598	2.600	2.602	2.604	2.606
6.8	2.608	2.610	2.612	2.613	2.615	2.617	2.619	2.621	2.623	2.625
6.9	2.627	2.629	2.631	2.632	2.634	2.636	2.638	2.640	2.642	2.644
7	2.646	2.648	2.650	2.651	2.653	2.655	2.657	2.659	2.661	2.663
7.1	2.665	2.666	2.668	2.670	2.672	2.674	2.676	2.678	2.680	2.681
7.2	2.683	2.685	2.687	2.689	2.691	2.693	2.694	2.696	2.698	2.700
7.3	2.702	2.704	2.706	2.707	2.709	2.711	2.713	2.715	2.717	2.718
7.4	2.720	2.722	2.724	2.726	2.728	2.729	2.731	2.733	2.735	2.737
7.5	2.739	2.740	2.742	2.744	2.746	2.748	2.750	2.751	2.753	2.755
7.6	2.757	2.759	2.760	2.762	2.764	2.766	2.768	2.769	2.771	2.773
7.7	2.775	2.777	2.778	2.780	2.782	2.784	2.786	2.787	2.789	2.791
7.8	2.793	2.795	2.796	2.798	2.800	2.802	2.804	2.805	2.807	2.809
7.9	2.811	2.812	2.814	2.816	2.818	2.820	2.821	2.823	2.825	2.827
8	2.828	2.830	2.832	2.834	2.835	2.837	2.839	2.841	2.843	2.844
8.1	2.846	2.848	2.850	2.851	2.853	2.855	2.857	2.858	2.860	2.862
8.2	2.864	2.865	2.867	2.869	2.871	2.872	2.874	2.876	2.877	2.879
8.3	2.881	2.883	2.884	2.886	2.888	2.890	2.891	2.893	2.895	2.897
8.4	2.898	2.900	2.902	2.903	2.905	2.907	2.909	2.910	2.912	2.914
8.5	2.915	2.917	2.919	2.921	2.922	2.924	2.926	2.927	2.929	2.931
8.6	2.933	2.934	2.936	2.938	2.939	2.941	2.943	2.944	2.946	2.948
8.7	2.950	2.951	2.953	2.955	2.956	2.958	2.960	2.961	2.963	2.965
8.8	2.966	2.968	2.970	2.972	2.973	2.975	2.977	2.978	2.980	2.982
8.9	2.983	2.985	2.987	2.988	2.990	2.992	2.993	2.995	2.997	2.998
9	3.000	3.002	3.003	3.005	3.007	3.008	3.010	3.012	3.013	3.015
9.1	3.017	3.018	3.020	3.022	3.023	3.025	3.027	3.028	3.030	3.032
9.2	3.033	3.035	3.036	3.038	3.040	3.041	3.043	3.045	3.046	3.048
9.3	3.050	3.051	3.053	3.055	3.056	3.058	3.059	3.061	3.063	3.064
9.4	3.066	3.068	3.069	3.071	3.072	3.074	3.076	3.077	3.079	3.081
9.5	3.082	3.084	3.085	3.087	3.089	3.090	3.092	3.094	3.095	3.097
9.6	3.098	3.100	3.102	3.103	3.105	3.106	3.108	3.110	3.111	3.113
9.7	3.114	3.116	3.118	3.119	3.121	3.122	3.124	3.126	3.127	3.129
9.8	3.130	3.132	3.134	3.135	3.137	3.138	3.140	3.142	3.143	3.145
9.9	3.146	3.148	3.150	3.151	3.153	3.154	3.156	3.158	3.159	3.161

x	0	0.1	0.2	0.3	0.4	0.5	0.6	0.7	0.8	0.9
10	3.162	3.178	3.194	3.209	3.225	3.240	3.256	3.271	3.286	3.302
11	3.317	3.332	3.347	3.362	3.376	3.391	3.406	3.421	3.435	3.450
12	3.464	3.479	3.493	3.507	3.521	3.536	3.550	3.564	3.578	3.592
13	3.606	3.619	3.633	3.647	3.661	3.674	3.688	3.701	3.715	3.728
14	3.742	3.755	3.768	3.782	3.795	3.808	3.821	3.834	3.847	3.860
15	3.873	3.886	3.899	3.912	3.924	3.937	3.950	3.962	3.975	3.987
16	4.000	4.012	4.025	4.037	4.050	4.062	4.074	4.087	4.099	4.111
17	4.123	4.135	4.147	4.159	4.171	4.183	4.195	4.207	4.219	4.231
18	4.243	4.254	4.266	4.278	4.290	4.301	4.313	4.324	4.336	4.347
19	4.359	4.370	4.382	4.393	4.405	4.416	4.427	4.438	4.450	4.461
20	4.472	4.483	4.494	4.506	4.517	4.528	4.539	4.550	4.561	4.572
21	4.583	4.593	4.604	4.615	4.626	4.637	4.648	4.658	4.669	4.680
22	4.690	4.701	4.712	4.722	4.733	4.743	4.754	4.764	4.775	4.785
23	4.796	4.806	4.817	4.827	4.837	4.848	4.858	4.868	4.879	4.889
24	4.899	4.909	4.919	4.930	4.940	4.950	4.960	4.970	4.980	4.990
25	5.000	5.010	5.020	5.030	5.040	5.050	5.060	5.070	5.079	5.089
26	5.099	5.109	5.119	5.128	5.138	5.148	5.158	5.167	5.177	5.187
27	5.196	5.206	5.215	5.225	5.235	5.244	5.254	5.263	5.273	5.282
28	5.292	5.301	5.310	5.320	5.329	5.339	5.348	5.357	5.367	5.376
29	5.385	5.394	5.404	5.413	5.422	5.431	5.441	5.450	5.459	5.468
30	5.477	5.486	5.495	5.505	5.514	5.523	5.532	5.541	5.550	5.559
31	5.568	5.577	5.586	5.595	5.604	5.612	5.621	5.630	5.639	5.648
32	5.657	5.666	5.675	5.683	5.692	5.701	5.710	5.718	5.727	5.736
33	5.745	5.753	5.762	5.771	5.779	5.788	5.797	5.805	5.814	5.822
34	5.831	5.840	5.848	5.857	5.865	5.874	5.882	5.891	5.899	5.908
35	5.916	5.925	5.933	5.941	5.950	5.958	5.967	5.975	5.983	5.992
36	6.000	6.008	6.017	6.025	6.033	6.042	6.050	6.058	6.066	6.075
37	6.083	6.091	6.099	6.107	6.116	6.124	6.132	6.140	6.148	6.156
38	6.164	6.173	6.181	6.189	6.197	6.205	6.213	6.221	6.229	6.237
39	6.245	6.253	6.261	6.269	6.277	6.285	6.293	6.301	6.309	6.317
40	6.325	6.332	6.340	6.348	6.356	6.364	6.372	6.380	6.387	6.395
41	6.403	6.411	6.419	6.427	6.434	6.442	6.450	6.458	6.465	6.473
42	6.481	6.488	6.496	6.504	6.512	6.519	6.527	6.535	6.542	6.550
43	6.557	6.565	6.573	6.580	6.588	6.595	6.603	6.611	6.618	6.626
44	6.633	6.641	6.648	6.656	6.663	6.671	6.678	6.686	6.693	6.701
45	6.708	6.716	6.723	6.731	6.738	6.745	6.753	6.760	6.768	6.775
46	6.782	6.790	6.797	6.804	6.812	6.819	6.826	6.834	6.841	6.848
47	6.856	6.863	6.870	6.877	6.885	6.892	6.899	6.907	6.914	6.921
48	6.928	6.935	6.943	6.950	6.957	6.964	6.971	6.979	6.986	6.993
49	7.000	7.007	7.014	7.021	7.029	7.036	7.043	7.050	7.057	7.064
50	7.071	7.078	7.085	7.092	7.099	7.106	7.113	7.120	7.127	7.134
51	7.141	7.148	7.155	7.162	7.169	7.176	7.183	7.190	7.197	7.204
52	7.211	7.218	7.225	7.232	7.239	7.246	7.253	7.259	7.266	7.273
53	7.280	7.287	7.294	7.301	7.308	7.314	7.321	7.328	7.335	7.342
54	7.348	7.355	7.362	7.369	7.376	7.382	7.389	7.396	7.403	7.409
55	7.416	7.423	7.430	7.436	7.443	7.450	7.457	7.463	7.470	7.477
56	7.483	7.490	7.497	7.503	7.510	7.517	7.523	7.530	7.537	7.543
57	7.550	7.556	7.563	7.570	7.576	7.583	7.589	7.596	7.603	7.609
58	7.616	7.622	7.629	7.635	7.642	7.649	7.655	7.662	7.668	7.675
59	7.681	7.688	7.694	7.701	7.707	7.714	7.720	7.727	7.733	7.740
60	7.746	7.752	7.759	7.765	7.772	7.778	7.785	7.791	7.797	7.804
61	7.810	7.817	7.823	7.829	7.836	7.842	7.849	7.855	7.861	7.868
62	7.874	7.880	7.887	7.893	7.899	7.906	7.912	7.918	7.925	7.931
63	7.937	7.944	7.950	7.956	7.962	7.969	7.975	7.981	7.987	7.994
64	8.000	8.006	8.012	8.019	8.025	8.031	8.037	8.044	8.050	8.056
65	8.062	8.068	8.075	8.081	8.087	8.093	8.099	8.106	8.112	8.118
66	8.124	8.130	8.136	8.142	8.149	8.155	8.161	8.167	8.173	8.179
67	8.185	8.191	8.198	8.204	8.210	8.216	8.222	8.228	8.234	8.240
68	8.246	8.252	8.258	8.264	8.270	8.276	8.283	8.289	8.295	8.301
69	8.307	8.313	8.319	8.325	8.331	8.337	8.343	8.349	8.355	8.361
70	8.367	8.373	8.379	8.385	8.390	8.396	8.402	8.408	8.414	8.420
71	8.426	8.432	8.438	8.444	8.450	8.456	8.462	8.468	8.473	8.479
72	8.485	8.491	8.497	8.503	8.509	8.515	8.521	8.526	8.532	8.538
73	8.544	8.550	8.556	8.562	8.567	8.573	8.579	8.585	8.591	8.597
74	8.602	8.608	8.614	8.620	8.626	8.631	8.637	8.643	8.649	8.654
75	8.660	8.666	8.672	8.678	8.683	8.689	8.695	8.701	8.706	8.712
76	8.718	8.724	8.729	8.735	8.741	8.746	8.752	8.758	8.764	8.769
77	8.775	8.781	8.786	8.792	8.798	8.803	8.809	8.815	8.820	8.826
78	8.832	8.837	8.843	8.849	8.854	8.860	8.866	8.871	8.877	8.883
79	8.888	8.894	8.899	8.905	8.911	8.916	8.922	8.927	8.933	8.939
80	8.944	8.950	8.955	8.961	8.967	8.972	8.978	8.983	8.989	8.994
81	9.000	9.006	9.011	9.017	9.022	9.028	9.033	9.039	9.044	9.050
82	9.055	9.061	9.066	9.072	9.077	9.083	9.088	9.094	9.099	9.105
83	9.110	9.116	9.121	9.127	9.132	9.138	9.143	9.149	9.154	9.160
84	9.165	9.171	9.176	9.182	9.187	9.192	9.198	9.203	9.209	9.214
85	9.220	9.225	9.230	9.236	9.241	9.247	9.252	9.257	9.263	9.268
86	9.274	9.279	9.284	9.290	9.295	9.301	9.306	9.311	9.317	9.322
87	9.327	9.333	9.338	9.343	9.349	9.354	9.359	9.365	9.370	9.375
88	9.381	9.386	9.391	9.397	9.402	9.407	9.413	9.418	9.423	9.429
89	9.434	9.439	9.445	9.450	9.455	9.460	9.466	9.471	9.476	9.482
90	9.487	9.492	9.497	9.503	9.508	9.513	9.518	9.524	9.529	9.534
91	9.539	9.545	9.550	9.555	9.560	9.566	9.571	9.576	9.581	9.586
92	9.592	9.597	9.602	9.607	9.612	9.618	9.623	9.628	9.633	9.638
93	9.644	9.649	9.654	9.659	9.664	9.670	9.675	9.680	9.685	9.690
94	9.695	9.701	9.706	9.711	9.716	9.721	9.726	9.731	9.737	9.742
95	9.747	9.752	9.757	9.762	9.767	9.772	9.778	9.783	9.788	9.793
96	9.798	9.803	9.808	9.813	9.818	9.823	9.829	9.834	9.839	9.844
97	9.849	9.854	9.859	9.864	9.869	9.874	9.879	9.884	9.889	9.894
98	9.899	9.905	9.910	9.915	9.920	9.925	9.930	9.935	9.940	9.945
99	9.950	9.955	9.960	9.965	9.970	9.975	9.980	9.985	9.990	9.995

Square roots for $10 \leq x \leq 100$

(1.2.1) (a) $x^2 - y^2$ (b) $x^2 + 2xy + y^2$ (c) $x - x^3$

(1.2.2) (a) $6x^2 + 7xy - 20y^2$ (b) $2x^3 - 5x^2 + 4x - 1$

(1.2.3) (a) $x^3 + 3x^2y + 3xy^2 + y^3$ (b) $1 - p^3$

(1.2.4) (a) $1 + 4x + 6x^2 + 4x^3 + x^4$ (b) $1 + x^2 + x^4$

(1.2.5) (a) $2a^3 - 2ab^2 + ba^2 - b^3$ (b) $1 - x^4$

(1.2.6) $x^5 - 4x^4 + 5x^3 - x^2 - 3x + 2$

(1.2.7) (a) $2c$ (b) $-a + 2b$ (c) b

(1.3.1) $(xz + yx)/(yz)$

(1.3.2) $(yz + xz + xy)/(xyz)$

(1.3.3) 1

(1.3.4) $a = 14/5$

(1.3.5) (a) 25% (b) $66\frac{2}{3}$% (c) $(100x/y)$%.

(1.3.6) (a) 1/5 (b) 85/100 = 17/20 (c) $3/x$ (d) 15%

(1.3.7) (a) \$140 (b) $x/4$

(1.3.8) (a) 1.91 (b) $(4x + 7y)/100$ (c) $(xy/100)$ %

(1.3.9) (a) $700/11 \approx 63.6\%$ (b) $400/11 \approx 36.4\%$

(1.3.10) $(3x + 8y)/11$ %

(1.3.11) \$99,000.

(1.3.12) (a) a^{13} (b) a^7 (c) a^{-6}

(1.3.13) (a) 100 (b) 1,000,000 (c) 1

(1.4.1) $x = 2, y = 1$

(1.4.2) $x = 9/22, y = -5/22, z = 1/11$

(1.4.3) $x = 3/2, y = 0$

(1.4.4) eqtn2-eqtn1 $\Rightarrow 1 = 0$

(1.4.5) $x = 0, y = 3/2$, or $x = 1, y = 1$

(1.4.6) $x = 3 - 2y$; $x = -7$

(1.4.7) (b) (c) only

(1.5.1) (a) x (b) x^3y (c) bc (d) x^2 (e) t^{13}

(1.5.2) (a) 5/3 (b) $(x + 1)/(x + 3)$ (c) 1/4

(1.5.3) (a) $x^4/2$ (b) $(\sqrt{x^2 + 6x + 3})/(x + 3)$

(1.5.4) ABC

(1.5.5) $x = 22; y = 13$

(1.5.6) $x = (d - b)/(a - c)$

(1.5.7) $x = -13/10$

(1.5.8) $x = -4$

(1.5.9) $x = 1, 2, 3.$

(1.5.10) $x = -10/29$

(1.5.11) $R = \pm 4$

(1.5.12) $R = 4/\sqrt{h}$

(1.5.13) $x = 3\ y = 4$ or $x = -3, y = -4.$

(1.5.14) $x = 1/(s - t - 1)$

(1.5.15) $y = 7/2; x = 9/4; z = 2$

(1.5.16) (a) $V = KT/P$ (b) $P = KT/V$

(1.5.17) $R = \sqrt{GMm/F}$

(1.5.18) (a) $t = (-u \pm \sqrt{u^2 + 2as})/a$, (b) 22

(1.5.19) 3, 6

(1.5.20) $L = 2W/(3W/2 - 2)$

(1.5.21) $(5x^2 - 2)/(2x^2 - x - 1)$

(1.5.22) $x = -4/3$

(1.5.23) $x = a$

(1.5.24) R = radius ; D = diagonal length; $D/R = \sqrt{2\pi}$

(1.5.25) $-1/11$

(1.5.26) 1

(1.5.27) 5/4

(1.5.28) .150

(1.5.29) (a) $7.08m$ (b) $0.87c$ (c) $0.99995c$

(1.5.30) (a) $R = V/I$, (b) $I = V/R$

(1.5.31) $m_1 = m_2(u - v_2)/(v_1 - u)$

(1.5.32) (a) $v = uf/(u - f)$, (b) $v = 12.5cm$

(c) approximately 10 cm for very large u.

(1.5.33) $x = 3b - a, y = a - 2b$

(1.5.34) (a) $\sqrt{2} + 1/2 + 12$

(b) $\sqrt{a} + 1/a + 3a^2$

(c) $c + 1/c^2 + 3c^4$

(d) $\sqrt{a + b} + 1/(a + b) + 3(a + b)^2$

(e) $\sqrt{y + 1/y} + 1/(y + 1/y) + 3(y + 1/y)^2$

(1.5.35) y^3

(1.5.36) $(a^4 + 4a^2 + 1)/(a^3 + a)$

(1.5.37) (c)

(1.5.38) wrong

(1.5.39) wrong: (a) (b) (c) (f)

(1.6.1) (a) 1 (b) $(y - 3)/2$

(1.6.2) (a) 2 (b) $a^{1/3} = ^3\sqrt{a}$

(1.6.3) (a) 81 (b) t^2 (c) impossible

(1.6.4) converts Yen to dollars

(1.6.5) $f^{-1}(y) = y^{1/3}/2$

(1.6.6) graph is mirror image

(1.6.7) (a) .8 (b) .1 (c) .6 (d) .4 (e) .15

(1.6.8) $f^{-1}(y) = 1/(y - 1)$, $f^{-1}(2) = 1$, $f^{-1}(1) = 1/0 = \infty$ (undefined)

(1.6.9) (a) $\sqrt{2} + 1/2 + 12$

(b) $\sqrt{a} + 1/a + 3a^2$

(c) $c + 1/c^2 + 3c^4$

(d) $\sqrt{a + b} + 1/(a + b) + 3(a + b)^2$

(e) $\sqrt{y + 1/y} + 1/(y + 1/y) + 3(y + 1/y)^2$

(1.6.10) $f^{-1}(y) = (y - 6)/4$

(1.6.11) (a) 23000 (b) 32500

(1.7.1) (a) 17 (b) 20

(1.7.2) 84

(1.7.3) 58

(1.7.4) (a) 65 miles (b) $65t$ miles (c) 3 hours

(1.7.5) 29

(1.7.6) proof

(1.7.7) 8π

(1.7.8) $72.1t$ miles

(3.1.1) 99, 100, 101

(3.1.2) 10/9 liters of liquid A.

160/27 liters of liquid B.

(3.1.3) $L = 15$ meters, $W = 30$ meters.

(3.1.4) P = perimeter and area= $P^2/16$

(3.1.5) after 6 hours at $7pm$

(3.1.6) 100 hours.

(3.1.7) $A = (4/h) + 4\sqrt{2h}$ where h = height

(3.1.8) R = radius in meters; T = time in hours. $R = \sqrt{6250T/\pi}$. $T = 1000000\pi/6250 \approx 503$ hours

(3.2.1) W = width of the field in meters.

cost is $C = (40000/W) + 40W$.

(3.2.2) $6, 8, 16$

(3.2.3) V = volume; A = surface area

(a) $6V^{2/3}$ (b) $(A/6)^{3/2}$

(3.2.4) (a) $150m$ (b) $75mph$ (c) 5 hrs (d) $50t$ miles
(e) $200/v$ hours (f) tv miles (g) x/v hrs (h) y/t mph
(i) $(va - ub)$ miles

(3.2.5) (a) $6m$ (b) vA/h

(3.2.6) \$10

(3.2.7) (a) $21, 34$ (b) 13

(3.2.8) $(x - yv)/s$

(3.2.9) $2 : 45pm$ ie. $1 + 1/4$ hrs later.

(3.2.10) 1/3 pints of 1% plus 1/6 pints of 4%

(3.2.11) ≈ 50050 seconds ≈ 13.9 hrs

(3.2.12) $2\pi 4000/24 \approx 1047mph$

(3.2.13) (a) $5\frac{5}{7}$ gallons (b) $(x/35)$ gallons
(c) $x/[50 - 3(v - 55)/5]$ gallons if $v > 55$
$x/50$ gallons if $v < 55$

(3.2.14) 4080

(3.2.15) (a) $64m^3$ (b) $(A/6)^{3/2}$

(3.2.16) (a) $81m^2$ (b) $(L/4)^2$ m^2

(3.2.17) (a) $8 + 4\sqrt{2}$ cm (b) $\sqrt{A}(2 + \sqrt{2})$ cm

(3.2.18) (a) $49mm$ (b) $L(1 + x)/(2\sqrt{x}) - L$ mm.

(3.2.19) 3/7 liters from A and 4/7 liters from B

(3.2.20) (a) $(1/2)$ mins (b) $1/4(v - 70)$ hrs

(3.2.21) 40 miles closer to SF

(3.2.22) $1500mph$

(3.2.23) (a) $81/156250 \approx .0005184$ mg (b) $2.592x$ mg

(3.2.24) the last square has $2^{63} \approx 10^{19}$ grains of sand.

(3.2.25) H = height; L = length W = width
Cost in \$ $= 3LW + 6LH + 6WH$.

(3.2.26) W = width in feet. $Cost = 20W^2 + 160/W$.
surface area$= (8/W) + 2W^2$

(3.2.27) W = width; height
(a) area$= WH + \pi(W^2/8)$
(b) height$= (2/W) - \pi W/8$

(3.2.28) R = radius in cm.
Surface Area $= 2\pi R^2 + 48/R$

(3.2.29) 12 cows and 28 chickens.

(3.2.30) x = length of a dividing fence
total length$= 4x + (4000/x)$

(3.2.31) It takes 4 hours to fly from LA to NY.

(3.2.32) 1600 bacteria.

(3.2.33) M = the number of machines set up
Total Money $= 200M(20 - M)$

(3.2.34) N = number of nickels; D = number of dimes;
Q = the number of quarters.
$NDQ = N(N/2)(18 - (3/2)N)$

(3.2.35) H = height in meters above the ground.
temp in $°C = 22 - (H/100)$

(3.2.36) T = the number of trees per acre.
number apples $= 5040T - 72T^2$

(3.2.37) $25/4$

(3.2.38) $1600\pi/3$ meters per minute.

(3.2.39) It passes 17 planes.

(3.2.40) *** GRAPH ***

(3.2.41) V = volume; Surface Area $= 6(^3\sqrt{V})^2$

(3.2.42) 2

(3.2.43) P = price of one ticket ; T = number tickets
sold; M = total profit.
$M = T(P - 100)$ and $T = 30000 - 100P$.
(a) $M = T(200 - (T/100))$
(b) $M = (30000 - 100P)(P - 100)$

(3.2.44) (a) $Y = \sqrt{3}X/2$ (b) Area $= \sqrt{3}X^2/4$

(3.2.45) H = aquarium height in meters.
Total Cost $= 3[(20/H) + 4\sqrt{20H}] + 16(\sqrt{20/H}) + 8H$

(3.2.46) $625/9\pi \approx 22$ days

(3.2.47) L =length of the prism; X = side length of the
triangle; T = time;
(a) Volume $= \sqrt{3}LX^2/4$
(b) $\sqrt{AT/L}3^{1/4}$

(3.2.48) P = price of one ticket
total income $= (1300 - 250P)P$

(3.2.49) W = width of the poster in cm.
Printed Area $= (W - 8)((500/W) - 10)$

(3.2.50) L = length of RECTANGULAR portion of the
field; D = diameter of the semi-circular ends.
Total Area $= \pi(D/2)^2 + D(500 - \pi D/2) = 500D - \pi D^2/4$

(3.2.51) It takes 8 times as much energy at 20mph compared to 10 mph

(3.2.52) (a) $y = 20 + 2x$ (b) $x = (y/2) - 10$
(c) -20 is same in both.

(3.2.53) 10000 seconds

(3.2.54) $6\pi \approx 19$ feet. Most people guess a very big
number.

(4.1.1) ≈ 51333 cm/min

(4.1.2) ≈ 14 Km per liter

(4.1.3) ≈ 70000 mph or $\approx 2.67 \times 10^{11}$ cm per day

(4.1.4) $\approx 2.2 \times 10^{26}$ grams

(4.1.5) $24,000,000$

(4.1.6) (a) $59\frac{3}{8}$ (b) $5/4$ mph
(c) 12 minutes (d) $720/11 \approx 65.5$ mph

(4.1.7) (a) December 31'st about $1\frac{3}{4}$ hours before midnight (b) about 1 minute before midnight.

(4.1.8) (a) $\approx 1.8 \times 10^{13}$ miles (b) ≈ 1450 miles
(c) ≈ 34 million miles (d) ≈ 125 feet

(4.6.1) 8 times as much

(4.6.2) 1/2 times as strong

(4.6.3) 3 times longer

(4.6.4) 125 times the volume

(4.6.5) 27 times longer

(4.6.6) how much heat an animal needs to produce per
gram of body weight. How much food an animal needs
to eat per gram of body weight

(4.6.7) formula has dimensions of area BUT it should
have dimensions of volume

(4.6.8) A/B is ms^{-2}; AB is meters

(4.6.9) Yes

(4.6.10) 3 times wider

(4.6.11) gets bigger by factor of (a) 4 (b) 8 (c) 16

(4.6.12) Block B is 8/3 times denser than block A

(4.6.13) (b) duck speeds up.

(5.1.1) x_n is nearly 0 when n is very big....

(5.1.2) (a) 4 (b) 0 (c) ∞ (d) 3 (e) 1/2

(5.1.3) 10%

(5.1.4) 33%

(5.1.5) 2

(5.1.6) 1

(5.1.7) .11%

(5.2.1) increase of 1/42

(5.2.2) 1.361

(5.2.3) $3, 2.1, 2.01, 2.001$

(5.2.4) $h^2 + 5h$

(5.3.1) (a) $1 + 2 + 3 + 4 + 5$ (b) $a_1 + a_2 + a_3 + a_4$ (c) $(a_2 + b_2) + (a_3 + b_3) + \cdots + (a_6 + b_6)$ (d) $f(-2) + f(-1) + f(0) + f(1) + f(2)$ (e) $3(x_1 + x_2 + x_3)$ (f) $c_{10} + c_{11} + \cdots + c_{15}$ (g) $(a_2 + b_2)^2 + (a_2 + b_3)^2 + \cdots + (a_5 + b_5)^2$ (h) $1/1! + 1/2! + 1/3! + 1/4! + 1/5!$ (i) $x^1/1 + x^2/2 + x^3/3 + x^4/4$ (j) $a_1 + a_2 + a_3 + a_4 + a_5 + a_6$ (k) $2^1 + 2^2 + 2^3 + 2^4 + 2^5$ (l) $(-1)^1 + (-1)^2 + (-1)^3 + (-1)^4 + (-1)^5$ (m) $-x + x^2 - x^3 + x^4$

(5.3.2) (a) $\frac{a_1}{a_1^2 + 1} + \frac{a_2}{a_2^2 + 1} + \cdots + \frac{a_{1000}}{a_{1000}^2 + 1}$

(b) $(a_2 - a_1) + (a_3 - a_2) + \cdots + (a_{101} - a_{100})$

(c) $(1/N)[x_1 + x_2 + \cdots + x_N]$

(d) $x/1! + x^2/2! + x^3/3! + \cdots + x^{200}/200!$

(5.3.3)

$$(a) \sum_{k=1}^{100} x_k \quad (b) \sum_{k=1}^{20} (a_k + 2b_k) \quad (c) \sum_{i=1}^{70} x_i$$

(5.3.4)

$$\frac{1}{1000} \sum_{i=1}^{1000} a_i$$

(5.3.5) cancel terms

(5.3.6) (c) 5050

(5.3.7) $(a_1 + a_2)(b_1 + b_2) \neq a_1 b_1 + a_2 b_2$

(5.3.8)

$$(a) \frac{1}{500} \sum_{i=1}^{500} x_i \quad (b) \frac{1}{N} \sum_{k=1}^{N} y_k \quad (c) \frac{1}{11} \sum_{p=10}^{20} a_p^2$$

(5.3.9) (a) 200 (b) 14 (c) 18

(6.1.1) (b) $y = (-3/5)x + 3$

(6.1.2) $y = (4/5)x - 13/5$

(6.1.3) $y = -1$

(6.1.4) $y = [(b - a)/2]x + (2a - b)$

(6.1.5) $y = 2x - 5$

(6.1.6) $y = x/3$

(6.1.7) $y = -x + 3$

(6.1.8) $y = x - 3$

(6.1.9) $y = (2/5)x - 2$

(6.1.10) $x = 13/5;\ \ y = -7/5$

(6.1.11) $(-9/5, -22/5)$

(6.1.12) $(-1, -1)$

(6.1.13) $2 : 30pm$

(6.1.14) $1 : 20pm$

(6.1.15) (a) Elvis cheaper for < 250 miles

(b) slope is price per mile

(6.1.16) (a) x=(9/5)y+32 (b) y=(5/9)x-160/9

(c) $29.4°C$ (d) $85°F$ (e) $-40°C = -40°F$

(6.1.17) (b) 150 Km North and 200 Km west of A.

(c) 300 Km per hour.

(6.1.18) $y = (11/2) - (5/2)x$

(6.1.19) (a) $-4/3$ (b) $(2, 0)$ (c) $(0, 10/3)$

(6.1.20) $(-21/11, 14/11)$

(6.1.21) (b) $(y_1 - y_0)/(x_1 - x_0) = (y_2 - y_1)/(x_2 - x_1)$

(c) yes

(6.1.22) (a) $(13/2, 11)$ (b) $(16/3, 9)$ (c) $(4, 47/7)$

(d) $(19/4, 8)$

(6.2.1) (a) 40 (b) 10 (c) $5a$ (d) $5a^2$ (e) 0 (f) $5a + 5b$

(g) $5/w$

(6.2.2) yes

(6.2.3) $9 billion

(6.2.4) mass $= K\, d^3$

where $d =$ diameter and $K =$ constant.

(6.2.5) Rate disease spreads $= KD(P - D)$ where $K =$ constant.

(6.2.6) (a) $T = KPV$ where $K =$ constant

(b) $P = T/KV$

(6.2.7) $3cm$

(6.2.8) (a) cost $= KMD$ where $M =$ mass; $d =$ distance and $K =$ constant (b) $30000

(6.2.9) $C =$ total cost; $F =$ fuel cost; $H =$ hourly pay; $S =$ speed; $t =$ time (a) $t = D/S$ (b) $F = KSD$ and crew cost$= HD/S$

(c) $C = KSD + H(D/S)$ (d) $600

(6.2.10) $F = GM_1 M_2/r^2$ where $F =$ force; M_1, M_2 are two masses and $r =$ distance between masses.

(6.2.11) No

(6.2.12) 1 year

(6.4.1) 7 billion

(6.4.2) (a) 2.2 (b) 2.236 (c) 1.6%

(6.4.3) (a) 33.044 feet (b) year 3025

(7.13.1) (a) .8351 (b) 3.7782 (c) 3.8351 (d) 2

(e) $-.2848$ (f) impossible (g) 2.9494 (h) impossible

(i) -3.1649

(7.13.2) (a) 5.1912 (b) 2.5795 (c) 3.2043 (d) 3.7576

(e) -1.8654

(7.13.3) (a) -0.6021 (b) -1.699 (c) 1.8713 (d) .1412

(e) -2.8282

(7.13.4) (a) 30.10 (b) 4 (c) 1.1273 (d) -5.038

(e) .6152 (e) $-.6152$ (f) -12.399

(7.13.5) (a) 3.38 (b) 41.7 (c) 316000 (d) 100 (e) 10^{20}

(f) 4090 (g) .00316 (h) .000134 (i) .002 (j) .03

(7.13.6) (a) 155000 (b) 379.5 (c) 1620 (d) 5721
(e) .0136

(7.13.7) (a) .25 (b) .02 (c) 74.33 (d) 1.38 (e) .00149

(7.13.8) (a) 1.26×10^{30} (b) 10000 (c) 13.4(d).00000916
(e) 4.12 (f) .243 (g) 3.99×10^{-13}

(7.13.9) (a) $log(600)$ (b) $log(2/3)$ (c) $log(32)$
(d) $log(x^2)$ (e) $log(x^5)$ (f) $log(x^{-2})$ (g) $log(xy)$
(h) $log(y^x)$ (i) $log(a^{x+y})$ (j) 0 (k) $log(2^{a-b})$

(7.13.10) (a) $log(x) + log(y)$ (b) $3log(x) + 2log(y)$
(c) $(1/2)log(x)$ (d) $log(x + y)$ (e) $log(x) + log(log(y))$

(7.13.11) $x = log(.3010) \approx -.5214$

(7.13.12) $x = log(.6) - 3 \approx -3.2218$

(7.13.13) $x = log(.6020)/log(2) = -.732$

(7.13.14) $x = log(y)$

(7.13.15) $x = log(5)/(log(2) - log(3))$

(7.13.16) $x = -ylog(3)/log(5)$

(7.13.17) (a) 32million (b) $t = 1/log(2) \approx 3.32 \approx$ April 1993

(7.13.18) (a) $1070 (b) $1144.90 (c) $1225.04
(d) $1967.15 (e) $1000(1.07)^x$ (f) after 102 years in 2092

(7.13.19) (a) 5 grams (b) $80^{(}17/50) \approx 63.2$ grams
(c) 50 years (d) $50log(5/8)/log(1/2) \approx 33.9$ years

(7.13.20) (a) $x2^t$ (b) $3 = 2^t$
(c) $t = log(3)/log(2) = 1.585$ hours

(7.13.21) (a) $R = 500$, $K = 10log(2)/log(6) \approx 3.88$
(c) 3.88 years

(7.13.22) 19.43 years

(7.13.23) $log(6.7) \approx .82$, slope$\approx .074$

(7.13.24) (a) .81 (b) .17 (c) 1.34 (d) -3.3 (e) 5

(7.13.25) (a) 3.2 (b) 2.2 (c) 480 (d) .005 (e) 10^{1000}

(7.13.26) ≈ 26500

(7.13.27) ≈ 736

(7.13.28) ≈ 8.89

(7.13.29) 36.7

(7.13.30) $C = log(2) \approx .3010$

(7.13.31) $t = [log(P/Q)/log(3)] - a$

(7.13.32) $t = [log(a/c)/log(d/b)]$

(7.13.33) $10^{ylog(4)} \approx 10^{.6020y}$

(7.13.34) $2^{a/log(2)}$

(7.13.35) $10^{-x} \to 0$ as $x \to \infty$

(7.13.36) $1 - 2^{-x} \approx 1$ when x large

(7.13.37) $x = 4$

(7.13.38) $x = 15.23$

(7.13.39) $40/log(2) \approx 133$ years

(7.13.40) $5730log(5)/log(2) \approx 13325$ years old

(7.13.41) $x = 10^{100000000}$

(7.13.42) (a) 1020 (b) 1.07×10^9 (c) 1.26×10^{30}

(7.13.43) 15100

(7.13.44)

(7.13.45) (a) $100\sqrt{10} \approx 141$ (b) 30 years

(7.13.46) $t =$ number years after 1900.
$100(2^{t/10}) = 10^6(2t + 4)$

(7.13.47) 0.435

(7.13.48) 337500

(7.13.49) 8/3

(7.13.50) (a) $x = ln(7)/ln(3)$ (b) $x = -ln(100)/ln(2)$
(c) $x = 3$ (d) $x = ln(30)/ln(7)$ (e) $x = 6$

(7.13.51) (a) $x = ln(10)$ (b) $x = 4$ (c) $x = ln(7)/ln(4)$
(d) $x = ln(6)/(ln(3) - ln(2))$

(8.1.1) (a) 50 miles per hour (b) 50 miles per hour (d)
50 miles per hour (e) graph (f) 50 miles per hour

(8.1.2) (a) 1.35 (b) 2.2 (c) 3.0

(8.1.3) (a) -11 (b) -13 (c) -14 (d) during first half
hour.

(8.1.4) 1500 gallons per hour.

(8.2.1) 30mph

(8.2.2) **GRAPH**

(8.2.3) (a) 6. (b) between $t = 0.8$ and $t = 1$ (c) 6

(8.2.4) $(1.000001)^2 - (1)^2)/(1.000001 - 1) = 2.000001$

(8.2.5) ** Table **

(8.2.6) (a) $^\circ F$ per hour. (b) at 7:00pm the temperature
is decreasing at a rate of 3 degrees per hour. (c) When
$f'(t)$ is positive, temperature is increasing (i.e. it's getting hotter). when $f'(t)$ in negative, temperature is
decreasing.

(8.2.7) (a) in 1995 there is no change in the company's
total profit. (b) millions of dollars per year. (c) If the
sign is negative, the company's total profit is decreasing
- this means the company is losing money. If the sign
is positive, the company is making money.

(8.2.8) 485 million.

(8.2.9) after 7 years, the balance is increasing at a rate
of 300 dollars per year.

(8.4.1) essay

(8.4.2) (a) $-3m^3$ per minute. (b) $-2.01m^3$ per minute.
(c) $-2m^3$ per minute. (d) $-2 - h\ m^3$ per minute.
(e) $-2m^3$ per minute. The volume of water in the tub
is **decreasing**

(8.4.3) (a) 15 grams (b) $t = 30$ minutes. (c) the ice
cube is gaining mass.

(8.4.4) (a) 32 feet per $second^2$ (b) 32 feet per $second^2$
(c) 80 feet per $second$

(8.4.5) (a) 2.001 (b) 4.001 (c) 6.001 (d) 8.001 (f) points
lie on a straight line (g) 2×1.5 ie 3 (h) 3.001

(8.4.6) (a) slope= 3 (b) slope= -2 (c) 1.5

(8.4.7) (a) $x < .5$ also $x > 1.6$ (b) derivative= 0 at .5
and 1.6 (c) edges of graph (d) 1

(8.4.8) graph

(8.4.9) (a) day 163 (Summer Solstice June 21) (b) day
357 (shortest day Dec 21) (c) day 95 (spring equinox)
(d) day 255 (vernal equinox) (e) 3 minutes

(8.4.10) (a) 5 (b) 5.2 (c) 10.4 (d) $C \times 5.2$.

(8.4.11) (a) 2.73 (b) 7.42 (c) 54.9.

(8.4.12) (a) .2 (b) .236 (c) .25 (d) .25 (e) .25

(8.4.13) (a) .162 (b) .223 (c) .246 (d) .249 (e) .25

(8.5.1) (a) $area/length = length$ (b) when the circle
has radius 3, then its area changes about 6π for a unit
change in the radius length.

(8.5.2) (a) feet per mile. (b) after 10 miles of flight, the plane is climbing at the rate of 500 feet per mile of flight. (c) $h'(x)$ is negative during the descent and landing. (d) the aircraft is underground!

(8.5.3) (a) gallons per meter (b) when the reservoir height is 30 meters, an increase in height of 1 meter increases volume by $\approx 5 \times 10^7$ gallons.

(8.5.4) (a) For a price of 50 dollars, 20,000 items are sold. (b) items per dollar (c) a small increase in price **reduces** the number of items sold by 200 times the small increase in price.

(8.5.5) (a) if no one is inoculated, 10^6 people will get the disease. (b) if 50% of people are inoculated, inoculating 1% more people will decrease the number who get sick by 10^4. (c) if 85% are inoculated only 100 will get the disease.

(8.6.1) (a) 7.2. (b) 7.02. (c) The tangent line approximation is better for smaller increases. So (b) is more accurate.

(8.6.2) $7.28x - 0.48$; gives $10^{0.6} \approx 3.89$ error is 2.5%

(8.6.3) (a) 1/4 (b) 2.1. (c) .11 percent. (d) 3.25 (e) 8.33 percent. (f) The tangent line approximation is most accurate for smaller changes

(8.6.4) graph

(8.6.5) (a) $f'(x) = a$. (b) Ca. (c) $C \times f'(x)$

(8.6.6) (a) a , c (b) $(a+c)x+(b+d)$, $a+c$ (c) $f'(x)+g'(x)$

(8.7.1) (a) $3x^2$ (b) $8x^3$ (c) $28x^3 - 6x$ (d) $-x^{-2}$

(8.7.2) (a) -7 (b) $18x + 12$ (c) 0

(8.7.3) (a) $4x^3 + 4x$ (b) $6x^{-3} + 3$ (c) $-4x^{-5}$

(8.7.4) (a) $3x^2 + 12x + 11$ (b) $2ax + b$ (c) 1

(8.7.5) $1 + x + x^2/2! + x^3/3! + x^4/4!$

(8.7.6) do it!

(8.7.7) $-2at + b$

(8.7.8) $y = 57x - 143$

(8.7.9) (b) $6x$ (c) 6

(8.7.10) $4\pi r^2$ surface area of sphere

(8.7.11) $2\pi r \ell$ surface area of pipe

(8.7.12) $2x$

(8.7.13) (a) $-9x^{-4}$ (b) $-(2/3)x^{-5/3}$ (c) $(1/2)x^{-1/2}$ (d) $-12.6x^{-5.2}$ (e) $1 - x^{-2}$

(8.7.14) (a) 20 days (b) falls $5m$ per day (c) $2000\pi \ m^3$

(8.7.15) $y = 3x - 1$

(8.7.16) $y = 5x - 8$

(8.7.17) $y = 0$

(8.8.1) (a) $5e^{5x}$ (b) $5e^{15} \approx 16345087$ (c) 5 (d) $25e^{5x}$

(8.8.2) $18e^{-3x}$

(8.8.3) any multiple of e^{4x}

(8.8.4) any multiple of e^{2x} or e^{-2x}

(8.8.5) calculate

(8.8.6) approximations are 1.105 error .015%, 1.345 error 0.36%, 1.945 error 3.4%

(8.8.7) calculate

(8.8.8) (a) c (b) ∞

(8.8.9) $y = 6x + 3$

(8.8.10) $10e^2$ mg/hour

(8.8.11) (a) $(-1/24)38e^{-1/8}$ (b) 60 (c) 98

(8.9.1) calculate

(8.9.2) calculate

(8.9.3) calculate

(8.10.1) (a) $20 + 10t$ (b) $40 \ m/s$ (c) 10 seconds (d) $-2 + \sqrt{24} \approx 2.9$

(8.10.2) (a) 0 (b) 0 (c) $1, .5, 0, -.5$ (d) $t = 15$ yrs

(8.10.3) (a) 6 (b) $y = 6x - 9$ (d) $(0, -9)$

(8.10.4) price rises for first 2 days. Sell when $t = 2$.

(8.10.5) $4400\pi \ m^2$ per hour

(8.10.6) $288\pi \ cm^3/sec$

(8.10.7) (a) $x = -3$ (b) $x > -3$ (c) $x > -3$

(8.10.8) graph

(8.10.9) graph

(8.10.10) after 3.4 months

(8.11.1) $y = 3x - 2$

(8.11.2) $y \approx 1 + 2.6x$

(8.12.1) (a) 0 (b) 2 (c) $6x$ (d) $n(n-1)x^{n-2}$ (e) $2x^{-3}$ (f) $80x^3 - 18x + 2$ (g) 0

(8.12.2) 12

(8.12.3) $f''(2) = 44 \quad f''(-1) = 8$

(8.12.4) graph

(8.12.5) (a) water is rising at rate of 1 foot per day on day 30. But this rate of rise is dropping by 1/2 ft per day. (b) falling by 1/2 ft per day

(8.12.6) (a) $2000m$ (b) $-30m/s$ (c) $-10 \ ms^{-2}$ (d) $t = 20$ (e) speed is $200 \ m/s$

(8.12.7) less than 2%: concave **down** means slope **decreases** with time

(8.12.8) $3e^{3x} \quad 9e^{3x}$

(8.12.9) $-e^{-x}, e^{-x}, -e^{-x}, e^{-x}, \cdots$

(8.12.10) (a) $x < -2$ and $x > 3$ (b) $x > 1/2$

(8.12.11) (a) $30 \ m/s$ (b) $t = 1$ (c) $-h(4) = 40m$ (d) $45m$

(8.12.12) (a) left (b) speeding up (c) increasing

(8.12.13) (a) 2 (b) 48

(8.12.14) (a) $f'(5)$ is positive and equal to $0.4 \times f(5)$. (c) $f''(5) > 0$

(8.13.1) $x = 5/2$ max is $f(5/2) = 33/4$

(8.13.2) min is 2 when $x = 1$

(8.13.3) square $125' \times 125'$

(8.13.4) $1000 \ m^2$

(8.13.5) maximum when $x = 2a$ is $4a^2$.

(8.13.6) $\$800\sqrt{10} \approx \$2529$

(8.13.7) $\$240$

(8.13.8) $80\sqrt{10} \approx 253$

(8.13.9) 10 machines

(8.13.10) $\$200$

(8.13.11) $\$1690$

(8.13.12) $20cm$

(8.13.13) $125000/\pi$

(8.13.14) $(1/2, 1/\sqrt{2})$

(8.13.15) $ln(10)$ years

(8.13.16) $2\ in$

(8.13.17) July 11'th.

(8.13.18) $833.33\ ft^3$

(8.13.19) $6.50

(8.13.20) $40\ m$

(9.1.1) (a) 6 (b) -20 (c) 10 (d) 2 (e) 0 (f) 4

(9.1.2) $a = 13/3$

(9.1.3) (a) $-1/2$ (b) $3/2$ (c) 1

(9.1.4) $t = 5 - \sqrt{2}$

(9.1.5) population increase between 1970 and 1980 is 1200 million

(9.1.6) oven temperature increased $120°C$ in 20 mins

(9.1.7) same area above and below graph between $x = 0$ and $x = 5$

(9.1.8) (a) $\int_0^{24} f(t)\ dt$ (b) $\int_0^7 f(t)\ dt = 200000$

(c) $\int_0^T f(t)\ dt = 400000$

(9.1.9) $a^2 + a$

(9.1.10) (a) 3/8   4% (b) 11/32   0.28%

(9.1.11) (a) $\approx .66$ (b) $\approx .55$ (c) ≈ 1.1

(9.1.12) graph is 10 times higher

(9.1.13) 6.56

(9.1.14) essay

(9.2.1) (a) 3 (b) 0 (c) 19/3

(9.2.2) 5/8

(9.2.3) 62

(9.2.4) $216^{1/3} = 6$

(9.2.5) (a) 137/60 (b) 19/3 (c) $(a/3) + b$

(9.2.6) (a) $a^3/3$ (b) $(1 - a^4)/4$ (c) $[(b^3 - a^3)/3] + (b - a)$

(9.2.7) (a) $(1/7)(e^{14} - 1)$ (b) $(1/ln(10))(10^5 - 10^{-1})$

(c) $(5/2ln(2)) + (1/3)$

(9.2.8) 171/4

(9.3.1) (a) $(2/3)x^3 - 5x + C$ (b) $x^6/6 + C$ (c) $x^3/3 + C$

(9.3.2) $y^4/4 + y + C$

(9.3.3) (a) $x^4 + x^3 + x^2 + x + C$ (b) $(x^3/3) + 3x^2 + 9x + C$

(c) $a(x^3/3) + b(x^2/2) + C$

(9.3.4) (a) $(-1/3)e^{-3x} + C$ (b) $(1/ln(2))2^x + C$

(c) $(1000/ln(10))10^x + C$

(9.3.5) $(x^5/5) + C$

(9.3.6) $x^3 + 7$

(9.3.7) $x^4/4 - x + 11/4$

(9.3.8) $f(x) = C$ =constant

(9.3.9) $e^{2x}/2 + C$

(9.3.10) e^{7t}

(9.3.11) $e^x + x^4 + 41$

(9.3.12) $h'(t) = K(15 - h(t))$

where K = constant of proportionallity

(9.3.13) $p(t) = Ae^{5t}$

$p(t) = 2000e^{5t}$ after 5 hours $2000e^{25}$

(9.3.14) $200ln(2)$

(9.4.1) $\frac{1}{1000} + \frac{4}{1000} + \frac{9}{1000} + \cdots + \frac{100}{1000}$

(9.4.2) $(1/10)[f(1/10) + f(2/10) + f(3/10) + \cdots + f(10/10)]$

(9.4.3) .385; error= 15.6%

(9.4.4) width= 1/5; left= 6/5; right= 7/5

(9.4.5) essay

(9.5.1) 17

(9.5.2) 3; looking at the straight line graph the average height is half way along and is thus 3

(9.5.3) average is 0. Half the graph is above and half is below the x-axis. These signed areas cancel out.

(9.5.4) average= 7/3; max= 4; min= 1

(9.5.5) 292 megawatts per hour

(9.5.6) (a) $(e^{100} - 1)/100$ (b) $ln[(e^{100} - 1)/100] \approx 15.3$

(9.5.7) $3 + (1/3000)$. Function is close to 3 over all interval.

(9.6.1) 309 acre feet

(9.6.2) 125 meters. Go Rat!

(9.6.3) $100\pi\ cm^3$

(9.6.4) decreases $162°C$

(9.6.5) $1/16 = .0625$ miles

(9.6.6) (a) 40/3 gallons (b) $60^{1/3} \approx 3.91\ hours$

(9.6.7) $\approx$ $0.60

(9.6.8) (1) 10 gallons (b) $t = 5$ hrs

(9.6.9) $0.45J$

(9.6.10) $83/3 = 27.66\ ms^{-1}$

(9.6.11) calculate

(9.6.12) $23/4\ cm^3$

(9.6.13) (a) $50 + 40e^{-1/2} \approx 74°C$ (b) $50 + 40e^{-t}$

(c) $ln(2) \approx 0.69$ hours

(9.6.14) $20 + \sqrt{24400} \approx 176$ minutes

(9.6.15) (a) graph (b) $12m/s$ (c) $t + t^2$ (d) $\int_0^5 (t + t^2)\ dt$

(10.2.1) (a) x^9 (b) y^7 (c) $y^{-2}x^{-6}$ (d) $a^{18}b^{-6}$

(e) $(3/32)x^{-3}$

(10.2.2) (a) $x^3 + 2x^2y - xy^2 - 2y^3$ (b) $1 - 8x^3$

(c) $-4LW - 8W^2$ (d) $a + b$

(10.2.3) (a) 4 (b) $a^2 - 2a + 1$ (c) y^2 (d) $9y^2 - 6y + 1$

(e) y^{-2}

(10.2.4) $x = 2$ or $x = 3$

(10.2.5) $x = 2$ or $x = 5$ or $x = -7$

(10.2.6) $x = 2$ and $y = 5$

(10.2.7) infinitely many solutions eg $x = 5$, $y = 0$ or $x = 2$, $y = 2 \cdots$

(10.2.8) $x = -7/20$

(10.2.9) $x = (4 - b)/(a - 3)$

(10.2.10) $x = 22/5$

(10.2.11) $x = -3/5$

(10.2.12) $W = (2a + b)/(4a + 3b)$

(10.2.13) $a = (1 - 8b)/14$

(10.2.14) (a) $x = 4$ (b) $x = -3$ (c) $x = 2$

(10.2.15) $x = 17/2$

(10.2.16) $x = (3b - 2k)/(2c - 3)$

(10.2.17) $x = 21$, $y = -29$ or $x = 5$, $y = 3$

(10.2.18) $x = (-5/(6y)) + (3y/2)$

(10.2.19) $(x^2 + 8x + 8)/(x^2 - 4)$

(10.2.20) $a = \pm\sqrt{c^2 - b^2}$

(10.2.21)

(a) plug in $x = 1$, $a = 1$, $b = 1$ get $2 = 1/2$

(b) plug in $a = 1$, $b = 1$, $x = 2$ get $1 = 3/2$

(c) plug in $x = 4$ get $3 = 5/3$

(d) plug in $a = b = 1$ get $1/2 = 2$

(10.2.22) (a) 9 (b) a^2 (c) $1/x^2$ (d) $(x+2)^2$

(10.2.23) (a) 2 (b) 1 (c) $y^{1/3}$

(10.2.24) (a) 3 (b) -2 (c) $4/3$ (d) $-2/3$

(10.2.25) 1

(10.2.26) $y = 5x - 13$

(10.2.27) $y = x/2$

(10.2.28) $y = (-x/3) + (11/3)$

(10.2.29) $(-3/2, -5/2)$

(10.2.30) $(3, 9)$ and $(-2, 4)$

(10.2.31) (a) 3 (b) $14x$ (c) $15x^4 - 6x^2 + 3$

(d) $4x - 5$ (e) $-15x^{-6}$ (f) 0

(10.2.32) $f(2) = 5$

(10.2.33) $x > 3$

(10.2.34) $18x$

(10.2.35) $4x^3$, $12x^2$, $24x$

(10.2.36) 8 times

(10.2.37) concave up

(10.2.38) (a) 2.6021 (b) 1.6990 (c) -2.6990

(10.2.39) (a) 0 (b) $-2log(y)$ (c) $(1/2)log(x)$

(10.2.40) (a) 20 (b) x^2 (c) 10^x (d) a^x

(10.2.41) (a) $e^{ln(37)}$ (b) $e^{ln(x)}$ (c) $e^{xln(10)}$ (d) $e^{xln(2)}$

(10.2.42) (a) $7e^x$ (b) $6e^{2x}$ (c) $ln(10)e^{xln(10)}$

(d) $ln(2)e^{xln(2)}$

(10.2.43) (a) $ln(10)10^x$ (b) $7ln(2)2^x$

(10.2.44) (a) $(x^3/3) + C$ (b) $7x + C$ (c) $(x^5/5) - x^3 + 5x + C$ (d) $-x^{-2} + C$ (e) $ln|x| + C$

(10.2.45) 14

(10.2.46) 8/3

(10.2.47) cross at $(0, 0)$ and $(1, 1)$. area$= 1/6$

(10.2.48) 8/3

(10.2.49) $e^3 - e$

(10.2.50) 4/3

(10.2.51) $x =$ side length. radius$= x/\sqrt{\pi}$

(10.2.52) 2500

(10.2.53) height$= h$. surface area$= 4h^2 + (15/h)$

(10.2.54) $30/(4\pi)$ cm per hour

(10.2.55) (a) 15 hours (b) $5log(20)/log(2) \approx 21.6$ hours

(10.2.56) 8pm

(10.2.57) (a) at 11am temperature is 17^oF

(b) ^oF per hour

(c) at 11am temperature is increasing at the rate 3^oF per hour

(10.2.58) $f'(1) \approx 0.8$

(10.2.59) first 10 mins 5200 micrograms per minute last 10 mins 2800 micrograms per min.

(10.2.60) 15 minutes

(10.2.61) $x = \pm\sqrt{2}$

(10.2.62) 6

(10.2.63) population in 1995 is 4250000. In 1985.

(10.2.64) (a) $3 + t + t^2$ (b) after 4 seconds

(c) $41\frac{1}{3}$ meters

(10.2.65) 16/5 meters

(10.2.66) $(\sqrt{204} - 2)/4 \approx 3.07$ hours

(10.2.67) $5110/ln(2) \approx 7370$

(10.2.68) $p(t) =$ number of tonnes pollution t years after 1900. $p'(t) = 10^7 \times 2^{t/10}$

$5110/ln(2) \times 10^7 \approx 737 \times 10^8$.

(10.2.69) $(200/3)(1 - e^{-0.9}) \approx 30$ Joules.

(10.2.70) $(1, 1)$ and $(-1, 1)$.

(11.0.1) $W = 4$ and $L = 8$ or $W = 8$ and $L = 4$

(11.0.2) $200mph$

(11.0.3) 2 hours

(11.0.4) 3 and 5

(11.0.5) 8, 18

(11.0.6) 3.5 hours

(11.0.7) Car A $60mph$ Car B $50mph$

(11.0.8) D ends after 9 hours

(11.0.9) 40/3 grams of W and 20/3 grams of X

(11.0.10) $(8/3)\%$

(11.0.11) A ends with $(13a/64) + (7b/32) + (3c/16)$

B ends with $(19a/64) + (9b/32) + (5c/16)$

C ends with $(a/2) + (b/2) + (c/2)$

player wants to go last.

(11.0.12) 2100 field zone and 3500 end zone

(11.0.13) adult price \$6 and child price \$3: total \$65

(11.0.14) 5.6 grams of 75% and 8.4 grams of 80%

(11.0.15) (a) 50 hours (b) $t = 1000x/(80 - 20x)$

(d) $t = -250$ it never reaches $5mg$ per liter because the incoming concentration is less than this, and is further diluted by the pure water initially there.

(11.0.16) 80 ppm

(11.0.17) 38.88 miles

(11.0.18) after 12 minutes and after 20 minutes

(11.0.19) perimeter$= 2\pi\sqrt{A/\pi}$ where $A =$ area

(11.0.20) $4L^3 - 52L + 72 = 0$

(11.0.21) radius$= 5$ cm height$= 3$ cm

(11.0.22) diameter$= 20/\sqrt{h}$ where $h =$ height

(11.0.23) 47

(11.0.24) (a) $2 + (5/8)$ hours $= 2$ hours 37.5 minutes

(b) 180 miles from the start

(11.0.25) for $t \leq 2$ answer is $2t$ cm

for $t > 2$ answer is $(4t - 4)$ cm

(11.0.26) (a) $(t/100)\%$ (b) after 100 days

(11.0.27) $t \leq 1/2$ $700 - 200t$ miles

$1/2 \leq t \leq 2$ $800 - 400t$ miles

$2 \leq t \leq 2 + (1/2)$ $400(t - 2)$ miles

$2 + (1/2) \leq t \leq 3$ $200t - 300$ miles

(11.0.28) (b) for A $10000t$ for B $20000(t - 10)$

(c) 1920 (ie $t = 20$) (d) 1923 (ie $t = 70/3$)

(11.0.29) (a) $A(t) = 10 \times 2^{t/12}$ $B(t) = 5 \times 2^{(t-48)/6}$

when $t \geq 48$ (b) $10 \times 2^{t/12} = 5 \times 2^{(t-48)/6}$

(c) $t = 108$ ie midnight Jan 5

(11.0.30) day 37 (between 35 and 39 is acceptable)

(11.0.31) (a) first 6 years (b) first 3 years

(11.0.32) (a) $100 2^{t \log(1.4)/(10 \log(2))}$

(b) $10 \log(2)/\log(1.4) \approx 20.6$ years (c) ≈ 41.2 years

(11.0.33) $1/0.72 \approx 1.39$ hours

(11.0.34) (a) 4 Km from start

(b) 1 hour (c) 4 Km from start

(11.0.35) 163.6 miles past Salt Lake City

(11.0.36) 13/8 hours

(11.0.37) 1 hour 40 mins

(11.0.38) $\pi \times 10^4/21 \approx 1496 m^3/hour$

(11.0.39) 5000π liters

(11.0.40) 20 seconds

(11.0.41) (a) $1/(180\pi)$ meters/min (b) 900π mins

(11.0.42) (a) 1/8 (b) $-\pi \ m^3/hr$

(11.0.43) $ln(5/2)/.081 \approx 11.3$ days

(11.0.44) \$35/6

(11.0.45) (b) $2 \times 10^6 \times e^{-2} \approx 270000$ trees per year

(c) $t = -ln(1/2) \approx 0.69$ years ≈ 8.3 months

(11.0.46) $p(t) = 10^6[9.4 + 3.45 sin\left(\frac{2\pi}{11.4}(t - 7.8)\right)]$

(11.0.47) $2^{t/40} = 2 \times 10^8/100600$

hence $t \approx 305$ years

(11.0.48) after 6 hours

(11.0.49) (a) $\sqrt{68}$ (b) $\sqrt{(a-2)^2 + 4a^2}$

(c) $x = 2/5 \quad y = 4/5$

(11.0.50) $x = 5/2 \quad y = \sqrt{5/2}$

(11.0.51) 195 inches

(11.0.52) (a) 12 Km from A the road is 400m higher than A

(b) 12 Km from A the road is descending as you move away from A at rate of $60m$ per Km.

(c) $394m$

(11.0.53) explain

(11.0.54) $V = 2x - xy$

(11.0.55) $(b/4) - (c/8)$

(11.0.56) 9/2 hours

(11.0.57) $y = (10 - 500x)/(3 - x)$

(11.0.58) (a) $16 - t - t^2/2$ (b) $\sqrt{17} - 1$ hours

(11.0.59) $60x^2 + 1600x^{-1}$

(11.0.60) 2760 miles from LA

(11.0.61) 8000/9 miles from LA

(11.0.62) 1.5 liters of oil

(12.1.1) (a) $0.5 \pm .02$ (b) $0.87 \pm .03$ (c) $0.71 \pm .02$

(d) $.34 \pm .01$ (e) $0.94 \pm .03$

(12.1.2) (a) $7, 1/\pi, \pi$ (b) $\sqrt{2}, 5/2, 2/5$ (c) $3, 2, 1/2$

(d) $1, 1/(6\pi), 6\pi$

(12.1.3) (a) $sin(t + \pi)$ leads by π

(b) $sin(2t + \pi)$ leads by π

(c) $sin(t + 3)$ leads by 2

(12.1.4) (a) $5cos(5t)$ (b) $6cos(2t - \pi)$

(c) $-21 sin(3t - 12)$ (d) $-2cos(4 - t)$

(12.1.5) graph

(12.1.6) graph

(12.1.7) graph

(12.1.8) (a) $-cos(x) + C$

(b) $-(1/2)cos(2x) + C$ (c) $(1/3)sin(3x) + C$

(12.1.9) 1

(12.1.10) 1

(12.1.11) (b),(a),(c)

(12.1.12) (d) $3sin((3\pi)(t - .47))$

(e) $1.5sin(\pi(t - 1.6)) + 2$ (f) $1 + sin(2\pi t)$

(12.1.13) $80\pi \ cm^3$

(12.1.14) (b) 1981 and 1986 (c) 1983 and 1988

(d) increasing by $5000/\pi$ rabbits per year

(12.1.15) greatest when $t = 6$. least when $t = 18$. 4800 megawatt hours.

(12.2.1) (a) $30x + 7$ (c) 15

(12.2.2) (a) $5x^4 + 12x^2 - 21$ (b) $18x^5 - 45x^4 - 3x^{-2}$

(c) $-4x^{-5}$ (d) $10x^9 + 36x^5 + 18x$

(12.2.3) (a) $2x sin(3x) + 3x^2 cos(3x) + 21cos(3x)$

(b) $(x^2 + 2x + 1)e^x$ (c) $ln(x) + 1$

(d) $3cos(3x)ln(x) + (1/x)sin(3x)$

(12.2.4) (a) $e^t[ln(t) + (1/t)]$

(b) $2t sin(t) + (t^2 + 1)cos(t)$ (c) $[cos(t)]^2 - [sin(t)]^2$

(d) $(6t - 2)sin(7t) + (21t^2 - 14t)cos(7t)$

(e) $2e^{2t}sin(3t) + 3e^{2t}cos(3t)$

(12.2.5) (a) $(x - 5)f(x)$ (b) $f(x) + (x - 5)f'(x)$

(c) -1000 (d) If price is increased from \$15 by \$1 to \$16 profit is reduced by about \$1000.

(12.2.6) (a) $n(x)p(x)$ (b) $n'(x)p(x) + n(x)p'(x)$

(12.2.7) (a) $K'(x) = g'(x)h(x) + g(x)h'(x)$

(b) $f'(x)K(x) + f(x)K'(x)$

(c) $f'(x)g(x)h(x) + f(x)g'(x)h(x) + f(x)g(x)h'(x)$

(d) $3x^2$ (e) $3[f(x)]^2 f'(x)$

(12.2.8) $(x^2 + 1)^2(6x), \quad 3[sin(x)]^2cos(x)$

(12.2.9) $4[f(x)]^3 f'(x)$

(12.2.10) Know $p' = .05p$ and $I' = .03I$ so

$(pI)' = p'I + pI' = .05pI + .03pI = .08pI$ this is 8% of total income.

(12.2.11) 5/4 square miles per year

(12.2.12) calculate

(12.2.13) do it

(12.2.14) (a) 400 Kg (b) increasing (c) 4 Kg/year

(12.2.15) \$32.5 billion per year

(12.2.16) average profit per computer \$5000. Profit per computer falls \$250 per year.

(12.2.17) 28 cm per hour

(12.2.18) $11000\pi \approx 34558$ Kg per year

(12.2.19) do it

(12.2.20) do it

(12.2.21) (a) $1/(2x + 1)^2$ (b) $[(1 + e^x) - xe^x]/(1 + e^x)^2$

(c) $-2x/(1 + x^2)^2$

(12.3.1) (a) $x = -5/2$ local min (b) $x = -1$ local min and $x = -3$ local max (c) critical points at $x = -1$. 2'nd derivative test fails (d) $x = \sqrt{2}$ local min $x = -\sqrt{2}$ local max (e) $x = 0$ local min

(12.3.2) (a) $x = 2$ local min $f(2) = -14$ also $x = -2$ local max (b) $x = -1$ local min $f(-1) = 0$ (c) $x =$

−1. 2'nd derivative test fails (but it is a local min) (d) $x = -3$ local max $f(-3) = -3$ (e) $x = 1/2$ local max $f(1/2) = ln(1/2) - 1/2$

(12.3.3) graph. Impossible to have a function with two consecutive maxima (at $x = 2, 3$) without a minimum inbetween.

(12.3.4) Local min at $x = (3/2) + 2n$, local max at $x = (1/2) + 2n$ for every integer n.

(12.3.5) approximately same as for $sin(x)$

(12.3.6) draw a staircase with round corners

(12.3.7) (a) $x = -1$ $y = -16$ min (b) none

(c) $x = -3$ $y = 86$ max and $x = 2$ $y = 73$ min

(d) $x = -4$ $y = 9$ max and $x = -1$ $y = -18$ min.

(12.4.1) (a) $(8x + 3)\Delta x$ (b) 0.54 (c) 0.5416 (d) 0.3%

(12.4.2) $4\pi R^2 \Delta R$ = volume of a thin shell of thickness ΔR.

(12.4.3) $10.5°K$

(12.5.1) $A = 6$ and $K \approx 1.7$

(12.5.2) $e^{-t} sin(4\pi t)$

(12.6.1) table looks like:

x	$ln(1 + x)$	$x - x^2/2 + x^3/3$	diff
0.5	0.4055	0.4167	.0112

error when $x = 1$ is 2.8%

(12.6.2) $(x/5) + (6/5)$.

$\sqrt{4.5} \approx 2.1$ error .02

$\sqrt{5} \approx 2.2$ error .036

(12.6.3) tangent line $y = (x/4) + 1$.

$\sqrt{4.5} \approx 2.125$ error .005.

$\sqrt{5} \approx 2.25$ error .014

(12.6.4) error= .01 = area of little square in top right corner

(12.6.5) .027% .17% .66%

(12.6.6) $-5x - 12$

(12.6.7) $1 + .0022x$ (a) 0.35% (b) 0.44%

(12.6.8) (a) time to double 17.67, 9.01, 6.11, 4.67 years.

(b) Rule of 72 gives 18, 9, 6, 4

(c) % error 1.9%, .07%, 1.9%, 3.6%

(13.1.1) t = number years. $m(t)$ = number of moths after t years. $c(t)$ = number cacti after t years. $m'(t)$ and $c'(t)$ ar relevant.

(13.1.2) height of plane and derivative of height

(13.2.1) (a) second order (b) third order (c) 5'th order (d) first order

(13.2.2) (a) $t^4 + t^2/2 + C$ (b) $t^5/5 + t^3/6 + Ct + K$

(c) $t^6/30 + t^4/24 + At^2/2 + Bt + C$ (d) $3t^2/4 + t + C$

(13.2.3) $4t^3/3 + 2t^2 + t + 3$

(13.2.4) $t^3 + 3t + 2$

(13.2.5) $t^2 + C$

(13.2.6) $(1/4)e^{2t} + At + B$

(13.2.7) (a) $m'(t) = 3t + 2$ (b)$3t^2/2 + 2t + C$

(c) 175 grams.

(13.2.8) $h(t)$ = height in meters after t seconds.

(a) $h'' = 20 + 4t$

(b) gen soln. $h = 10t^2 + 2t^3/3 + Ct + K$

(c) particular soln $h = 10t^2 + 2t^3/3$

(d) Height after 10 secs 5000/3 m.

(e) Velocity after 10 seconds 400 m/s.

(13.2.9) (a) $B'(t) = 200 + 10t$ (b) 90000

(13.2.10) $y = 2t^3 - t^2 + At + B = 2t^3 - t^2 + t + 3$.

(13.3.1) graph

(13.3.2) graph

(13.3.3) $y = 3e^{4t}$ solution $\to \infty$

(13.3.4) $y = -8e^{-t/2}$ solution $\to 0$

(13.3.5) $k = 3/2$. $y(6) = 2e^{9/2}$.

(13.3.6) (a) $m' = 0.2m$. (b) $m(t) = Ae^{0.2t}$

(c) $m(t) = 20e^{-0.6}e^{0.2t}$ (d) $20e^{-0.6} \approx 10.97$ grams

(13.3.7) - sign because mass is decreasing. $ln(0.5)/(-.04) \approx 17.3$ days

(13.3.8) $k = -ln(3/20)/10 \approx 0.19$

(13.3.9) $k = ln(1/2)/100 \approx -.0069$

$m(t) = m_0 e^{t[ln(1/2)/100]} \approx m_0 e^{-.0069t}$.

$100ln(0.15)/ln(0.5) \approx 273.7$ years

(13.3.10) $A = -3$. $k = -0.5$

(13.3.11) (a) $m' = km$ (b) $m = 10e^{t/2}$ (c) $2ln(6)$ days

(13.4.1) (1) (a) oscillates between 0.5 and 1.5

(b) gets bigger and bigger, goes to ∞

(c) increases towards 1 but never gets there.

(d) oscillates larger and larger amounts

(e) oscillates less and less approaching a value of 0.4 as $t \to \infty$.

(f) decreases towards limiting value of -0.2

(g) decreases towards $-\infty$

(h) increases towards ∞

(i) decreases rapidly towards 0.

(j) decreases linearly for ever towards $-\infty$

(13.6.5) b,e,f

(13.7.1) $y = 1 + Ae^{-5t}$. $y \to 1$ as $t \to \infty$

(13.7.2) (a) $y' = 4(7/4 - y)$ (b) $y' = (3/2)(2 - y)$

(13.7.3) $2 - e^{-0.3t}$

(13.7.4) $6 + 4e^3 \approx 86$

(13.7.5) $f(t) = (7/3) + Ae^{-3t}$

(13.7.6) describe

(13.7.7) $y(t) = 15/2$ constant

(13.7.8) $x(t) = 100 - 10e^{-3t}$.

$t = ln(5)/ - 3 \approx -0.536$

(13.7.9) (a) 12.8 secs (b) 16.6 secs (c) 25.6 secs

(13.7.10) $y' = k(30 - y)$ $k = -ln(5/7)$.

$ln(2/7)/ln(5/7) \approx 3.7$ days.

(13.7.11) (a) $20 + 70e^{ln(1/7)/2} \approx 46.38°$ C

(b) $20ln(6/7)/ln(1/7) \approx 1.58$ mins.

(13.8.1) sketch $y = 10/(9e^{-t} + 1)$ and $y = 10/(9e^2 e^{-t} + 1)$

(13.8.2) sketch

(13.8.3) one increases more slowly than the other (which? why?)

(13.8.4) sketch

(13.8.5) $k = 4/5$ $y(1) = 10/(e^{-4/5} + 1) \approx 6.9$

(13.8.6) (a) $p' = (p/25)(1 - (p/10^{10}))$

(b) $-25ln(1/9) \approx 54.9$ years

(c) $-25ln(9) \approx -54.9$ years

(13.8.7) (a) $m' = (25/49)m(1 - m/100)$

(b) $(49/25)ln(49) \approx 7.63$ days

(c) $100/(49e^{-250/49} + 1) \approx 77$ gms

(13.9.1) 2'nd row: 1.33, 1.76, 2.35

4'th row: 1.5%, 3.3%, 4.5%

(13.9.2) $y(0.4) \approx 1.4$

(15.0.1) (a) $6xy^4$ (b) $12x^2y^3$ (c) $4xy - 3y^3 + 28x^3y^5$

(15.0.2) (a) $\frac{\partial}{\partial x}(x/y) = 1/y \quad \frac{\partial}{\partial y}(x/y) = -xy^{-2}$

(b) $\frac{\partial}{\partial x}(3x + 7y + 2) = 3 \quad \frac{\partial}{\partial y}(3x + 7y + 2) = 7$

(c) $\frac{\partial}{\partial x}(2x - 3y)^2 = 8x - 12y$

$\frac{\partial}{\partial y}(2x - 3y)^2 = -12x + 18y$

(d) $\frac{\partial}{\partial x}(x^{-2}y^{-2}) = -2x^{-3}y^{-2}$

$\frac{\partial}{\partial y}(x^{-2}y^{-2}) = -2x^{-2}y^{-3}$

(15.0.3) (a) $\frac{\partial}{\partial x}(e^{2x-3y}) = 2e^{2x-3y}$

$\frac{\partial}{\partial x}(e^{2x-3y}) = -3e^{2x-3y}$

(b) $\frac{\partial}{\partial x}(sin(xy)) = ycos(xy)$

$\frac{\partial}{\partial y}((sin(xy)) = xcos(xy)$

(c) $\frac{\partial}{\partial x}(e^{2xy}) = 2ye^{2xy} \quad \frac{\partial}{\partial x}(e^{2xy}) = 2xe^{2xy}$

(d) $\frac{\partial}{\partial x}(xe^y) = e^y \quad \frac{\partial}{\partial y}(xe^y) = xe^y$

(15.0.4) (a) $e^y(2x + 2)$ (b) $e^x sin(3y)$

(c) $3e^{-2y}cos(3x)$ (d) $sin(2y + x) + xcos(2y + x)$

(15.0.5) $f_x = 6x \quad f_y = 8y$

(a) 6   (b) 16   (c) 0.

(15.0.6) $\frac{\partial p}{\partial x}$ is the rate of change of pressure as the mass of air in the tire increases

$\frac{\partial p}{\partial T}$ is the rate of change of pressure as the temperature of the air in the tire increases.

(15.0.7) f_x is the rate of increase of the investment as the amount invested increases.

f_y is the rate of increase of the investment as the length of time the money is invested increases.

$f_x(3000, 5) = 1.6$ means that increasing the amount invested from \$3000 by \$1 will increase the investment by \$1.6 in 5 years time.

$f_y(3000, 5) = 300$ means increasing the investment period from 5 years to 6 years will (approximately) increase the investment by \$300 with an initial investment of \$3000.

(15.0.8) $V = x^2h \quad S = 2x^2 + 4xh$

(a) $V_x = 2xh \quad V_h = x^2 \quad S_x = 4x + 4h \quad S_h = 4x$

(b) $S_x(2, 3) = 20$ (c) unit of length

(d) surface area increases at a rate of 20 units of area per unit increase of length of side of base, when base is 2 and height is 3.

(15.0.9) (a) $F_m = GM/R^2 \quad F_R = -GmM/R^3$

(b) gravitational force **decreases** as distance of rocket from Earth increases.

(15.0.10) (a) $(x - 20)y$

(15.1.1) $z = 2x - 4y$

(15.1.2) $z = -5x + (3y/2) + 2$

(15.1.3) $(0, 0, 5), \quad (1, 0, 7) \quad (0, 1, 2)$ and many more

(15.2.1) (a) $h(-2, 4) = -90$

(b) $h_y(-2, 4) = -22$ falls 22 feet per mile.

(15.2.2) $A = WL$ (a) $A_W = L$ so 3 cm^2 per cm

(b) $A_L = W$ so 5 cm^2 per cm

(c) length (d) 2 cm per cm

(15.2.3) $R = 2h$ where R = radius and h = height

(15.2.4) $z = 4x - 2y - 3$

(15.2.5) $z = 3x - 2y$

(15.3.1) 13.116

(15.3.2) 72.014

(15.3.3) (a) 226 (b) $50\frac{2}{3}$

(15.3.4) $.13 \ cm^2$ error is 0.15%

(15.3.5) $\Delta x/2$

(15.3.6) $f_t(30, 5) = -5/3$ and $f_x(30, 5) = -3/2$

$47 + (1/6)$

(15.3.7) (d) 3/2 hours

(15.4.1) $x = 1/3$ and $y = -4/3$

(15.4.2) $x = 4/11$ and $y = 1/11$

(15.4.3) (a) cost is $10WL + 80(W^{-1} + L^{-1})$

(b) For minimum cost, length and width both 2, height is 1/2

(15.4.4) $x = 40 \quad y = 50$

Index

(-2.5,0) (4,7 7

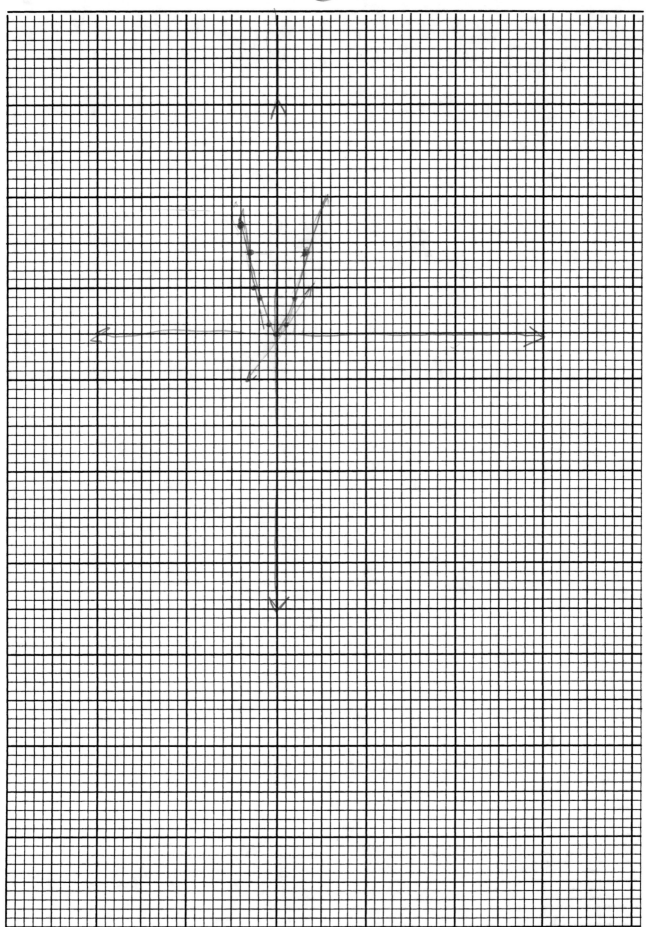

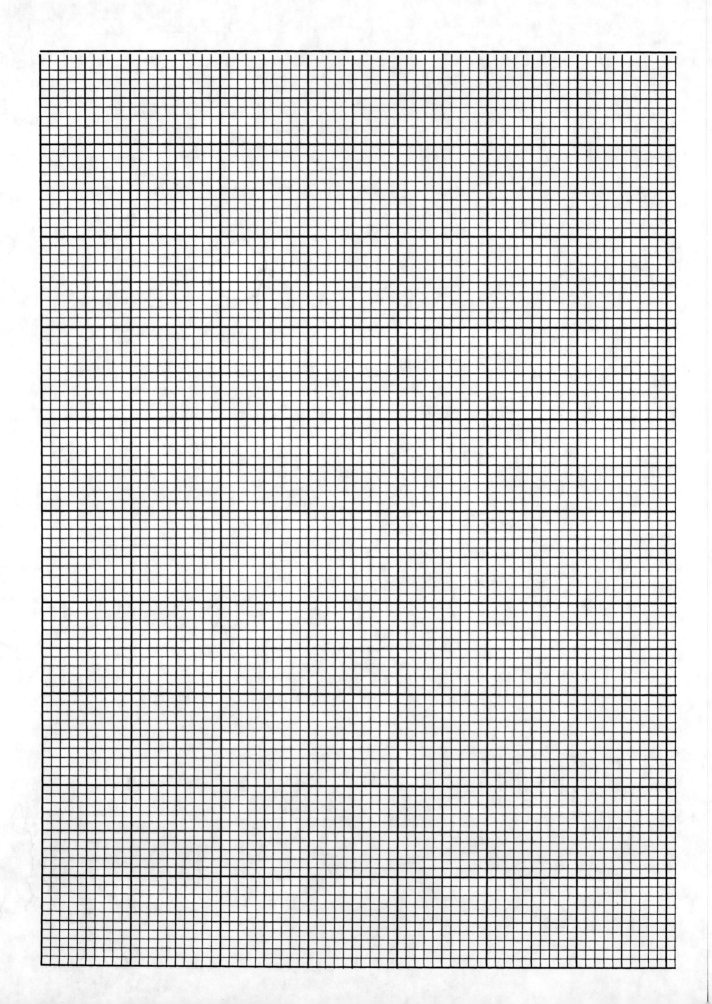

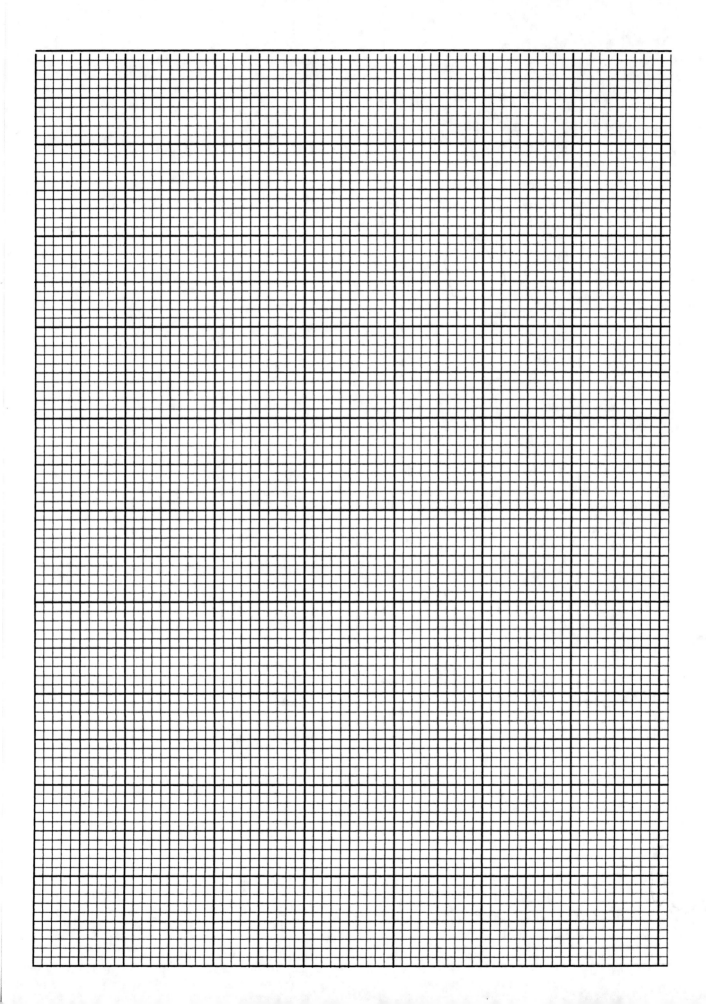

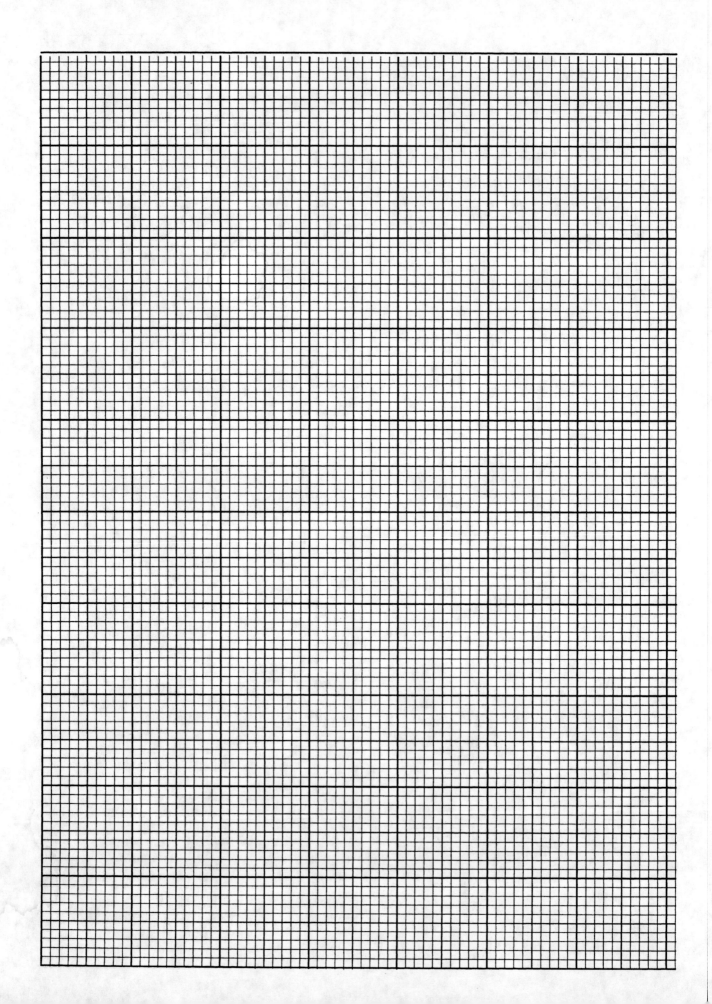

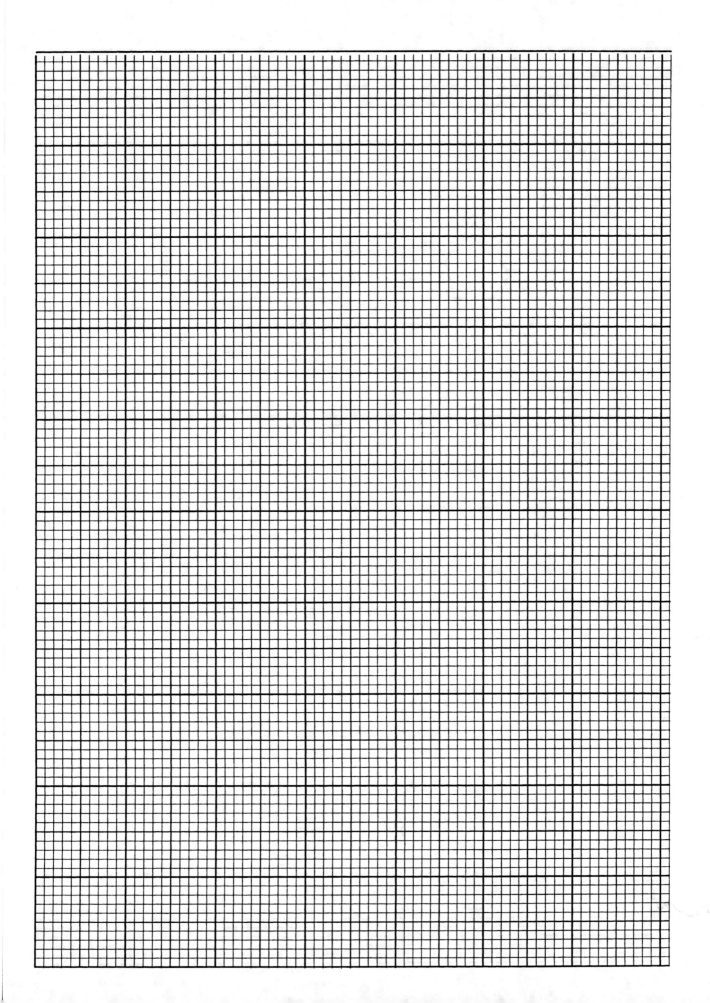

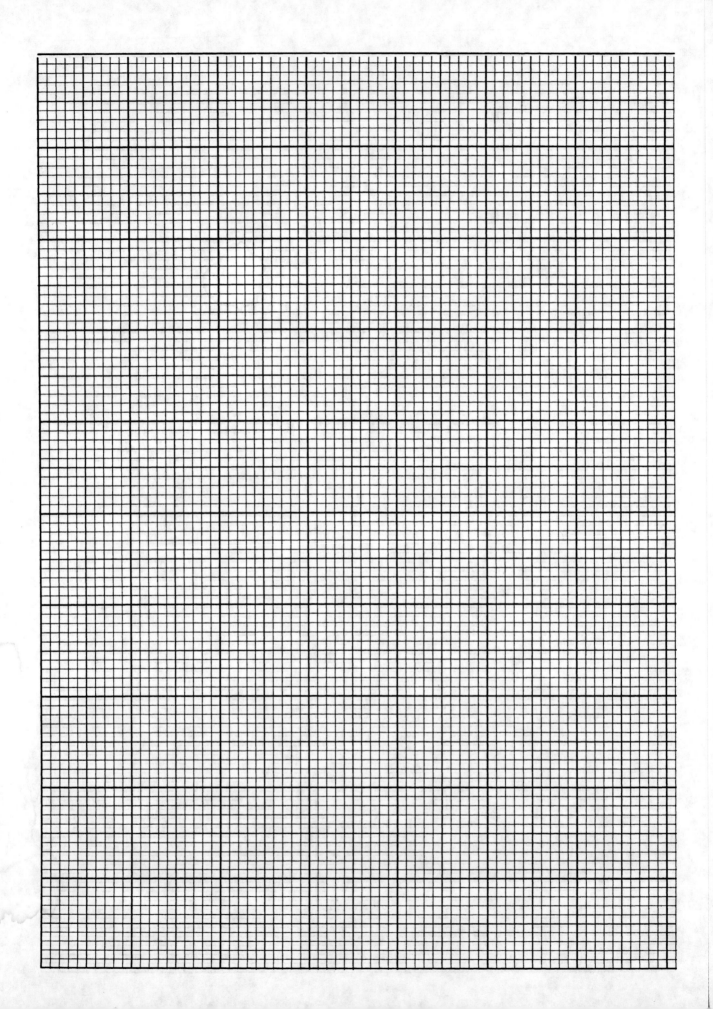

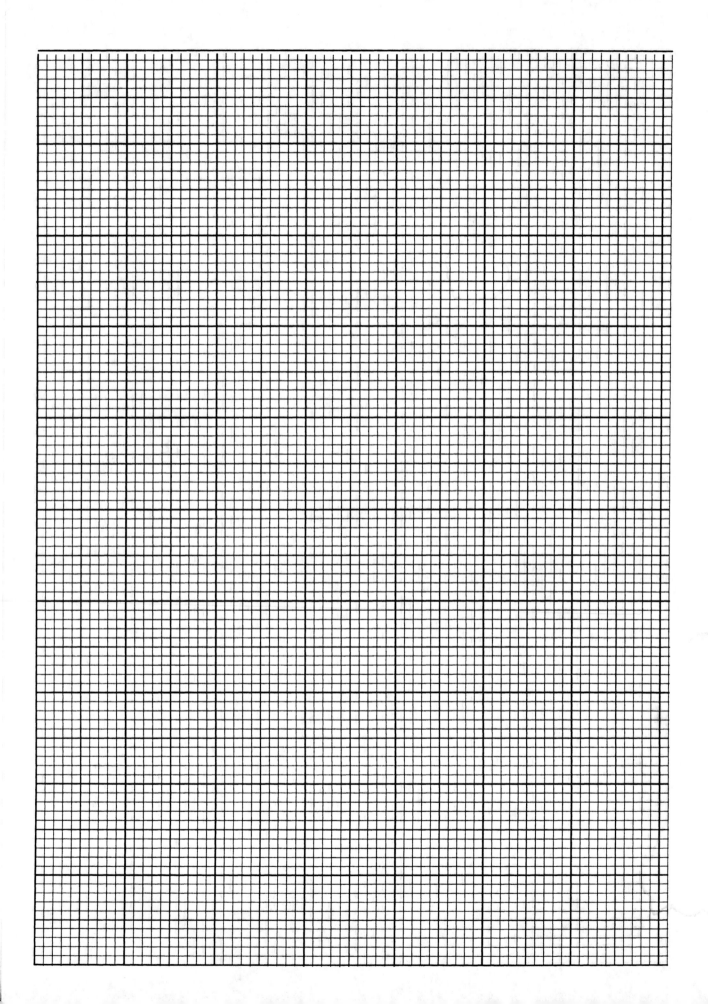

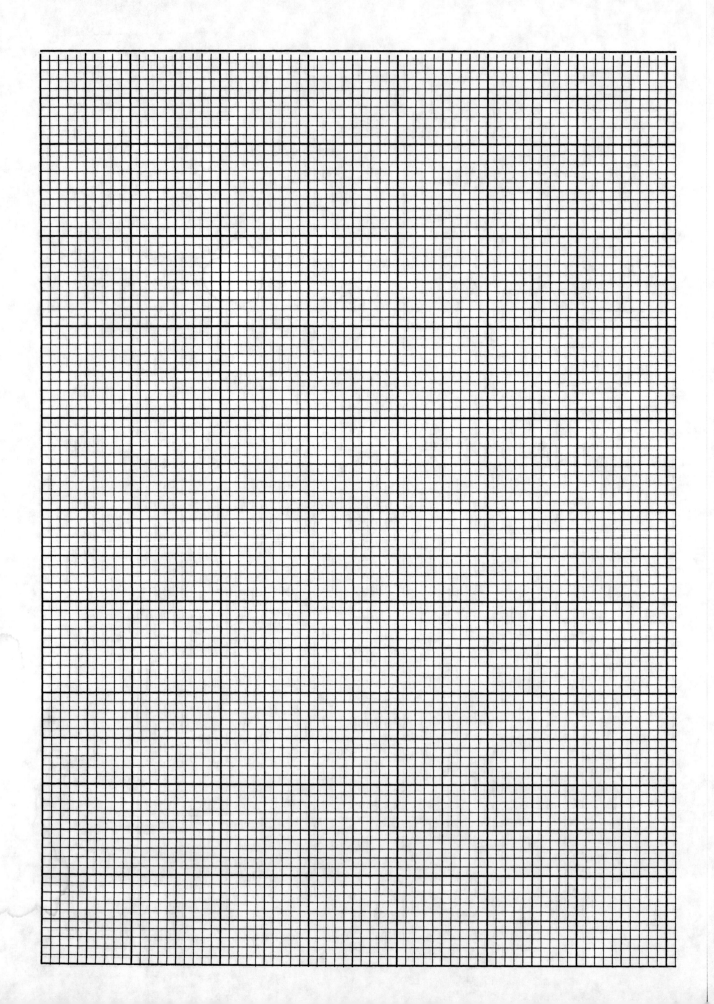

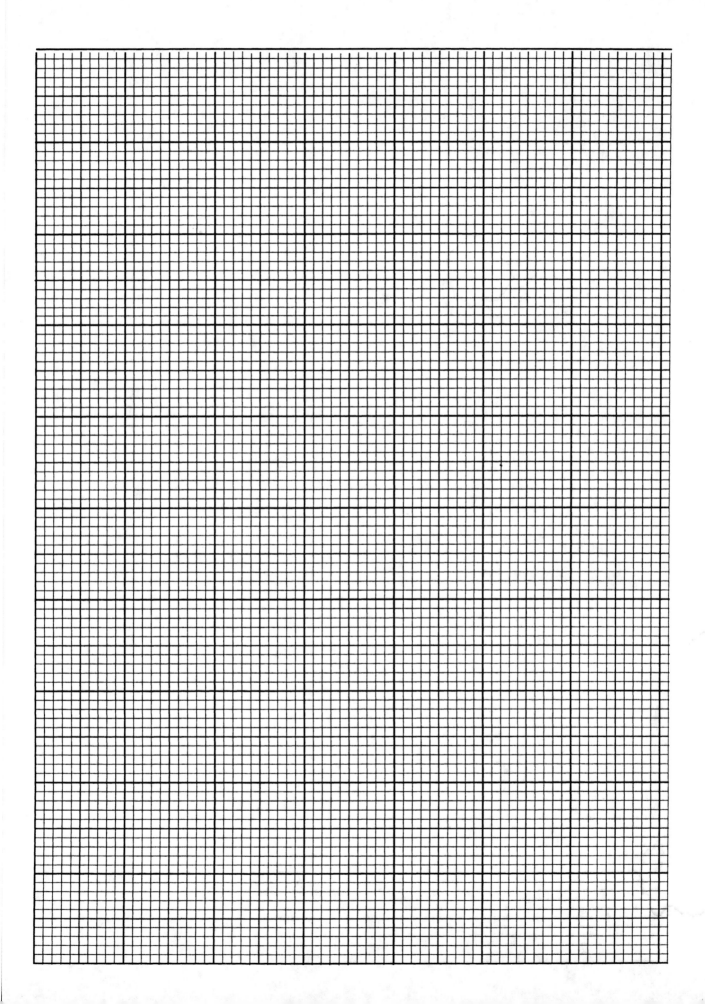

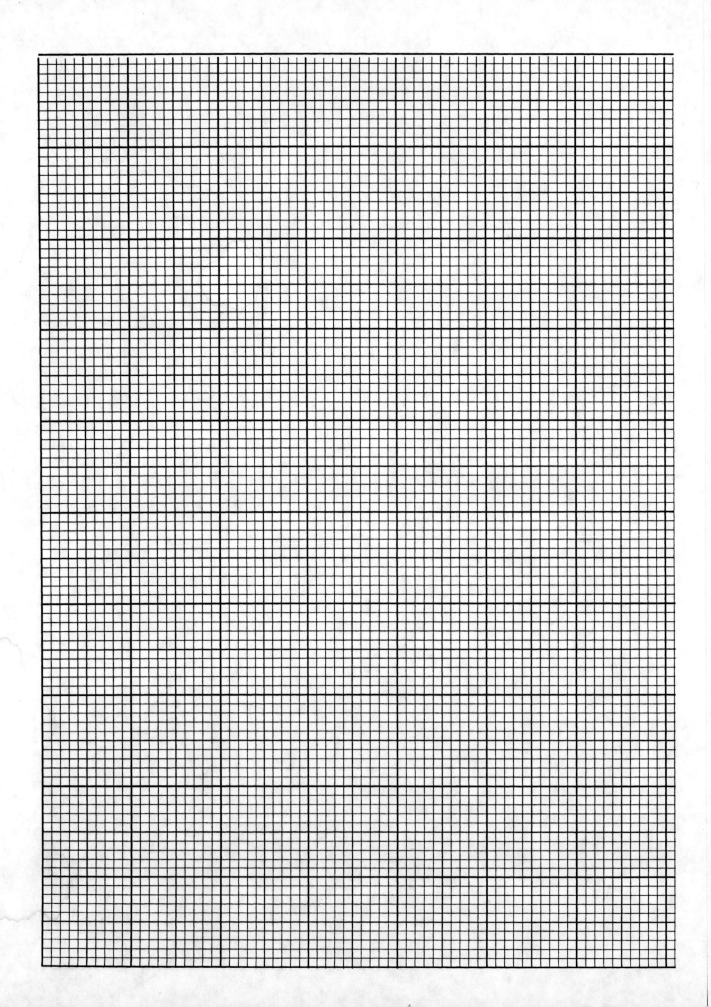

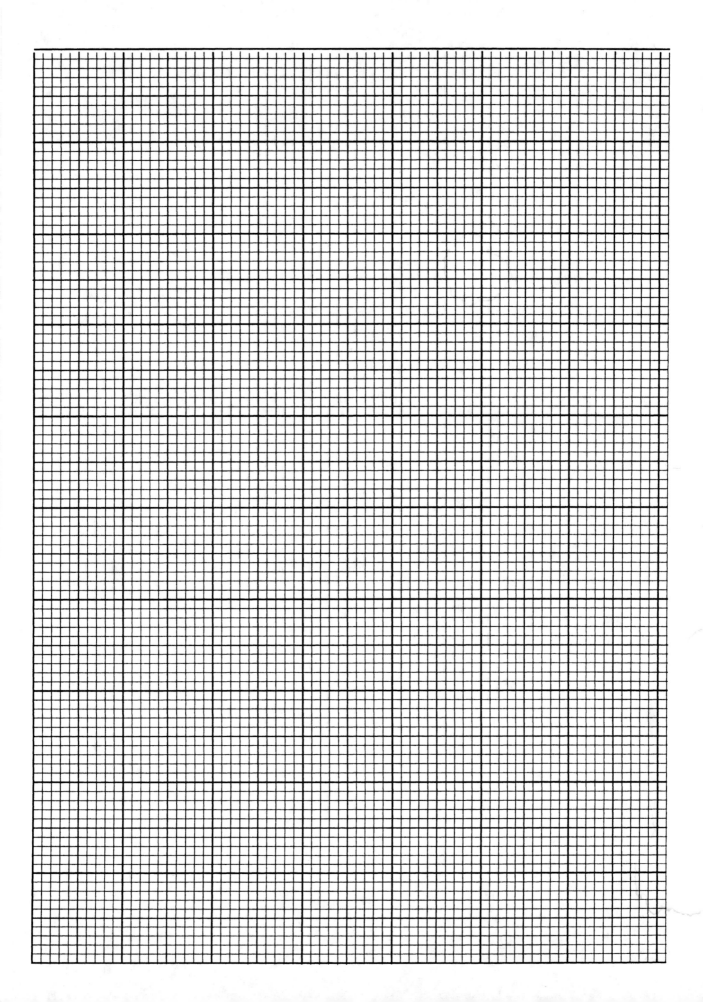

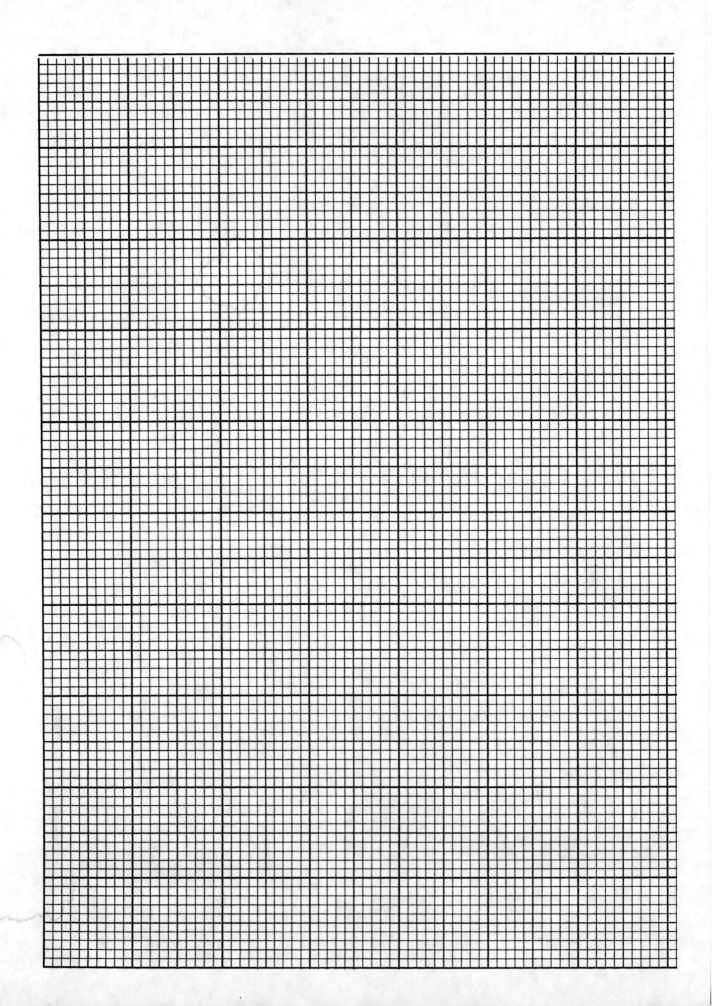

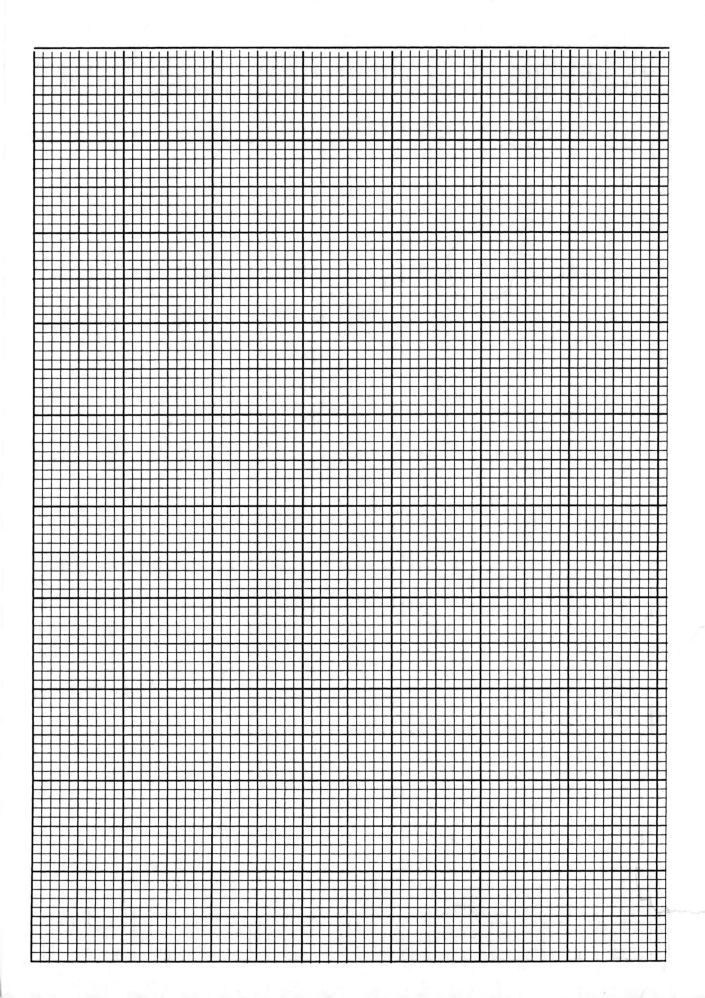

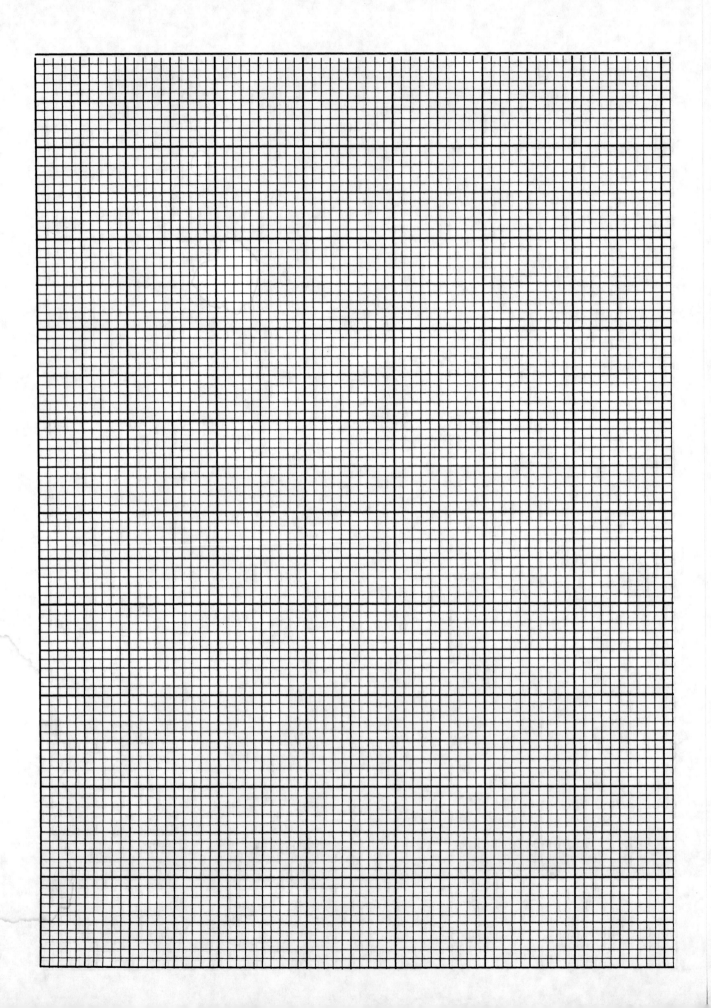

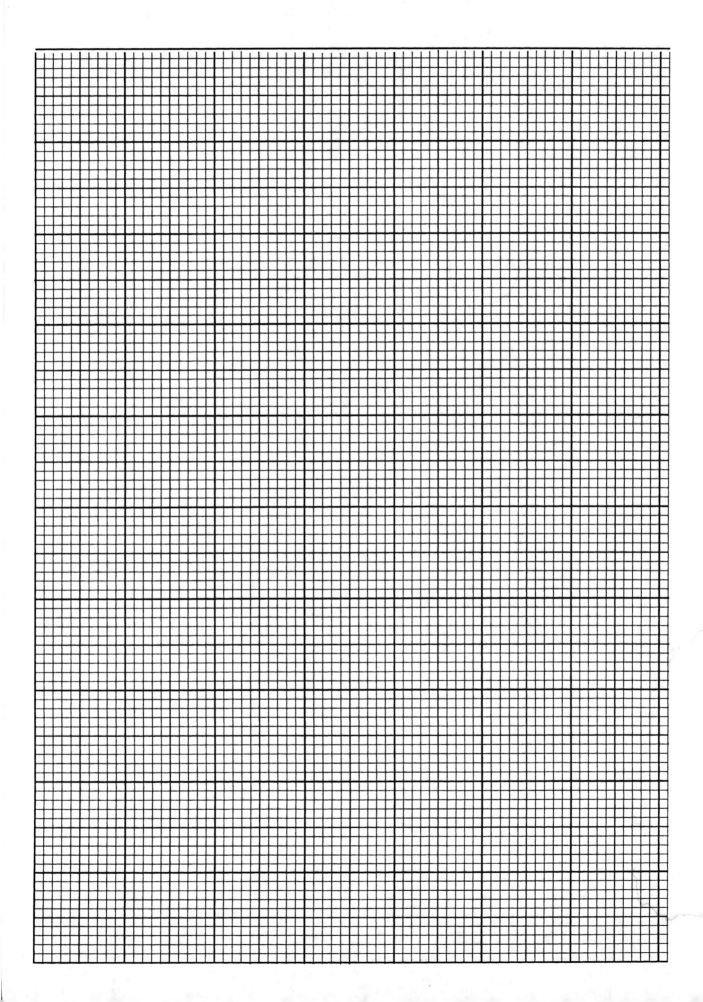

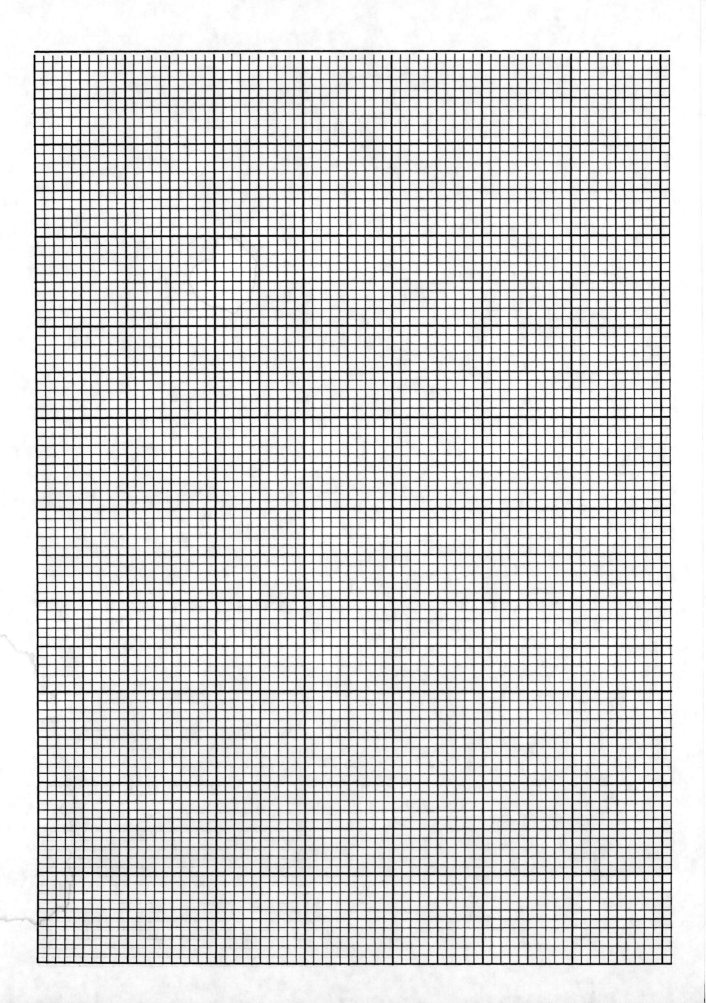

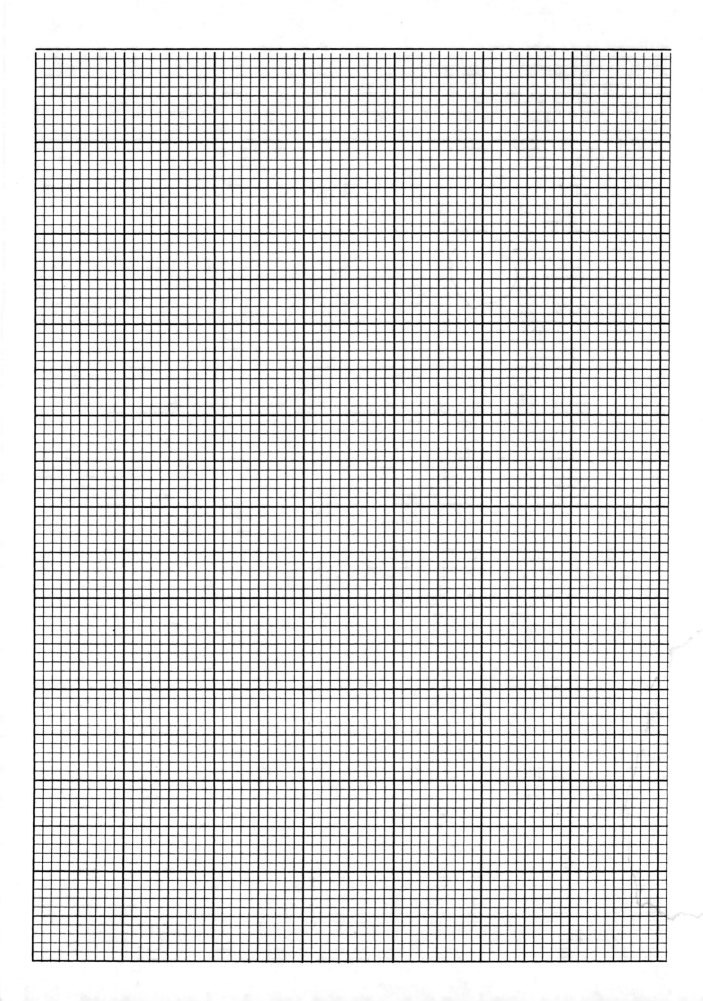

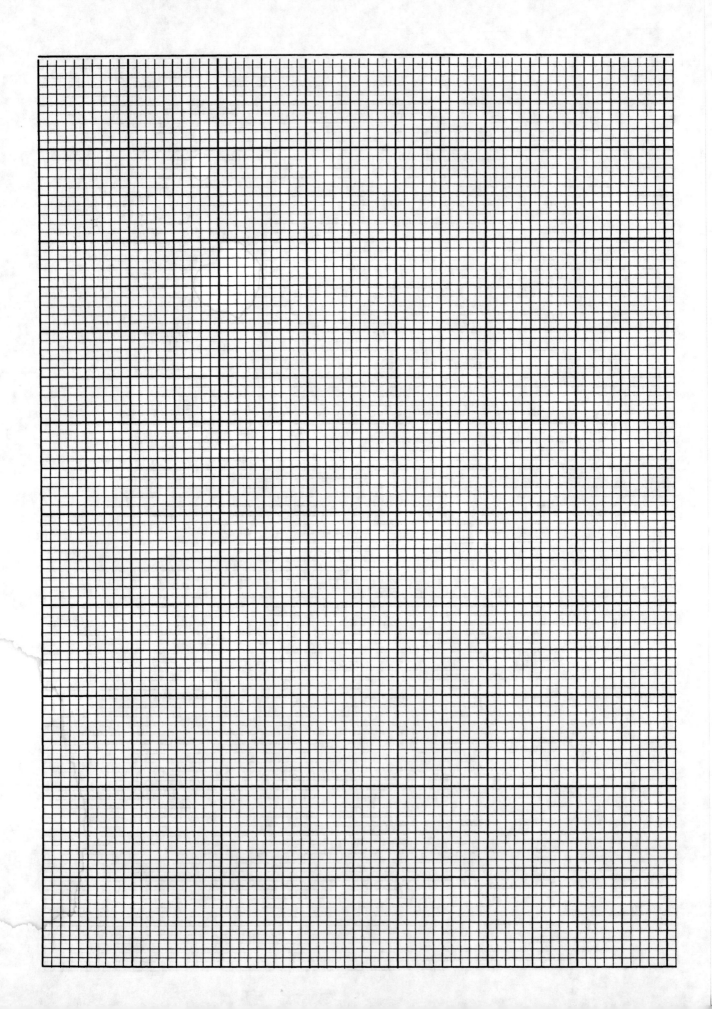

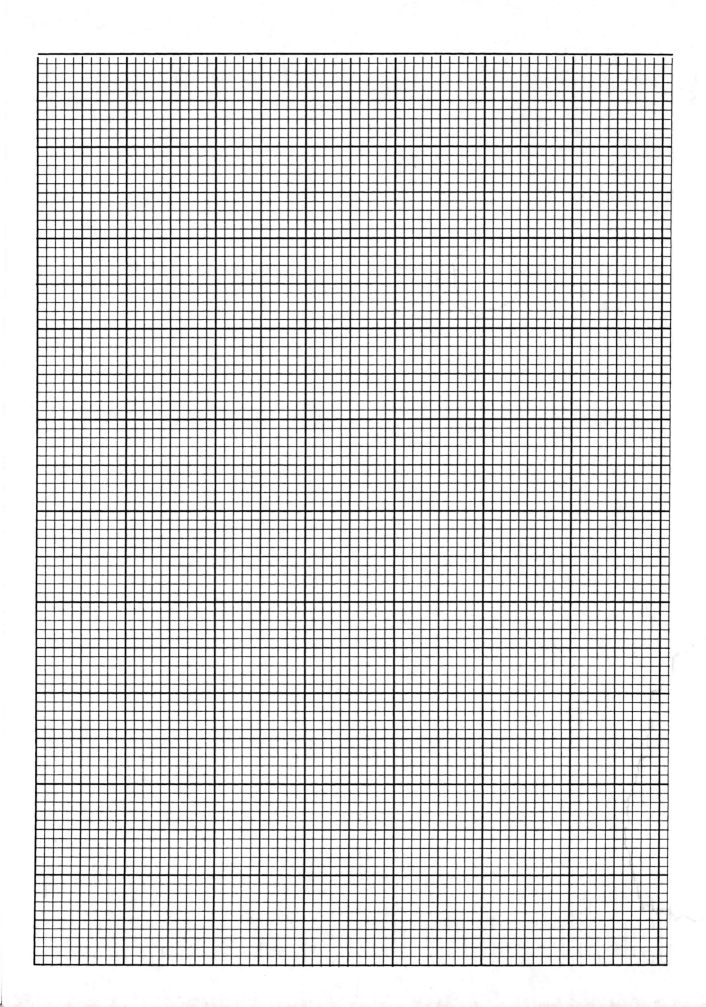

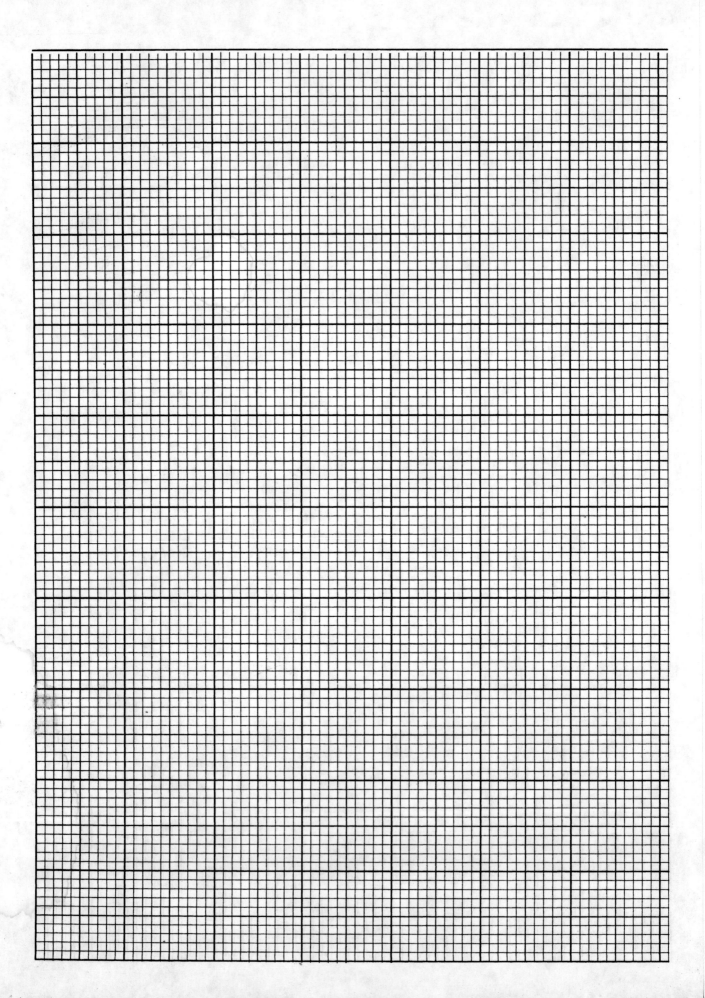